D0757685

OPTICAL
TRANSFORMS

OPTICAL TRANSFORMS

H. Lipson

*The University of Manchester Institute
of Science and Technology,
Manchester, England*

1972

ACADEMIC PRESS · LONDON · NEW YORK

ACADEMIC PRESS INC. (LONDON) LTD.
24/28 Oval Road,
London NW1

United States Edition published by
ACADEMIC PRESS INC.
111 Fifth Avenue
New York, New York 10003

Library of Congress Catalog Card Number: 78-170763
ISBN: 0-12-451850-8

PRINTED IN GREAT BRITAIN BY
WILLIAM CLOWES AND SONS LIMITED, LONDON, COLCHESTER AND BECCLES

Contributors

J. E. BERGER   Centre for Crystallographic Research, Roswell Park Division of Health Research Inc., Buffalo, New York, U.S.A.

B. CHAUDHURI   Department of Physics, University of Gauhati, Gauhati, Assam, India

W. P. ELLIS   University of California, Los Alamos Scientific Laboratory, Los Alamos, New Mexico 87544, U.S.A.

G. HARBURN   Physics Department, University College, Cathays Park, Cardiff, S. Wales

J. A. LAKE   The Rockefeller University, New York, New York 10021, U.S.A.

H. LIPSON   The University of Manchester Institute of Science and Technology, Manchester, England

S. G. LIPSON   Department of Physics, Technion, Haifa, Israel

J. SHAMIR   Department of Electrical Engineering, Technion, Haifa, Israel

D. SHECHTMAN   Department of Materials Engineering, Technion, Haifa, Israel

C. A. TAYLOR   Department of Physics, University College, Cathays Park, Cardiff, S. Wales

B. J. THOMPSON   The Institute of Optics, University of Rochester College of Engineering and Applied Science, Rochester, New York 14627, U.S.A.

Preface

Optics is one of the oldest branches of physics, and also one of the most persistent. In spite of repeated claims that its role is finished, because it does not directly lead to the study of the structure of the atom, optical apparatus and ideas continue to provide a necessary basis for many new concepts, which help both physics itself and the other branches of science of which physics is a necessary part. When microscopy was supposed to be a finished subject, Zernike invented the phase-contrast microscope, which has been of enormous use to the biologist; when light sources were thought to have reached their limit, Schawlow and Townes invented the laser, which now has applications in almost every branch of science; and when the theory of image formation was thought to be a closed subject, Leith and Upatnieks following up the much earlier work of Gabor, invented their improved version of holography, which is having a remarkable influence in technologies of all sorts.

This book deals with one particular part of optics—the production and use of optical transforms. Basically, optical transforms are Fraunhofer diffraction patterns, the study of which had been for a long time regarded as of only limited interest. Their application to X-ray diffraction problems, however, brought the realization that the diffraction pattern of an object sometimes provided information in a more direct form than an image of the object itself, and gradually it has become evident that this property can be useful in subjects other than X-ray diffraction.

I have therefore tried to gather together the experience of a number of workers who have made different uses of optical transforms. Each has presented his ideas in his own way in a separate chapter of the book. Although each author was given a brief outline of what all the others were writing, there is inevitably some overlapping, but I have thought it worthwhile not to eliminate this so that each chapter is complete in itself. For the same reason, references are collected at the end of each chapter, instead of being in a consolidated list at the end of the book.

The final chapter is different from the rest. Whereas in general my aim has been to include only those subjects that have, over the years, shown their value in established researches, in the last chapter some new and rather tentative projects are described. Most are still in embryonic form, but I hope that their inclusion will encourage the reader to think about the possibility of the application of diffraction methods to the solution of his own problems. In spite, therefore, of its rather unfinished appearance, it is possible that Chapter 11 will prove to be at least as important as the other chapters.

I hope, then, that the book will stimulate others who may find that optical ideas may have a place in their work that they had not previously suspected. The history of physics abounds in unexpected inter-relations between its various branches, and if this book induces any further instances of the applications of optics it will have served the purpose for which it was intended.

JANUARY 1972

H. LIPSON

Contents

1. Basic Principles

H. LIPSON

2. Coherence Requirements

B. J. THOMPSON

3. Determination of Crystal Structure

B. CHAUDHURI

CHAPTER 1

Basic Principles

H. LIPSON

The University of Manchester Institute of Science and Technology,
Manchester, England

I. HISTORICAL INTRODUCTION

A. The Wavelength Problem

Many physical instruments have developed more or less accidentally, with little theoretical understanding of their basic principles. As the need for improvements has arisen, however, so theoretical investigations have become necessary, and from these more detailed designs have developed; ultimately quite complicated instruments have appeared, seeming perhaps almost completely

1

unrelated to their humble origins. Theory may direct practice into unexpected channels, or may even indicate that limits may exist which preclude further practical developments.

Of no field is this more true than that of optics. From about 1600, when Leeuwenhoek's apprentice invented the first telescope, to the present day, with its gigantic astronomical telescopes, there has been tremendous progress, in which theory and practice have complemented each other. The microscope has taken even greater steps. From Hooke's early instruments to the electron microscope, which can show nearly atomic detail, there has been a series of steps forward which cover a range of resolution of about 10^5. Progress has not however been continuous; at certain stages, indeed, it has looked as if an impasse had been reached. In trying to find ways round these impasses, however, new ideas had been introduced and these have led to the remarkable progress in microscopy of which the electron microscope is the culmination.

This book is concerned with microscopy in its broadest sense. The first microscopes were simple lenses and it is amazing what discoveries were made with them—even with a drop of honey in a circular hole. But it was soon found that combinations of lenses were easier to use and gave better results, and even in the early days Huygens had worked out in detail the theory of the eyepiece that is still named after him. Ways to minimize or eliminate aberrations were discovered and with precise means for measuring the refractive indices and dispersive powers of glasses (in this study the name of Fraunhofer is particularly prominent) quite complicated lens systems of superb performance could be devised. With improved workmanship there appeared to be no limit to what could be accomplished.

In 1873, however, Abbe introduced his wave theory of image formation and showed that even with perfect lenses resolutions of less than about half the wavelength of light could not be achieved. It seemed, then, that Nature had set a limit of the order of 0·2 μm on what the microscopist could distinguish with his instrument. The only way round the limitation would be to see if some radiation with a shorter wavelength could be used in the image-forming process.

Ultra-violet radiation was an obvious suggestion; microscopes were built for the near ultra-violet, but difficulties of designing lenses for an invisible radiation and of using the instrument solely by photography made the relatively small gain in resolution hardly worth while. No other possibilities seemed to exist.

When, in 1912, came the discovery that X-rays—discovered by Röntgen in 1895—were a radiation with a wavelength of the order of 1 Å; Laue, Friedrich and Knipping diffracted the radiation by means of a crystal of copper sulphate and Bragg's interpretation of the diffraction pattern of rock salt enabled him to establish a scale of wavelength to an accuracy of about one per cent. Could a microscope be built to make use of this new radiation?

The answer was "No"; X-rays could not be appreciably refracted and so no lenses could be made, and, although they can be reflected from crystal planes, the task of making a mirror accurate enough to match the wavelength of the rays was out of the question. Although reflecting X-ray microscopes have been made, they cannot give resolution even as good as the light microscope.

In 1924 a more hopeful possibility emerged; de Broglie put forward his theory of the wave nature of moving particles, and it was soon verified experimentally for low-voltage electrons (Davisson and Germer, 1927) and for high-voltage electrons (Thomson, 1927). Electrons with energies of 50 KeV have a wavelength of about 0·05 Å and since they can be deflected by electric or magnetic fields, it should be possible, by designing suitably shaped electrodes or pole pieces, to produce focused beams and hence images of objects. The limit of resolution with such short wavelengths should be as much as any physicist could desire; the wavelength is a great deal smaller than atomic separations, which are round about 2 Å, and no detail smaller than this can be envisaged.

So electron microscopes were built and their limitations were explored. It was soon found—theoretically and practically—that corrected electron lenses could not be made and that to obtain a good image the lenses must be "stopped down" very considerably; beam angles θ of a fraction of a degree had to be used and the value of the limit of resolution, given by the expression $0\cdot6\lambda/\sin\theta$, was very much larger than $\lambda/2$. Soon however the resolution of light microscopes was reached and surpassed and it seemed likely that progress to 1Å would be reached. But in fact, resolutions have not reached this value. Below about 10 Å difficulties of design and electrical control mount up, and even the most sanguine claims do not reach atomic separations.

Moreover, the electron microscope is by no means a generally useful instrument like a light microscope; electron beams can be produced only in a vacuum and can pass only through extremely thin specimens. The electron microscope has produced results of immense importance, particularly in biology, but it has not reached the stage—nor is it likely to do so—of the general applicability of the light microscope.

Other suggestions based on de Broglie's theory have also been made. Protons, for example, can be used, but these heavy particles cannot compete with electrons and the proton microscope has not proved to be of great utility. We therefore seem to be in difficulties. X-rays would be usable, since they can pass through air, but they cannot be focused; electrons can produce images but only of specialized objects, and the ultimate resolution of electron waves cannot be exploited. What can we do?

B. *Abbe's Theory*

In order to answer this question, we must look into Abbe's theory more closely in order to see whether we can circumvent the problems that it raises. There are many ways of expressing this theory, but the most direct is that propounded by Zernike (1946), who states that the process of image formation in an optical instrument consists of two stages of diffraction: the incident light is diffracted by the object and the diffraction pattern formed when this diffracted light is brought to a focus in the image (Fig. 1). This statement of Abbe's theory emphasizes the importance of the illumination system: it is not merely required to throw light on to the object; it provides a necessary part

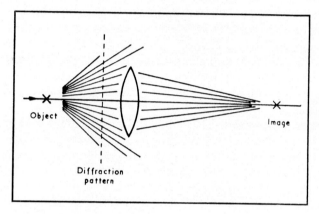

Fig. 1. Part of the light diffracted by an object is collected by a lens which refracts it to produce an interference pattern which is the image.

of the image-forming process and considerable care must be devoted to its construction if the highest possible resolution is to be obtained.

To understand these ideas it is simplest to assume that the object is illuminated by a plane parallel wave of wavelength λ, that is, by completely coherent light. (This concept will be discussed in more detail in Chapter 2.) The light will be scattered in all directions by the object (Fig. 1), the wave scattered in any particular direction being the sum of the waves scattered with different phase relations from different parts of the object. The amplitudes and phases of the scattered waves are functions of the direction of scattering; if we confine ourselves, to begin with, to a one-dimensional object, we may write for the wave ψ at an angle α (Fig. 2).

$$\psi = \int_{-\infty}^{\infty} f(x) \exp(-ikx \sin \alpha)\, dx \qquad (1.1)$$

where $f(x)$ represents the amplitude of the wave scattered by a point distant x from the axis, and k is $2\pi/\lambda$ (Lipson and Lipson, 1969). This expression arises because the path difference between the wave scattered from 0 and that scattered from P (Fig. 2) in $x \sin \alpha$. It is convenient to put $k \sin \alpha = u$ and then

$$\psi(u) = \int_{-\infty}^{\infty} f(x) \exp(-iku) \, du. \tag{1.2}$$

$\psi(u)$ is said to be the *Fourier transform* of $f(x)$.

As we can see, ψ is in general a complex quantity, that is, it has both an amplitude and a phase. The amplitude is that of the light scattered in the direction θ, and the phase is the phase of the wave relative to that scattered by the point 0. Phases are, of course, always relative to some standard. Only

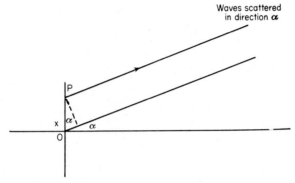

Fig. 2. Optical path difference between the waves scattered at an angle α by two points separated by a distance x.

the intensity is observable and this is equal to $|\psi|^2$ or to $\psi\psi^*$ where ψ^* is the complex conjugate of ψ.

In practice, the object will not be illuminated by a plane parallel wave; indeed, such illumination is quite undesirable and it is best to have incoherent illumination, that is, waves falling on the object from a large variety of directions. If we take one such wave falling at an angle β on the object, we then have that

$$\psi(u, \beta) = \int_{-\infty}^{\infty} f(x) \exp\{-ikx(\sin \alpha + \sin \beta)\} \, dx \tag{1.3}$$

$$= \psi(u) \exp\{-ikx \sin \beta\}.$$

The wave scattered in a particular direction is now the sum of a large number of waves and the problem becomes too complicated to deal with unless $f(x)$

is an extremely simple function. It is for this reason we have considered only the simple case of plane parallel illumination.

To form an image, we need to bring all the waves scattered from one point x in the object to a single point in space, each point in the object space having a corresponding point in image space. This is the function of the optical instrument. Now the condition for an image point to exist is that all the waves should arrive at the point in the same phase, that is, that all the optical paths between the object point and image point should be equal. In the back focal plane F of the lens (Fig. 3) all the waves that are parallel to each other will come to a focus; therefore in this plane the function $|\psi|^2$ will be observed. Now since the relative phases between object points O and image points I' are the same and may be made zero, it is clear that the phase change between

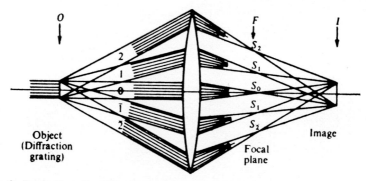

Fig. 3. Production of a diffraction pattern in the back focal plane of a lens. For clarity, the object is taken to be a diffraction grating, but the principle applies to any object.

O and F is equal and opposite to the phase change between F and I'. Thus the relation between the wave function in the focal plane of the lens to that in the image is the inverse of that between the wave function in the object and that in the focal plane of the lens. In other words, the image is the inverse Fourier transform of $\psi(u)$;

$$f'(x) = \int_{-\infty}^{\infty} \psi(u) \exp(iku)\, du, \qquad (1.4)$$

the inverse nature of the transformation being indicated by the omission of the minus sign in the exponential.

This is the mathematical equivalent to Zernike's statement, quoted at the beginning of this chapter, that an image is the diffraction pattern of the diffraction pattern of the object.

The statement is perhaps easier to understand if we introduce the concept of interference, rather than diffraction, for the second stage. We may regard diffraction as a process that occurs naturally, when a light wave is limited in some way, and interference as the deliberate bringing together of a number of separate waves. The light wave diffracted by the object can be regarded as a large number of separate waves that are diverging from each other; these are brought together by a lens so that they interfere with each other, and in one plane—the image plane—the relative phases are such as to produce an image of the object.

C. Optical Transforms

The theory can be easily extended to two-dimensional plane objects, perpendicular to the axis of the lens system. Each point in the object now has to be specified by a vector $\mathbf{r}$, the direction of the incident beam by a vector $\mathbf{s}_0$ and that of the diffracted beam by a vector $\mathbf{s}$. If the diffracting object is represented by the function $f(\mathbf{r})$, we have that the total scattered wave is

$$\psi = \int_{-\infty}^{\infty} f(\mathbf{r}) \exp \frac{2\pi i |\mathbf{r}|}{\lambda} (\cos \alpha - \cos \beta) \, dA.$$

where dA is the element of area around the point $\mathbf{r}$, and α and β are the angles of incidence and scattering respectively. If we give the vectors $\mathbf{s}$ and $\mathbf{s}_0$ moduli of $1/\lambda$, then Eq. (1.4) can be written as

$$\psi = \int_{-\infty}^{\infty} f(\mathbf{r}) \exp 2\pi i (\mathbf{r}.\mathbf{s} - \mathbf{r}.\mathbf{s}_0) \, dA \qquad (1.5)$$

or

$$\psi = \int_{-\infty}^{\infty} f(\mathbf{r}) \exp (2\pi i \mathbf{r}.\mathbf{S}) \, dA \qquad (1.6)$$

where $\mathbf{S} = \mathbf{s} - \mathbf{s}_0$ (Lipson and Taylor, 1958).

This general expression is valid for all angles of incidence, a necessary condition when we come later to consider three-dimensional diffraction. For the moment, however, we need to deal only with normal incidence, and Eq. (1.6) becomes

$$\psi(\mathbf{S}) = \int_{-\infty}^{\infty} f(\mathbf{r}) \exp (2\pi i \mathbf{r}.\mathbf{S}) \, dA \qquad (1.7)$$

which is the *Fourier transform* of the diffracting object $f(\mathbf{r})$. The intensity observed is $|\psi|^2$ or $\psi\psi^*$ (Section I.B). This is the *optical transform*.

D. X-ray Diffraction by Crystals

As we saw in Section I.A, the problem raised in the use of X-rays to examine crystals is that there is no way of producing an image by means of lenses;

only the first stage of the image-forming process can be carried out—the observation of the diffraction pattern. It is therefore necessary to try to deduce the nature of the diffracting object purely from this diffraction pattern.

At first sight, the task would seem to be almost impossible. One can, of course, recognize simple objects, such as a circular or rectangular hole, from their diffraction patterns, or even simple combinations of these, but it is hardly to be expected that one could deduce the arrangements of large numbers of atoms in such an indirect way. In fact, considerable success *has* been achieved in this art, largely because of one basic fact—that most solid matter is crystalline and hence consists of a relatively small group of atoms regularly repeated in three dimensions. The periods of repetition in a crystal can be deduced quite simply from its diffraction pattern, and the basic problem then is to find the number (often quite small) of atoms in the repeating unit, which is called the *unit cell*.

A single crystal behaves as a diffraction grating. Because it is three-dimensional, the conditions for diffraction are more complicated than those for a one-dimensional grating (Lipson and Cochran, 1966), but they can be briefly summarized in the following way. First, each order of diffraction is specified by three integers, in place of the single integer for a one-dimensional grating; these integers are represented by the symbols *hkl*. W. L. Bragg (1913) showed that each order could be regarded as a reflexion from a set of lattice planes (*hkl*) and for this reason the orders of diffraction are commonly called reflexions. This nomenclature is acceptable so long as it is realized that ordinary specular reflexion is not taking place.

Secondly, because of the three-dimensional nature of the diffraction, three conditions, known as Laue's equations (Lipson and Cochran, 1966), have to be obeyed simultaneously. Since there are only two variables, represented by two independent direction cosines giving the direction of the incident beam with respect to the crystal, it is unlikely that any order of diffraction will be produced if a monochromatic beam of X-rays is allowed to fall on a stationary crystal. Thus another degree of freedom must be introduced. It may be a variation in wavelength, resulting in the now-obsolete Laue method which makes use of the white radiation from an X-ray tube. But more commonly it is a rotation, angular oscillation or some other form of angular motion. Details will not be discussed here, but the latest forms of apparatus, introducing also a movement of the film, result in X-ray diffraction photographs in which indices *hkl* can be assigned to each spot quickly and unambiguously. The most recent apparatus dispenses with photography, and uses electronic means to adjust the crystal and measure the intensities of the various orders of X-ray diffraction.

The ultimate result of any of these methods is to present a complete diffraction pattern of a crystal in the form of the intensity of each possible order of diffraction. From this the atomic arrangement within each unit cell has to be derived.

W. H. Bragg (1915) suggested the basis of a general method of tackling the problem. He pointed out that a crystal, being a periodic structure, could be represented by a three-dimensional Fourier series and that the Fourier coefficients must be somehow related to the order of diffraction. The relationship is, in fact, quite simple: the Fourier coefficients are the *structure factors* of the orders of diffraction, that is, complex quantities whose moduli are the amplitudes of the orders of diffraction and whose phases are the relative phases of the waves. If these two quantities could be experimentally determined for each order of diffraction, the Fourier series could be summed and a mathematical picture of the complete crystal structure would result. The equation involved is

$$\rho(x, y, z) = \sum \sum \sum |F(h, k, l)| \cos \{2\pi(hx + ky + lz) - \alpha\} \qquad (1.8)$$

where $\rho(x, y, z)$ is the electron density at the point (z, y, z) in the unit cell, $|F(h, k, l)|$ is the amplitude of the order hkl, and α is the relative phase angle.

These phase angles cannot however be measured experimentally, and thus essential information is lacking. This defect is known as the *phase problem*. Considerable ingenuity has been devoted to overcoming this problem, the various methods being described in books such as that by Lipson and Cochran (1966), and Woolfson (1970) and great success has been achieved. But no complete solution has become available and each new crystal-structure problem represents a new challenge to those who devote their attention to it.

It would be inappropriate to devote any space in this book to a description of these methods. The importance of the Fourier method is simply that it is the second part of the Zernike image-forming process (Section I.B); observation of the diffraction pattern is the first part and the summation of the Fourier series is the second—the production of the image. From this point of view then, we must regard an optical instrument as a device for performing Fourier summations—a so-called Fourier transformer. It can carry out this process more quickly and in far greater detail—since it can produce a continuous function—than any digital computer can do.

However, it can operate only in two dimensions. For ordinary vision this limitation is not important; the retina is two-dimensional and thus can accept only two-dimensional representations of three-dimensional objects. But for crystal structures, it would appear that limitation would nullify the optical approach. In fact, it is not difficult to reduce crystal-structure determination to two dimensions and until relatively recently nearly all researches started in this way; W. L. Bragg (1929) verified his father's Fourier theory in two dimensions only using data from the crystal diopside, $CaMg(SiO_3)_2$. The principle of the idea is that one deals only with central sections of a three-dimensional diffraction pattern; this corresponds to the *projection* of the structure on to a plane parallel to the section chosen. If three principal sections are chosen, to give projections on to the faces of the unit cell, the complete crystal structure in three dimensions usually follows directly.

E. Ewald Sphere and Reciprocal Space

Because of the complexity of three-dimensional diffraction, one cannot easily find which orders of diffraction, if any, will be produced with a crystal in a given orientation with respect to the incident X-ray beam, nor the directions in which such orders will emerge. As the orientation of the crystal is changed, different orders of diffraction flash out and there is no logical sequence in which they follow each other. This problem was neatly solved by Ewald (1921) using a construction based upon what is called the *reciprocal lattice* of the crystal.

The reciprocal lattice is a set of points each of which represents an order of diffraction. The axes of this lattice are each perpendicular to two of the axes of the crystal lattice (for example, the reciprocal axis OA^* is perpendicular to the plane containing OB and OC, the crystal axes), the reciprocal-lattice translations a^*, b^* and c^* being given by the relations

$$a^* = \lambda/d_{100}, \quad b^* = \lambda/d_{010} \quad \text{and} \quad c^* = \lambda/d_{001},$$

where the d's are the spacings of the planes indicated. As we shall show later (Section I.H), the reciprocal lattice can be regarded as the diffraction pattern of the crystal lattice.

Ewald's construction for indicating diffraction conditions can most easily be shown in two dimensions. In a section of the reciprocal lattice (Fig. 4), we draw a line QO passing through the origin O in the direction of the incident beam. On this line as diameter we draw a circle with radius $1/\lambda$, having its centre at C. Then, if the circle passes through a reciprocal-lattice point, such as P, a diffracted beam will appear in the direction CP.

In three dimensions the circle becomes a sphere, known as the *sphere of reflexion* (if one is an X-ray crystallographer), the *sphere of observation* (if one is an X-ray physicist) or the *Ewald sphere* (if one's approach is historical). The chance that the surface of the Ewald sphere will pass through an ideal point is, of course, zero, and this is the reason why, to explore a finite part of the reciprocal lattice, the crystal must be made to pass through a finite range of orientation (Section I.D).

The reciprocal lattice exists rigorously only for a perfect infinite crystal. For a finite crystal the points become extended regions in space, and for an imperfect crystal detail appears *between* the reciprocal-lattice points. Then the whole of the volume becomes important and is known as *reciprocal space*. The Ewald-sphere construction is still valid for giving the surface in reciprocal space that can be explored by a parallel beam of monochromatic radiation. Because of the relationship between diffraction and Fourier transformation (Section III.B) reciprocal space is sometimes known as *Fourier space*.

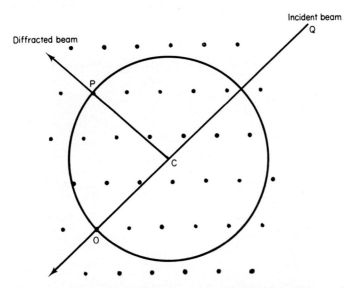

Fig. 4. Ewald construction showing when an order of diffraction will be produced. The spots represent reciprocal-lattice points and the circle represents a section of the Ewald sphere, based upon the incident ray QO; when this sphere passes through a point such as P, a diffracted ray is produced in the direction CP.

F. X-ray Microscope

W. L. Bragg (1939) followed up the approach described in Section I.D by producing an image of one projection of the diopside structure. For this he needed the set of orders of diffraction with one index zero, and he chose the $h0l$'s. He had a plate drilled with holes whose areas were proportional to the modulus of the structure factor for a particular order, and whose separation was inversely proportional to the spacing of the lattice planes, the direction of the separation being perpendicular to those planes. Two such holes would give a diffraction pattern consisting of Young's fringes, which would correspond to a Fourier component of the structure. From a complete set of holes, the diffraction pattern would be a complete representation of the structure, if the phases were also appropriately chosen. This was the difficulty, which will be discussed in more detail in Chapter 6. Bragg chose the $h0l$ section of the diffraction pattern because the positions of the Ca and Mg atoms were known from symmetry, and these would almost certainly cause the orders of diffraction to be all in the same phase. In other words, the phase problem (Section I.D) did not exist, and the diffraction pattern of the set of holes (Fig. 5) showed clearly the arrangement of the atoms projected on to the (010) face of the unit cell.

Bragg called his apparatus, which will be described in the next section, an X-ray microscope because it follows exactly the two operations by which a microscope produces an image. It is not however a true microscope in that

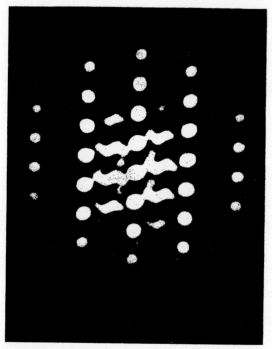

Fig. 5. Optical synthesis of structure of diopside, $CaMg(SiO_3)_2$, projected on (010), produced by diffraction from a reciprocal-lattice mask (Bragg, 1939).

the image is not produced in the same radiation that was diffracted by the specimen, and Buerger (1950) therefore called it a two-wavelength X-ray microscope. The development of this device and its application to more general problems will be discussed in Chapter 6.

G. Fly's Eye

The X-ray microscope, despite its importance in illustrating the basic principles of the subject, has not proved to be of great use in research. What would have been of more use would have been a device for producing quickly a representation of a central section of the diffraction pattern of a possible arrangement of atoms in a crystal; this would then be compared with the observed X-ray diffraction pattern to see whether the two bore any similarity to each other. A proposed arrangement of atoms could in this way be tested extremely quickly.

Bragg first suggested this idea; he used a pin-hole camera that had a two-dimensional array of pin-holes on a square lattice. A small source of light would then produce, on the plate of the camera, a similar array of images

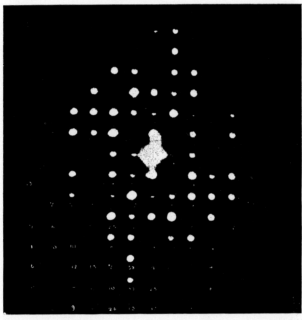

Fig. 6. Diffraction pattern equivalent to a reciprocal-lattice section of sodium benzyl-penicillin, produced by fly's-eye diffraction grating. The numbers represent observed structure amplitudes (Crowfoot *et al.*, 1949).

approximating to points. The eye of the fly has a set of regularly arrayed lenses and so produces a set of images on its retina; it was this resemblance that gave the device its name. If the lamp were moved to a new position, and a second exposure made on the same plate, a regular arrangement of pairs of spots would be produced. By moving the lamp to several positions, a two-dimensional diffraction grating with any chosen fine structure could be produced. A projection of a crystal structure could be represented and by giving different exposures different atoms could be simulated.

Bunn made use of this device in solving the structure of penicillin (Crowfoot *et al.*, 1949). One of his attempts is shown in Fig. 6. The structure was not found solely by these methods, but it provided evidence complementary to that given by other evidence and played a vital part in the success of the research.

Because of this success, other workers attempted to make the device more precise, by making arrays of lenses and by trying to simulate more accurately the scattering of different atoms; a darker patch is not necessarily the best representation of a heavier atom. But these methods did not develop into any important procedures, despite their elegance and simplicity.

H. Optical Transform

One of the reasons why the methods described in the last section proved sterile was that a new and better idea had developed. It is not necessary to produce a repetitive pattern in order to find the intensities in a zone of X-ray reflexions; a single unit gives all the information that is required. This may be seen in two dimensions by making a diffraction grating with a fine structure. In Fig. 7 the unit is an irregular set of six holes; these give the diffraction pattern shown in (a). If we put another similar unit parallel to it, the diffraction pattern (b) is similar but is crossed by Young's fringes; four units at the corners of a unit cell (c) give crossed fringes; a horizontal row gives finer fringes (d) and two such rows give crossed fringes again. A large number of rows, representing a two-dimensional diffraction grating, gives a set of sharp spots (f).

These sharp spots represent the diffraction pattern of the lattice of the diffraction grating; they lie on the reciprocal lattice (Section I.E) which will be dealt with later, in Chapters 3 and 10. The positions of the reciprocal-lattice points are clearly dependent only upon the unit-cell dimensions of the grating, but the intensities depend upon the structure of the unit of the grating. Thus, where the diffraction pattern (a) is weak, the resulting spots in the pattern (f) are also weak, and where (a) is strong, the spots in (f) are also strong. Thus, in order to find a set of intensities of orders of diffraction from a given grating, we need two separate pieces of information; these are the diffraction pattern of the unit—the optical transform (Section I.C)—and the relative positions of the orders of diffraction, or the reciprocal lattice. The second piece of information can be presented in the form of a lattice of points. If this is then superposed on the optical transform, the intensities of the various orders of diffraction can be read off directly.

In addition to being simpler to produce than the fly's eye diffraction pattern, the optical transform is more versatile. It is easy to see the effects of small changes of scale, or of changes of orientation of the diffracting unit. Basically, the extra information is contained in the space between the reciprocal-lattice points, and this is deliberately removed when devices such as the fly's eye are

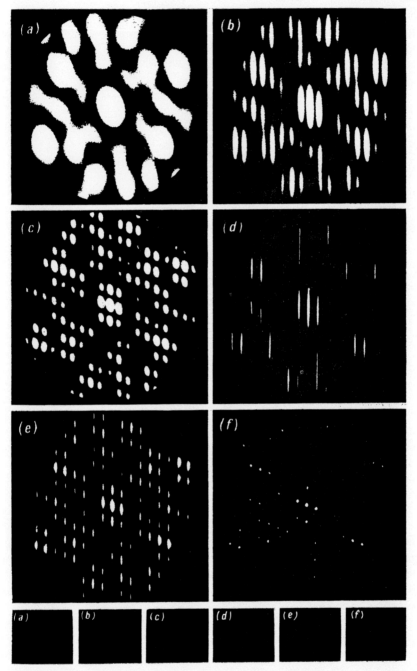

Fig. 7. Relationship between diffraction patterns of (a) single set of six holes; (b) two parallel sets; (c) four sets on corners of parallelogram; (d) row of sets based upon (b); (e) two parallel rows based upon (d); (f) extended lattice based upon (c).

used. Thus the optical-transform method has proved to be of much greater value than the other methods described, and has displaced them completely. The method as an aid to crystal-structure determination will be described in Chapter 3.

II. FRAUNHOFER DIFFRACTION

A. Experimental Conditions

It is implicit in the theory outlined in Section I.B that the object is illuminated in a plane parallel beam of light and that each point in the diffraction pattern corresponds to one direction in space. These conditions can be closely simu-lated by using a lens to produce a plane parallel beam from a point source

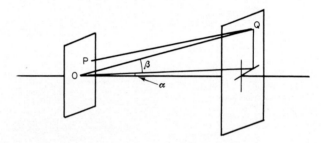

Fig. 8. Path difference between waves scattered from two points O and P in a plane object, interfering at a point Q with angular coordinates α and β.

and another lens to bring the beam to a focus; any point in the focal plane of the second lens then corresponds to the point of convergence of a plane wave travelling in some specific direction. The distribution of light in the focal plane is then the distribution of intensity in the optical transform of any object placed between the two lenses. The limitations of the apparatus are two-fold: first, the point source must be finite and therefore the beam incident on the diffracting object cannot be quite parallel; secondly, it follows that each point in the diffraction pattern cannot correspond to precisely one direction. In any real piece of equipment, therefore, it is necessary to gauge the size of these effects and to limit them to manageable proportions; methods will be described in Chapter 2.

Although the apparatus is quite simple, in order to obtain the best results from it several quite stringent conditions must be obeyed; these will be described in Section IV. Nevertheless, quite useful results can be obtained with relatively crude apparatus, and anyone who wishes to try out transform methods before deciding whether to indulge in more elaborate apparatus is advised to see what can be accomplished with small uncorrected lenses on an ordinary optical bench.

B. Theory

The theory of Fraunhofer diffraction is fully worked out in most textbooks on optics, such as that by Lipson and Lipson (1969). Here we shall give a simplified version which will be adequate for the purposes of this book.

Let the diffracting object be planar (Fig. 8) and illuminated by a plane wave travelling perpendicularly to it. Let a point P have coordinates (x,y). We require to find how the waves from all points P in the object add together in some other parallel plane at a large distance l away. Let Q be a point on such a plane. The coordinates of Q are $l\alpha$, $l\beta$, where α and β are the angular coordinates of the line OQ. Then the path difference $OQ\text{-}PQ$ can easily be seen to be equal to $2l(\alpha x + \beta y)$, since α and β are small. Thus the diffraction function ψ (Eq. 1.7) is

$$\psi(\alpha, \beta) = \int_{-\infty}^{\infty} f(x, y) \exp\{-ik(\alpha x + \beta y)\}\, dx\, dy. \qquad (1.9)$$

From this equation it is possible to work out the diffraction patterns of various objects. In particular, if the object is an opening, $f(x)$ is constant and we merely require to integrate the expression

$$\psi(\alpha, \beta) = \int \exp\{-ik(\alpha x + \beta y)\}\, dx\, dy \qquad (1.10)$$

over the area of the opening.

C. Some Examples

Only one opening—the rectangle—can, however, be treated purely analytically, simply because it is the only shape for which the two dimensions can be separately treated. If the sides of the rectangle are a and b, Eq. (1.10) becomes

$$\psi(\alpha, \beta) = \int_{-a/2}^{a/2} \exp(-ik\alpha x)\, dx \int_{-b/2}^{b/2} \exp(-ik\beta y)\, dy. \qquad (1.11)$$

This can easily be shown to become

$$\psi(\alpha, \beta) = ab\, \frac{\sin \tfrac{1}{2}k\alpha a}{\tfrac{1}{2}k\alpha a} \cdot \frac{\sin \tfrac{1}{2}k\beta b}{\tfrac{1}{2}k\beta b}. \qquad (1.12)$$

This expression has several important properties. First, it shows that the amplitude at the centre of the diffraction pattern is proportional to the area of the aperture—a result which is true for any uniformly illuminated aperture.

The intensity is thus proportional to the square of the area. Secondly, it shows that the diffraction pattern has a centre peak whose general shape is inversely related to the shape of the aperture, being more extended along the direction of the shorter side. Parallel to the edges of the aperture are sets of successively

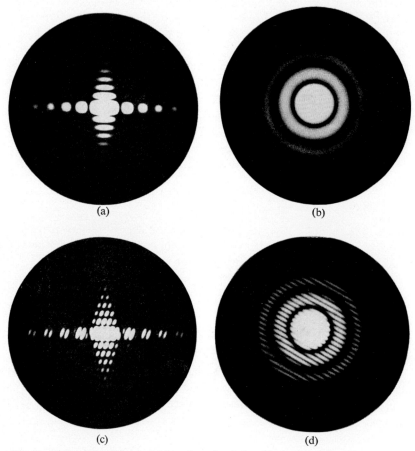

(a)   (b)

(c)   (d)

Fig. 9. Diffraction pattern of (a) rectangular hole; (b) circular hole; (c) two parallel rectangular holes (a); (d) two circular holes (b).

weaker peaks. All these results are shown in Fig. 9(a). There are other peaks in the diffraction pattern, but they are much weaker and are very difficult to observe.

The diffraction pattern of a circular hole is far more important than that of a rectangular hole, since it provides a basis for the theory of optical instruments. The full theory (e.g. Lipson and Lipson, 1969) is rather complicated, but the result can be expressed in terms of Bessel functions. The diffraction pattern is shown in Fig. 9(b); it contains a central peak, known as the Airy disc, surrounded by successively weaker rings.

If two similar apertures are placed parallel to each other, the diffraction pattern is the same as that of each separately, crossed by Young's fringes; the fringes are perpendicular to the separation of the apertures, and have spacings inversely proportional to the value of the separation. These results are illustrated in Figs. 9(c) and (d), which are the diffraction patterns of two rectangular and circular holes respectively; the latter holes are closer together, and thus show coarser fringes.

The main result that emerges from a general study of diffraction patterns is that diffraction detail is reciprocal to aperture detail. The bigger the aperture, the smaller the diffraction pattern; this is basically the reason for the high resolving power of large telescopes. The larger the separation of two apertures, the finer the detail that they produce. For this reason, the diffraction pattern of an aperture in real space is said to exist in reciprocal space (Section I.E). In particular, a crystal, which has a lattice in real space, has a reciprocal lattice (Section I.E) in diffraction space. The larger the unit cell, the finer the reciprocal lattice, but because atoms are all roughly the same size the extent of the total diffraction patterns of crystals does not vary greatly.

III. Image Formation

A. Transform as Fourier Analysis

It is worthwhile looking at the process of diffraction, and its relationship to Fourier transformation, in a more physical way than that described in Section I.C. In the same way that a mathematical function can be represented as the sum of an infinite number of sinusoidal curves, so we can regard a physical aperture as so composed: in the opaque parts the curves add to zero; in the transparent parts they add to produce the amplitudes of the waves transmitted. The process of diffraction can be considered as analysing the object into these components. A plane wave falls on the aperture and, at each angle of diffraction, there is interaction with only one of the components. If the component is large, a relatively high intensity will be diffracted; if it is small, the diffracted intensity will be smaller. Thus the Fraunhofer diffraction pattern can be regarded as an analysis of the object into its periodicities.

If the object is itself periodic—such as a diffraction grating—then it contains only a specific set of components, which we can regard as the fundamental and its harmonics. Thus only discrete diffractions occur—the orders of diffraction. If the diffraction grating were a sine curve, it would give only the two first orders of diffraction.

In two dimensions, the Fourier analysis covers direction as well as intensity. The presence of a strong peak at any particular point indicates a strong periodicity in the object in the direction indicated by the vector joining the strong peak to the centre of the pattern; the spacing of this periodicity is inversely related to the magnitude of this vector.

In theory, all diffraction patterns extend to infinity. In practice they decrease in intensity as the angle of diffraction increases, the actual range of observation depending upon the fineness of detail present in the object.

B. *Image as a Fourier Synthesis*

We have seen in Section I.B that an image is produced when the waves diffracted from the object are caused to interfere. This concept is easier to envisage if the object is a diffraction grating, since we then have only specific orders, each pair of which gives a set of Young's fringes; the addition of these fringes gives an interference pattern that resembles the object. The resemblance cannot be perfect because for perfection an infinite series is required, and only a limited set of interferences can be introduced into an optical system. The effects caused by this limitation will be discussed in Chapter 10.

A further limitation occurs because the intensity of a wave depends upon the square of the amplitude. Suppose that the optical instrument is capable of dealing only with a ratio of 100:1 in intensity; then waves with less than one-tenth of the maximum will not be included in the interference pattern. This is a serious limitation, which is known as the *termination of series* error in crystallography, since the higher orders are generally the weaker ones; the limitation, however, applies to all weak orders of diffraction.

Experimental methods of exploiting this object-image relationship, particularly concerning its application to crystal structures, will be described in Chapter 6.

C. *Relation to Information Theory*

The optical transform gives complete information about the object producing it. Again we may take the diffraction grating as an example. The zero order gives the information that the grating exists and its width indicates the total length of the grating. The first orders give the periodicity, but tell nothing about the nature of the diffracting elements. The succeeding orders give more and more information about the precise form of these elements, and the more orders that are included the more accurate will this form be defined. In other words, the transform gives the spectrum of *spatial frequencies* in the object.

These ideas apply also to a continuous transform. Provided that there are no phase changes, such transforms contain a strong peak at the centre, and the shape of this is inversely related to the general shape of the diffracting object. This effect can be seen in the diffraction pattern of the rectangular hole shown in Fig. 11(a), but it also applies to more complicated arrangements; for example, if a set of holes covers approximately a rectangular area, its transform will also have a central peak of the same dimensions as that of the rectangle. Moreover, other peaks in the transform will also have this same general shape. The intermediate parts of the transform give information about the separate components of diffracting object, and the outer parts give precise information about the nature of these components, subject to the limitations described in the previous section.

If the transform is altered in some way, its interference pattern will no longer be a true image of the object. It may therefore seem that there can be no reason for carrying out this operation. In fact, there is; it may be worthwhile sacrificing complete veracity in order to bring out some desired detail. For example, if an object consists of a perfectly transparent film of varying thickness no image will be visible; yet it will give a diffraction pattern because the waves transmitted through various parts will have different phases. The interference pattern will have constant intensity but varying phase, but it is possible to alter the transform in such a way that these phase variations become evident as variations in intensity. This method, introduced by Zernike, is called *phase-contrast*, and is described in more detail in Chapters 8 and 10.

Other devices, to be dealt with in these chapters, are also possible and are used in practice. It must always be remembered, however, that they must alter the image in some way that may be more or less important. False detail may be introduced and anyone using these methods must be on guard against interpreting unusual detail as being of physical significance unless it can be checked in some independent way.

IV. APPARATUS

A. Choice of Dimensions

As described by Taylor and Lipson (1964) there are several ways of producing optical transforms experimentally. They fall into two main classes. One uses very small objects and minimal optical equipment; the other makes use of objects a few centimetres across, and uses much more complicated optical apparatus. There is much to be said for the principle of concentrating one's efforts on the non-repetitive part of one's work, and therefore in this chapter we shall consider only a device, the optical diffracto-meter (Fig. 10), which needs great care in construction but which can be used to produce transforms of easily-made objects.

2

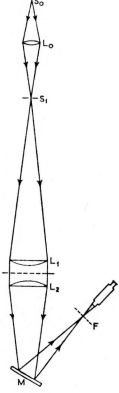

Fig. 10. Optical diffractometer. Light from the pin-hole S_1 is made parallel by the lens L_1 and is brought to a focus to F by means of the lens L_2 after reflexion at the plane mirror M. If an object is placed between the lenses, its diffraction pattern (optical transform) is seen centred on F.

Let us suppose that our object is several centimetres across, and will therefore give extremely small diffraction angles. A rectangle of side 50 mm would have its first zero (Fig. 9a) at an angle given by

$$\alpha = 5 \times 10^{-4} \text{ mm/50 mm}$$
$$= 10^{-5} \text{ rad.}$$

At a distance of 1 m, this angle would produce a separation of only 10^{-2} mm. This is extremely small, but is nevertheless easily visible with a good microscope, and can be recorded on fine-grain film. Thus lenses of about 1 m focal length are needed. For smaller diffracting objects, lenses of proportionately shorter focal length can be used, and thus simple experiments can be carried out on an ordinary optical bench. But for general work it is best to have an instrument that will cope with the most exacting requirements.

The pin-hole (Fig. 10) must clearly be at least as small as the detail to be observed if the collimating lens has the same focal length as the observing lens. There is no *a priori* reason why the focal lengths should be equal, but it is obviously convenient to order two similar lenses—astronomical-telescope objectives—when the optical diffractometer is being constructed.

B. Perfection of Lenses

Although much useful work can be carried out with ordinary lenses (Section II.A)—particularly with plano-convex lenses set for minimum spherical aberration—lenses corrected for spherical aberration must be used for transforms of the highest quality. These will normally be contact doublets, corrected also for chromatic aberration; this correction, however, is not needed since monochromatic light is to be used. The lenses should also have as few blemishes as possible; the high-density glass needed cannot apparently be made without some bubbles and inclusions of foreign material. For telescopes, these will reduce the aperture slightly but, for the diffractometer, if they obstruct a beam passing through a hole in a diffracting mask they can alter the diffraction pattern considerably. Absence of any such effect can always be tested by noting, first, that each diffraction pattern is centro-symmetric and, secondly, that it does not change as the mask is moved in its own plane.

Good lenses need good treatment. They should be supported in a framework that does not vary with temperature; for this reason the framework should be made of a single material, such as steel, so that there will be no differential thermal expansions. The framework should be supported by means of vibration-free mounts.

The adjustment of the lenses should be as accurate as possible; they should be coaxial and the pin-hole should be at the focus of the collimating lens. To the accuracy required, these requirements are not easy to satisfy; a procedure has been worked out by Taylor and Thompson (1958) and is described by Taylor and Lipson (1964). It depends upon the reflexions from the surfaces of the lenses, and takes about 1–2 hours to carry out completely. The method provides its own check. If a large pin-hole, about 1 mm, is used, several images are produced by reflexions from the six surfaces of the two doublets; in proper adjustment these can be seen to lie on a straight line, which is the axis. There will be some scatter, since no lenses are perfect, but this scatter should be quite small.

For visual work, all that is necessary is to focus the observing microscope (Fig. 10) so that the image of the pin-hole is as small as possible. [Ideally, one should also see the Airy-disc pattern (Section II.C) of the aperture of the lenses, but few lenses are good enough for this; the pattern observed should, of course, have circular symmetry]. For photographic work the focal plane of the observing lens should be located to an accuracy of the order of 0·01 mm. This rather surprising requirement can be shown to be necessary because, with lenses subtending an angle of about 4°, errors greater than this introduce unacceptable optical path differences.

C. *Illuminating Systems*

It is required that as much light as possible should be passed through the instrument. Because the light has to pass through a small pin-hole, and then through a diffracting mask of possibly only a small area, a powerful source is needed. A 250-watt compact-source mercury arc lamp has proved very suitable; it has a small arc and the light is concentrated in a few lines of which one, usually the yellow, can be extracted by means of a filter placed above the pin-hole (Fig. 10). In principle, the lamp should be placed as near as possible to the pin-hole and no condenser should be needed, but, because the lamp housing takes up some space, it is necessary to use a condensing lens to focus the arc on the pin-hole.

This leads us to consider the coherence properties of the illumination. We may, for example, have the Airy disc of the condenser falling on the pin hole, so that it is coherently illuminated; however this illumination would be very weak. Alternatively, we may try to form a perfect image of the arc on the pin-hole so that each part is separately illuminated and there is no coherence at all. This is the system that is aimed at in practice, although complete incoherence is impossible because of diffraction and lens aberrations. These matters are considered in more detail in Chapter 2.

The advent of the laser has, however, changed these considerations; it is now possible to have a very bright source which illuminates the pin-hole coherently. This extends considerably the scope of the optical diffractometer. Transforms can be projected on to screens and shown to groups of students instead of to individuals; diffraction patterns of small masks, previously requiring long exposures, can now be photographed quite quickly, and the adjustment of the diffractometer becomes easier because the laser beam can be seen. Apart from cost, the laser has every advantage over the compact-source lamp.

V. KINEMATIC AND DYNAMIC THEORIES OF DIFFRACTION

The theory of diffraction set out in Section I.D is essentially an over-simplification: the scattered orders of diffraction do not simply leave the crystal; they interact with it in the same way as the incident beam does. The fact that the theory has nevertheless been extremely successful is a result of the weakness of interaction of X-rays and most crystals; multiply-diffracted beams are so weak that, except in special circumstances, they can be neglected.

The theory of interaction of waves and crystals that ignores all but the primary interaction is called the *kinematic theory*. For particularly strong reflexions it breaks down and produces an effect known as *extinction*, which causes the intensities of such reflexions to be weaker, sometimes several times weaker, than kinematic theory would indicate. Various experimental and empirical procedures exist for correcting for extinction, and the effect is not a serious hindrance to producing accurate representations of crystal structures.

For electrons, however, the effect is much greater. The interaction of electrons with crystals is very strong, as evidenced by the fact that electron-diffraction patterns can be *seen* on a fluorescent screen. For the interpretation of electron-diffraction phenomena therefore, we require that the complete theory of interaction of electron waves with crystals should be understood; this is called *dynamic theory*. Although the general outlines of this theory are known (James, 1957) its application in practice is very difficult; for this reason the development of our understanding of electron-diffraction patterns lags well behind that of X-ray diffraction patterns.

For this reason also imaging in the electron microscope is to some extent suspect. The orders of diffraction from a crystalline specimen are not as simply related to the Fourier coefficients of the crystal as they are for X-rays, and therefore the interference pattern that they produce is not an exact image of the crystal. This point is discussed more fully in Chapter 10 (Section VIII.B).

REFERENCES

Abbe, E. (1873). *Archiv. f. Mikroskop. Anat.*, **9**, 413.

Bragg, W. H. (1915). *Phil. Trans. R. Soc. A*, **215**, 253.

Bragg, W. L. (1913). *Proc. Camb. phil. Soc.*, **17**, 43.

Bragg, W. L. (1929). *Proc. R. Soc. A*, **123**, 537.

Bragg, W. L. (1939). *Nature, Lond.*, **143**, 678.

Buerger, M. J. (1950). *J. appl. Phys.*, **21**, 909.

Crowfoot, D., Bunn, C. W., Rogers-Low, B. W. and Turner-Jones, A. (1949). "The X-ray Crystallographic Investigation of the Structure of Penicillin," Oxford University Press, Oxford.

Davisson, C. J. and Germer, L. H. (1927). *Phys. Rev.*, **30**, 715.

Ewald, P. P. (1921). *Z. Kryst.*, **56**, 129.

James, R. W. (1957). "Optical Principles of the Diffraction of X-rays," Bell, London.

Lipson, H. and Cochran, W. (1966). "The Determination of Crystal Structures," Bell, London.

Lipson, S. G. and Lipson, H. (1969). "Optical Physics," Cambridge University Press, Cambridge.

Taylor, C. A. and Lipson, H. (1964). "Optical Transforms," Bell, London.

Taylor, C. A. and Thompson, B. J. (1958). *J. Sci. Instr.*, **34**, 439.

Thompson, G. P. (1927). *Nature, Lond.*, **119**, 890.

Woolfson, M. M. (1970). "An Introduction to X-ray Crystallography," Cambridge University Press, Cambridge.

Zernike, F. (1946). "La Théorie des Images Optiques," CNRS, Paris.

CHAPTER 2

Coherence Requirements

B. J. Thompson

*The Institute of Optics, University of Rochester,
College of Engineering and Applied Science,
Rochester, New York, U.S.A.*

I. Introduction

This book deals largely with coherent light phenomena. The optical transform is the classical Fraunhofer diffraction pattern and, of course, diffraction and interference are coherent light phenomena; optical data processing is a coherent process of image formation by double diffraction; holography is a two-step coherent image-forming process. Hence it is not surprising that we have to ask the question, how coherent is the light? It was to answer this question (or the equivalent one, how incoherent is incoherent light?) that the whole theory of partially coherent light was developed. Verdet (1865) realized that even sunlight was not strictly incoherent. Strong correlation exists in the optical field produced by the sun over intervals on the order of 0·05 mm. The eye, of course, operating unaided in sunlight, does not resolve this distance and hence can be considered to be receiving an incoherent field. Fizeau and Michelson were also aware that the optical field produced by a star was not completely uncorrelated and designed interferometers to measure the star diameters from a measurement of the correlation. These early workers did not think in terms of the correlations in the light field but derived these results by an integration over the source. As we shall see, Michelson's (1890) concept of "visibility" is closely related to today's concept of "degree of coherence". Studies of the correlation existing in optical fields and the concepts of partially coherent light occupied the efforts of many researchers including such prominent persons as Von Laue, Berek, van Cittert and Zernike. Hopkins (1951, 1953) interpreted many of these earlier results and applied them to study of image formation with partially coherent light. The theory was further generalized by Blanc-Lapierre and Dumontet (1955) and by Wolf (1955). Since then the work has blossomed into detailed theoretical and experimental discussions. Excellent reviews can be found in a number of texts (Born and Wolf, 1967; Beran and Parrent, 1964; Troup, 1968). Recent reviews by Mandel and Wolf (1965) and Hopkins (1969) are also valuable reading.

A. *The Measurement of Intensity*

All detectors that are available to us—the eye, photocells, film, etc.—are responsive to the quantity we call intensity. Furthermore, we are familiar with the idea of linear superposition of intensities; this is the way that our eye works, since it is normally dealing with incoherent light. Hence we are quite used to handling optical systems that are linear in intensity. The quantity we call intensity is the long-time average of the square modulus of the optical field.

Let us define an optical field $V(x,t)$, where V is a complex quantity and a function of both a spatial coordinate x and a time coordinate t.

$$I(x,t) = V(x,t) V^*(x,t), \tag{2.1}$$

where the star denotes a complex conjugate. The measured intensity is then given by

$$I(x) = \langle I(x,t) \rangle = \langle V(x,t) V^*(x,t) \rangle, \tag{2.2}$$

where the sharp brackets denote a time average defined by

$$I(x) = \langle I(x,t) \rangle = \lim_{T \to \infty} \frac{1}{2T} \int_{-T}^{T} I(x,t)\, dt. \tag{2.3}$$

B. Addition of Optical Fields

We may often have to add two beams of light where the measured intensities of the individual beams are defined as

$$\begin{aligned} I_1(x) &= \langle V_1(x,t)\, V_1^*(x,t) \rangle, \\ I_2(x) &= \langle V_2(x,t)\, V_2^*(x,t) \rangle. \end{aligned} \tag{2.4}$$

1. Incoherent Superposition

If the two beams are incoherent with respect to each other then the intensities simply add and the resultant intensity $I_R(x)$ is given by

$$I_R(x) = I_1(x) + I_2(x), \tag{2.5}$$

and the system is linear in the intensity.

2. Coherent Superposition

Normally this is discussed as a part of interference and the statement is made that we must add the complex amplitudes of the two fields. The monochromatic separation of the space and time dependency is then invoked and the complex field $V(x,t)$ is written as

$$V(x,t) = \Psi(x) \exp(-2\pi i \nu t), \tag{2.6}$$

where $\Psi(x)$ is a complex amplitude dependent only upon the spatial variable x and the time-dependent part is separated in the exponential; ν is the frequency of the light. Thus when two optical fields that are coherent with respect to each other are added, the resultant complex amplitude, $\Psi_R(x)$, is determined by a summation of the individual amplitudes. Hence,

$$\Psi_R(x) = \Psi_1(x) + \Psi_2(x), \tag{2.7}$$

Finally the resultant intensity is given by

$$\begin{aligned} I_R(x) &= \Psi_R(x)\, \Psi_R^*(x) = [\Psi_1(x) + \Psi_2(x)]\,[\Psi_1(x) + \Psi_2(x)]^*, \\ &= I_1(x) + I_2(x) + 2\sqrt{I_1(x)\, I_2(x)}\, \cos \delta, \end{aligned} \tag{2.8}$$

where $I_1(x) = \Psi_1(x)\,\Psi_1^*(x)$, $I_2(x) = \Psi_2(x)\,\Psi_2^*(x)$ and δ is the phase difference between the two beams.

3. General Case

It is instructive to return to the general situation and determine the instantaneous optical field $V_R(x,t)$ given by

$$V_R(x,t) = V_1(x,t) + V_2(x,t). \tag{2.9}$$

The instantaneous intensity is then

$$I_R(x,t) = V_R(x,t) V_R^*(x,t), \tag{2.10}$$

and the intensity is

$$I_R(x) = \langle I_R(x,t) \rangle;$$
$$I_R(x) = \langle V_1(x,t) V_1^*(x,t) \rangle + \langle V_2(x,t) V_2^*(x,t) \rangle$$
$$+ \langle V_1(x,t) V_2^*(x,t) \rangle + \langle V_2(x,t) V_1^*(x,t) \rangle. \tag{2.11}$$

The first two time-averaged terms are simply the intensities $I_1(x)$ and $I_2(x)$ (Eq. 2.2). The remaining two terms are cross-correlation functions. If $V_1(x,t)$ and $V_2(x,t)$ both vary randomly with time and randomly with respect to each other then these last two terms in the expression for $I_R(x,t)$ are zero in the time average. Thus we have the incoherent addition of the two intensities according to Eq. (2.5).

If $V_1(x,t)$ and $V_2(x,t)$ vary randomly with time but the random variations are identical in each case then the final two terms become

$$\langle V_1(x,t) V_2^*(x,t) \rangle \rightarrow \Psi_1(x) \Psi_2^*(x)$$
$$\langle V_2(x,t) V_1^*(x,t) \rangle \rightarrow \Psi_2(x) \Psi_1^*(x) \tag{2.12}$$

and the coherent superposition result of Eq. (2.8) applies.

The general situation is, of course, when $V_1(x,t)$ and $V_2(x,t)$ vary randomly with time but with some partial correlation between them. We will need to derive an expression for this general case which will tend to the two limits. However, before going to a derivation and discussion of the general case it will be instructive to discuss the result of some conceptual experiments that will provide insight into the fundamental ideas of partially coherent light and illustrate the need for a careful study.

II. PARTIALLY COHERENT LIGHT

A. Basic Conceptual Experiments

An unknown distant source of finite size produces an optical field in a given plane. This field is examined by two small circular apertures P_1 and P_2, whose separation can be changed, in an otherwise opaque screen. A converging lens of known focal length is located behind this screen and a photographic or photoelectric record of the intensity distribution is obtained in its focal plane. Firstly with one aperture, P_1, alone the intensity distribution observed is that readily associated with the diffraction pattern of a circular aperture.

It can, therefore, be concluded that the amplitude distribution across the aperture P_1 is uniform and furthermore that the radiation across the aperture is essentially coherent.

The second aperture, P_2, gives a similar result, but when the two apertures are opened together, and are at their closest separation, two-beam interference fringes are observed which are formed by the division of the incident wavefront by the two apertures. At this closest separation the fringes are extremely clear (see Fig. 1A). As the separation of the apertures increases, the photographic record of the fringes are as shown in Fig. 1. The fringes almost disappear in D and E only to reappear in F through J, fade again at K, and reappear very faintly at O. Intensity plots corresponding to a typical sample of the photographic records are shown in Fig. 2. As the separation of P_1 and P_2 is increased these results show that:

1. the fringe spacing increases;
2. the intensities of the fringe minima are never zero;
3. the relative height of the maxima above the minima steadily decreases;
4. the absolute height of the maxima decreases and the height of the minima increases;
5. eventually the fringes disappear, at which point the resultant intensity is just twice the intensity observed with one aperture alone;
6. the fringes reappear with increasing separation of P_1 and P_2 but the fringes contain a central minimum not a central maximum.

The results 1–5 can be summarized by defining a quantity called visibility $\mathscr{V}$ (first introduced by Michelson for this very purpose).

$$\mathscr{V} = \frac{I_{max} - I_{min}}{I_{max} + I_{min}}. \qquad (2.13)$$

If this visibility function is plotted against the separation of the apertures P_1 and P_2 for the example given in Fig. 1, a curve similar to that shown in Fig. 3 results. For the closest separation the addition is approximately coherent and hence the actual wave disturbances have to be added as given by Eq. (2.8). For the separation of the two apertures that produced the results of Figs. 1(K) and (N) no resultant fringes are seen and the visibility is zero. The addition of the two beams is essentially incoherent and the resultant intensity is given by an intensity addition of the two independent intensities as in Eq. (2.5).

The same series of tests is carried out with the same apertures but on a different optical field. At first sight the experimental results appear identical but upon closer inspection the intensity distribution has a slightly different

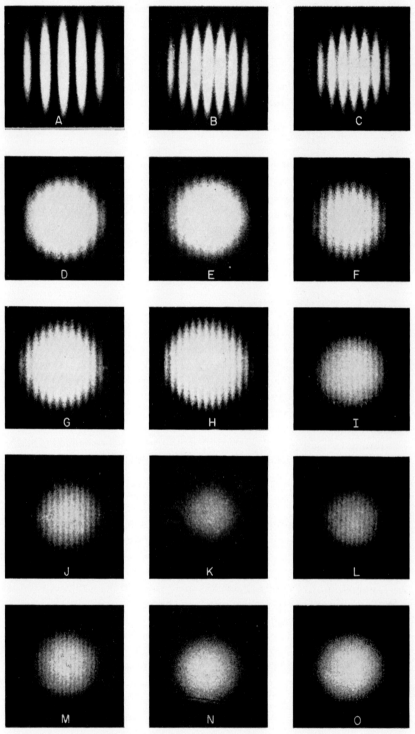

Fig. 1. Photographs of two-beam interference fringes produced by a given optical field, for varying separation of the two sampling apertures.

form. A typical result is shown in Fig. 4; (a) is the resultant intensity distribution for a given separation of the two apertures with this second field and (b) is the equivalent result with the first optical field. If the fringe visibility is computed from Fig. 4(a) the value of $\mathscr{V}$ varies with the position in the fringe field, gradually decreasing as the distance from the central fringe increases.

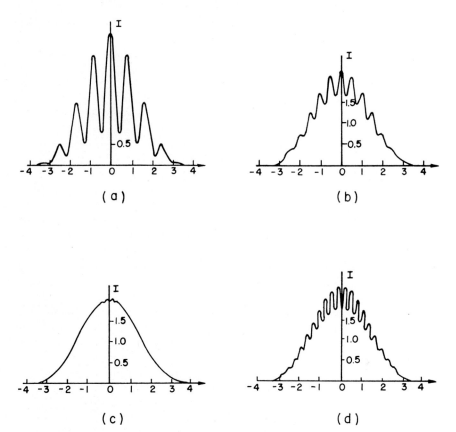

Fig. 2. Typical intensity distributions corresponding to photographic records similar to those shown in Fig. 1.

The fringe visibility as measured at the centre of the fringe field for a range of separations of the aperture at P_1 and P_2 would follow the visibility curve of Fig. 3. Hence we could conclude at this point that the change in visibility as a function of position in a given fringe field was caused by the path difference between the two interfering beams. A measurement of the spectral width of the incident light would show that it had a line width considerably broader than in the first experimental test.

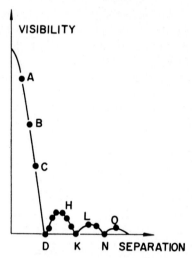

Fig. 3. Visibility, as a function of separation of the two apertures, for the fringes in Fig. 1.

There is an obvious connection between the results just described and the existence of so called white-light fringes in classical interference experiments; this type of fringe is seen only for extremely small path differences and the fringes rapidly disappear as the path difference between the two interfering

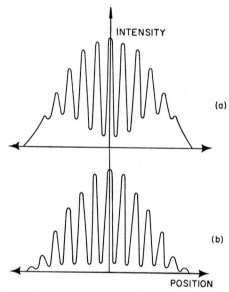

Fig. 4. The intensity distribution in the fringe pattern for two optical fields.

beams increases. This conclusion can be demonstrated by conducting a different test on the optical fields using a Michelson interferometer (Fig. 5).

The incident light is divided by the beam splitter *B*, one beam passing through compensating plate *C* to mirror 1 where it is reflected back along the same path and then passes through the beam splitter. The second beam initially goes through the beam splitter and to mirror 2 and is reflected and again reflected by the beam splitter, emerging parallel to the first beam. If mirrors 1 and 2 are perpendicular, circular fringes are seen when looking in towards the beam splitter from the emergent direction of the beams. If mirror 1 is slowly moved backwards, the fringe visibility as defined by Eq. (2.13) decreases

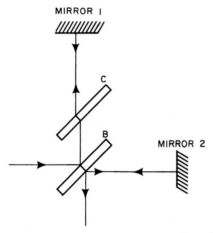

Fig. 5. A Michelson interferometer for determining the effect of optical path difference on fringe visibility.

as a function of the path difference introduced by moving the mirror. For the two optical fields under test, the second field produces fringe that rapidly lose visibility as the path difference is increased whereas the first optical field produces fringes that retain their visibility for much longer path differences.

The usual equations for coherent and incoherent additions of optical field are clearly inadequate for a correct description of the experimental results described here. It might be surmised that they are but particular limits of a more general expression that would describe these results. Hence we need to formulate a general law of addition of optical fields.

B. Two-Beam Interference

There are several methods of deriving the two-beam interference law with partially coherent light; we will use the one first given by Thompson and Wolf (1957) (see also Born and Wolf, 1969).

The light emitted by a finite-sized source S having a narrow but finite spectral width is allowed to fall upon an opaque screen containing two small apertures at points P_1, with coordinate $\mathbf{x}_1$, and P_2 with coordinate $\mathbf{x}_2$ (see Fig. 6). The complex amplitudes at these two apertures are $V(\mathbf{x}_1, t)$ and $V(\mathbf{x}_2, t)$ where these functions each satisfy the wave equation given by

$$\nabla^2 V(\mathbf{x}, t) = \frac{1}{c^2} \frac{\partial V^2(\mathbf{x}, t)}{\partial t^2}, \qquad (2.14)$$

where ∇^2 denotes the Laplacian operator in the coordinate $\mathbf{x}$ and c is the velocity of light.

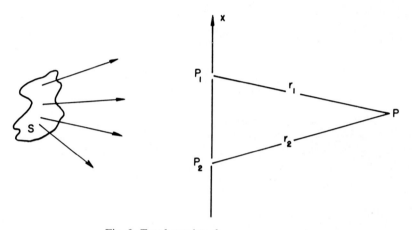

Fig. 6. Two-beam interference arrangement.

We wish to determine the resultant intensity as measured at a point P placed behind the screen. Let r_1 and r_2 be the distances $P_1 P$ and $P_2 P$ respectively; we will restrict our attention to small path differences and we will use the shorthand notation

$$V_s(t) = V(\mathbf{x}_s, t), \qquad s = 1, 2.$$

The points P_1 and P_2 may be regarded as centres of secondary disturbances and hence the complex amplitude at the point P is given by

$$V_p(t) = K_1 V_1(t - t_1) + K_2 V_2(t - t_2), \qquad (2.15)$$

where $t_1 = r_1/c$, $t_2 = r_2/c$ and c is the velocity of light in the medium. The factors K_1 and K_2 depend upon the size of the apertures and the relative position of P with respect to P_1 and P_2. The resultant intensity at P is the observable quantity and is the time-averaged measure given by

$$I_p = \langle V_p(t) V_p^*(t) \rangle \qquad (2.16)$$

the sharp brackets denoting the time average as defined by Eq. (2.3). Hence,

$$I_p = K_1 K_1^* \langle V_1(t - t_1) V_1^*(t - t_1) \rangle + K_2 K_2^* \langle V_2(t - t_2) V_2^*(t - t_2) \rangle$$
$$+ K_1 K_2^* \langle V_1(t - t_1) V_2^*(t - t_2) \rangle + K_2 K_1^* \langle V_2(t - t_2) V_1^*(t - t_1) \rangle. \quad (2.17)$$

For stationary fields the time origin can be shifted and $t_2 - t_1$ set equal to τ. Then Eq. (2.17) becomes

$$I_p = |K_1|^2 I_{p_1} + |K_2|^2 I_{p_2} + 2|K_1 K_2| \, Re \, \Gamma_{12}(\tau), \quad (2.18)$$

where Re denotes a real part and

$$\left. \begin{array}{l} I_{p_1} = \langle V_1(t) V_1^*(t) \rangle, \\ I_{p_2} = \langle V_2(t) V_2^*(t) \rangle, \end{array} \right\} \quad (2.19)$$

$$\Gamma_{12}(\tau) = \langle V_1(t + \tau) V_2^*(t) \rangle, \quad (2.20)$$

I_{p_1} and I_{p_2} are the intensities at the point P_1 and P_2 respectively. Because of the introduction of the complex notation I_{p_1} is actually twice the measurable intensity at the point P_1. With the same comment $|K_1|^2 I_{p_1}$ is the intensity at P due to aperture P_1 alone, which we shall denote by I_1. The term $\Gamma_{12}(\tau)$ is a cross correlation function and is termed the *mutual-coherence function* of the light vibrations at P_1 and P_2. This function can be normalized by setting

$$\gamma_{12}(\tau) = \frac{\Gamma_{12}(\tau)}{\sqrt{\Gamma_{11}(0) \Gamma_{22}(0)}} = \frac{\Gamma_{12}(\tau)}{\sqrt{I_{p_1} I_{p_2}}}, \quad (2.21)$$

where $\gamma_{12}(\tau)$ is called the *complex degree of coherence*. It will be noted that I_{p_1} is an autocorrelation function and may be written as $\Gamma_{11}(0)$ and hence may be called a *self-coherence function*.

Equation (2.18) then becomes

$$I_p = I_1 + I_2 + 2\sqrt{I_1 I_2} \, Re \, \gamma_{12}(\tau). \quad (2.22)$$

If the mean frequency of the illumination is ν and the mean wavelength λ then we may write

$$I_p = I_1 + I_2 + 2\sqrt{I_1 I_2} |\gamma_{12}(\tau)| \cos(\beta_{12}(\tau) + \delta_{12}),$$
$$\beta_{12}(\tau) = 2\pi\nu\tau + \arg\gamma_{12}(\tau),$$
$$\delta_{12} = 2\pi/\lambda(r_2 - r_1). \quad (2.23)$$

It must be noted that $0 \leqslant |\gamma_{12}(\tau)| \leqslant 1$ and that in the coherent limit $|\gamma_{12}(\tau)| = 1$ and in the incoherent limit $|\gamma_{12}(\tau)| = 0$.

Under the condition that the two interfering beams have equal intensity

$$I_p = 2I[1 + |\gamma_{12}(\tau)| \cos(\beta_{12}(\tau) + \delta_{12})].$$

It is interesting to compute the visibility, as defined by Eq. (2.13), of the interference fringe formed under these circumstances. We require I_{max} and I_{min} defined by

$$I_{max} = 2I[1 + |\gamma_{12}(\tau)|],$$
$$I_{min} = 2I[1 - |\gamma_{12}(\tau)|];$$

hence the visibility at the point P in the field is given by

$$\text{Visibility at } P = |\gamma_{12}(\tau)| = \mathcal{V}_p. \tag{2.24}$$

Thus the modulus of the complex degree of coherence is equal to the visibility of the fringes and is hence directly measurable. However, the visibility may vary continuously with τ as indicated in Fig. 4(a).

In general when $I_1 \neq I_2$ we may set $I_1 = I$ and $I_2 = \alpha I$ then

$$I_{max} = I[1 + \alpha + 2\alpha^{1/2}|\gamma_{12}(\tau)|],$$
$$I_{min} = I[1 + \alpha - 2\alpha^{1/2}|\gamma_{12}(\tau)|],$$

and the visibility at P is

$$\mathcal{V}_p = \frac{2\alpha^{1/2}}{1 + \alpha}|\gamma_{12}(\tau)|. \tag{2.25}$$

The visibility is now not a direct measure of the modulus of the complex degree of coherence, which can only be determined if the ratio of $I_1 : I_2$ is known. Clearly when $\alpha = 2.5$ an error of about 10% is made in assuming that the visibility is a measure of the modulus of the complex degree of coherence. However, if I_1 and I_2 are monitored separately then an accurate measure can be made.

C. Concepts of the Theory of Partial Coherence

In the last section many of the basic concepts of the theory of partially coherent light were included as part of the derivation of the fundamental two-beam interference. For clarity they will be summarized here together with their properties.

The key function in the theory is the mutual-coherence function $\Gamma_{12}(\tau)$ defined in general by

$$\Gamma_{12}(\tau) = \Gamma(x_1, x_2, \tau) = \langle V^*(\mathbf{x}_1, t) V(\mathbf{x}_2, t + \tau) \rangle, \tag{2.26}$$

where $V(\mathbf{x}, t)$ is the analytic signal associated with a Cartesian component of the electric field vector. The mutual-coherence function depends upon seven variables—six space variables and the time delay τ. The normalized form of the mutual-coherence function is termed the complex degree of coherence $\gamma_{12}(\tau)$ defined by Eq. (2.21). $\Gamma_{12}(\tau)$ and hence $\gamma_{12}(\tau)$ are both observable quantities and if the mutual-coherence function is known in one plane it can be determined in a second plane since the mutual-coherence function obeys the pair of wave equations:

$$\nabla_s^2 \, \Gamma_{12}(\tau) = \frac{1}{c^2} \frac{\partial^2 \, \Gamma_{12}(\tau)}{\partial \tau^2} \qquad (s = 1, \, 2). \tag{2.27}$$

The quantity normally detected is the intensity which is recovered from the mutual coherence function by setting $x_1 = x_2$ and $\tau = 0$:

$$I(x) = \Gamma(x_1, x_2, 0). \tag{2.28}$$

1. Quasi-Monochromatic Approximation

For many problems in optics, particularly when approximations to the coherent situation are sought, the light may be assumed to be quasi-monochromatic, i.e.,

$$\Delta v \ll v, \tag{2.29}$$

where Δv is the spectral width and v the mean frequency. In many experimental situations the conditions are such that

$$\tau \ll \frac{1}{\Delta v}, \tag{2.30}$$

which is the important consideration for making the quasi-monochromatic approximation. Under these conditions the space-dependent and time-dependent parts can be separated and

$$\Gamma(x_1, x_2, \tau) \approx \Gamma(x_1, x_2, 0) \exp(-2\pi i v \tau). \tag{2.31}$$

$\Gamma(x_1, x_2, 0)$ is called the *mutual-intensity function* and is simply written as Γ_{12}. The wave equations defined by Eq. (2.27) tend in the quasi-monochromatic limit to a pair of Helmholtz equations.†

$$\nabla_s^2 \, \Gamma_{12} + k^2 \, \Gamma_{12} = 0 \qquad (s = 1, \, 2), \tag{2.32}$$

where $k = 2\pi/\lambda$.

The corresponding normalized function $\gamma(x_1, x_2, 0)$ is the complex degree of coherence for quasi-monochromatic light and has no special name.

2. Spatial and Temporal Coherence

It is useful on occasion to separate out the space-dependent and time-dependent parts of the coherence function. *Temporal coherence* effects arise from the finite spectral width of the source radiation. Strictly monochromatic

† In exactly the same way that the function $V(x,t)$ satisfies the single wave equation (2.14) and tends to a Helmholtz equation in the coherent limit $V(x,t) = \Psi(x)\exp(-2\pi i v t)$ and $\nabla^2 \Psi(x) + k^2 \Psi(x) = 0$.

radiation is, of course, always coherent. Hence, we define a *coherence length* Δl and a *coherence time* Δt defined by

$$
\left.
\begin{aligned}
\Delta l &= \frac{c}{\Delta \nu} = \frac{\lambda^2}{\Delta \lambda}, \\[2mm]
\Delta t &= \frac{l}{\Delta \nu}.
\end{aligned}
\right\}
\tag{2.33}
$$

The term *spatial coherence* has been used to describe partial coherence effects arising from the finite size of an incoherent source. Hence for the equipath position for the addition of two beams, we define a coherence interval as the separation of two points such that $|\gamma_{12}(0)|$ is equal to a pre-chosen value. The pre-chosen value has to be specified. $|\gamma_{12}(0)| = 0 \cdot 88$ or $= 0$ are often used.

III. PROPAGATION OF PARTIALLY COHERENT LIGHT

The mutual-coherence function propagates according to two wave equations as shown in Eq. (2.27) and the mutual intensity propagates according to a pair of Helmholtz equations (see Eq. 2.32). Here we are interested in the existence of optical fields and how coherent the fields are. It is important to note that a field is coherent if the mutual intensity describing that field can be put into the product form

$$
\Gamma_{12} = \Psi(x_1) \Psi^*(x_2),
\tag{2.33}
$$

where

$$
\nabla^2 \Psi(x) + k^2 \Psi(x) = 0.
\tag{2.34}
$$

However, an incoherent field cannot exist in free space even though an incoherent source consistent with the above statements can be defined. Very often, however, a field can be considered essentially incoherent under certain

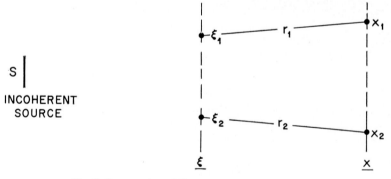

Fig. 7. Propagation of the mutual-intensity function.

experimental circumstances. For example, in image formation, if the degree of coherence in the field that illuminates the object is very much smaller than the impulse-response width of the imaging system as projected onto the object plane, then the image formation is essentially incoherent. A survey of image formation with partially coherent light discusses these effects in detail (Thompson, 1969).

A plane incoherent primary quasi-monochromatic source S illuminates a plane ξ (see Fig. 7). $\mathbf{x}_1$ and $\mathbf{x}_2$ are two points distant r_1 and r_2 from points $\boldsymbol{\xi}_1$ and $\boldsymbol{\xi}_2$, respectively. Each point on the source produces coherent illumination at the points $\mathbf{x}_1$ and $\mathbf{x}_2$ and hence the mutual intensity function in the x plane is given by

$$\Gamma(\mathbf{x}_1, \mathbf{x}_2) = \int_S \Psi(s, \mathbf{x}_1) \Psi^*(s, \mathbf{x}_2) \, ds, \qquad (2.35)$$

where $\Psi(s, \mathbf{x})$ is the contribution to the disturbance at $\mathbf{x}$ from a typical source point s. The disturbance $\Psi(s, \mathbf{x}_1)$ and $\Psi^*(s, \mathbf{x}_2)$ may be described by means of the Kirchhoff diffraction formulae as

$$\Psi(s, \mathbf{x}_1) = \frac{-iA}{2\lambda} \int \Psi(s, \boldsymbol{\xi}_1) \exp \frac{ikr_1}{r_1} \Lambda_1 \, d\boldsymbol{\xi}_1,$$

$$\Psi(s, \mathbf{x}_2) = \frac{-iA}{2\lambda} \int \Psi(s, \boldsymbol{\xi}_2) \exp \frac{ikr_2}{r_2} \Lambda_2 \, d\boldsymbol{\xi}_2, \qquad (2.36)$$

where A is a constant and the Λ's are the appropriate inclination factors. Hence on substitution from Eq. (2.36) into Eq. (2.35)

$$\Gamma(\mathbf{x}_1, \mathbf{x}_2) = \frac{-A^2}{4\lambda^2} \int_S \int \int \Psi(s, \boldsymbol{\xi}_1) \Psi^*(s, \boldsymbol{\xi}_2) \exp \frac{ik(r_1 - r_2)}{r_1 r_2} \Lambda_1 \Lambda_2^* \, d\boldsymbol{\xi}_1 \, d\boldsymbol{\xi}_2 \, ds. \qquad (2.37)$$

Integrating over the source gives

$$\Gamma(\mathbf{x}_1, \mathbf{x}_2) = \frac{-A^2}{4\lambda^2} \int \int \Gamma(\boldsymbol{\xi}_1, \boldsymbol{\xi}_2) \exp \frac{ik(r_1 - r_2)}{r_1 r_2} \Lambda_1 \Lambda_2^* \, d\boldsymbol{\xi}_1 \, d\boldsymbol{\xi}_2. \qquad (2.38)$$

Equation (2.38) shows that if the mutual intensity function over a given plane is known then the mutual intensity in any other plane can be determined without knowing the source characteristics. Even though the actual source may not be incoherent an equivalent incoherent source can be defined. The intensity distribution in the $\mathbf{x}$ plane is determined by setting $\mathbf{x}_1 = \mathbf{x}_2$ and hence

$$I(x) = \frac{-A^2}{4\lambda^2} \int \int \Gamma(\boldsymbol{\xi}_1, \boldsymbol{\xi}_2) \exp \frac{ik(r_1 - r_2)}{r_1 r_2} \Lambda_1 \Lambda_2^* \, d\boldsymbol{\xi}_1 \, d\boldsymbol{\xi}_2. \qquad (2.39)$$

A. The van Cittert–Zernike Theorem

An extremely useful result derived independently by van Cittert (1934) and Zernike (1938) relates to the mutual intensity function, and hence the degree of coherence, in the optical field produced by an extended incoherent quasi-monochromatic source, when the medium between the source and the field is homogeneous. This result can be derived directly from Eq. (2.38) by assuming the condition that the ξ plane is actually incoherent. Hence Eq. (2.38) becomes

$$\Gamma(\mathbf{x}_1, \mathbf{x}_2) = C \int I(\xi) \exp \frac{ik(r_1 - r_2)}{r_1 r_2} \, d\xi, \tag{2.40}$$

where the inclination factor has been included in the constant C and r_1 and r_2 now mean the distances from a single point in the ξ plane to the points $\mathbf{x}_1$ and $\mathbf{x}_2$ respectively. The complex degree of coherence is then

$$\gamma(\mathbf{x}_1, \mathbf{x}_2) = \frac{C}{\sqrt{I(\mathbf{x}_1) I(\mathbf{x}_2)}} \int I(\xi) \exp \frac{ik(r_1 - r_2)}{r_1 r_2} \, d\xi. \tag{2.41}$$

The integral in Eq. (2.41) is the same as the one that arises in the diffraction problem of a spherical wave illuminating a plane screen. However, the result first derived by van Cittert and later by Zernike has to be interpreted quite differently; as Wolf (see Born and Wolf, 1969) concisely described it

"Equation (2.41) implies that the complex degree of coherence, which describes the correlation of vibrations at a fixed point x_1 and a variable point x_2 in a plane illuminated by an extended quasi-monochromatic primary source, is equal to the normalized complex amplitude at the corresponding point x_1 in a certain diffraction pattern centred at x_2. This pattern would be obtained on replacing the source by a diffracting aperture of the same size and shape as the source, and on filling it with a spherical wave converging to x_2, the amplitude distribution over the wave-front in the aperture being proportional to the intensity distribution across the source."

B. Application of the van Cittert–Zernike Theorem

Further to interpret Eq. (2.41) we shall consider an incoherent source in the ξ, η plane and determine the coherence function in the x, y plane (see Fig. 8). Any two points in the x, y plane are $P_1(x_1, y_1)$ and $P_2(x_2, y_2)$. The distances from a typical source point to the point P_1 and P_2 are r_1 and r_2 respectively; r is the distance between the source and the x, y plane. Hence,

$$\left. \begin{aligned} r_1{}^2 &= r^2 + (x_1 - \xi)^2 + (y_1 - \eta)^2, \\ r_2{}^2 &= r^2 + (x_2 - \xi)^2 + (y_2 - \eta)^2. \end{aligned} \right\} \tag{2.42}$$

If, as is often the case, the dimensions of the source and the separation between P_1 and P_2 are small compared to the separation between the source and the plane containing P_1 and P_2 then,

$$r_1 \sim r + \frac{(x_1 - \xi)^2 + (y_1 - \eta)^2}{2r},$$

$$r_2 \sim r + \frac{(x_2 - \xi)^2 + (y_2 - \eta)^2}{2r}. \tag{2.43}$$

The degree of coherence is, then

$$\gamma_{12}(0) = \frac{\exp(ik\theta) \iint\limits_{s} I_s(\xi, \eta) \exp \dfrac{[ik(x_1 - x_2)\xi]}{r} \exp \dfrac{[-ik(y_1 - y_2)\eta]}{r}}{\iint\limits_{s} I_s(\xi, \eta) \, d\xi \, d\eta} d\xi \, d\eta,$$

and

$$\theta = \frac{(x_1^2 + y_1^2) - (x_2^2 + y_2^2)}{2r}. \tag{2.44}$$

$I_s(\xi, \eta)$ is, of course, the intensity distribution across the source. Equation (2.44) shows that, providing the dimensions of the source and the separation between P_1 and P_2 are small compared to the distance between the source and the plane

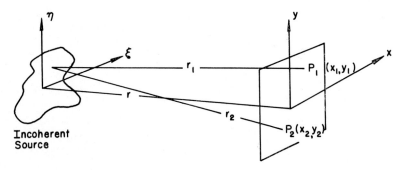

Fig. 8. Coordinate system for illustrating the van Cittert–Zernike theorem.

containing P_1 and P_2, the degree of coherence is the Fourier transform of the intensity function of the source suitably normalized. Hence all the knowledge of optical transforms becomes extremely valuable in determining the degree of coherence from specific incoherent sources.

1. *Incoherent Slit Source*

With the result given in Eq. (2.44) the degree of coherence arising from a uniform slit incoherent source can immediately be written down from knowledge of the solution of the equivalent diffraction problem. Hence,

$$\gamma_{12} = \exp{(ik\theta)} \sin\left(\frac{k(x_1 - x_2)\,a}{r}\right) \Big/ \left(\frac{k(x_1 - x_2)\,a}{r}\right), \qquad (2.45)$$

where $2a$ is the width of the slit. In the y direction there is no coherence if the slit is considered infinitely long. In fact the slit source is finite in the other direction and some coherence will result of similar form to Eq. (2.45) but the correlation distances will be much smaller.

2. *Rectangular Source*

The result for a uniform rectangular incoherent source of width $2a$ and length $2b$ is then

$$\gamma_{12} = \exp{(ik\theta)}\,\mathrm{sinc}\left[\frac{ka(x_1 - x_2)}{r}\right] \mathrm{sinc}\left[\frac{kb(y_1 - y_2)}{r}\right] \qquad (2.46)$$

where the notation $\mathrm{sinc}\,x = \sin x / x$ is used.

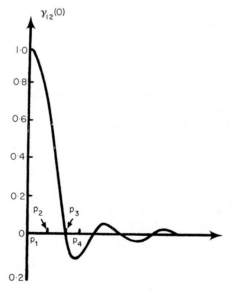

Fig. 9. The degree of coherence in the field produced by an incoherent circular source.

3. *Circular Source*

The situation that very often arises experimentally is the field produced by an incoherent circular source. The degree of coherence also has circular symmetry and

$$\gamma_{12} = \exp\left(ik\theta\right) 2J_1\left(\frac{k\alpha\rho}{r}\right)\bigg/\left(\frac{k\alpha\rho}{r}\right), \tag{2.47}$$

$$\alpha = [(x_1 - x_2)^2 + (y_1 - y_2)^2].$$

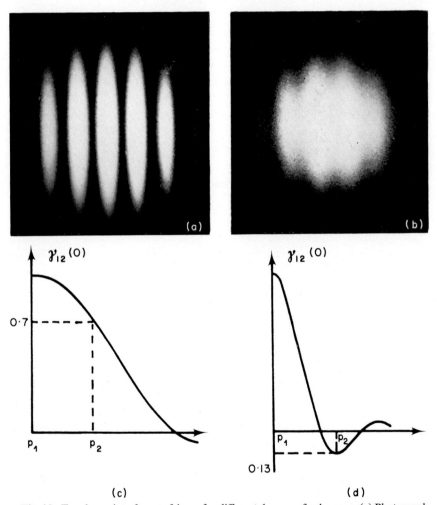

Fig. 10. Two-beam interference fringes for different degrees of coherence. (a) Photograph of the high-contrast fringes formed with $\gamma_{12}(0) = 0.7$ as indicated in Fig. 10(c). (b) Photograph of the low-contrast fringes (showing a central minimum) formed with $\gamma = -0.13$ indicated in Fig. 10(d).

The degree of coherence between the points in the field of an incoherent circular source as a function of the separation of the two points is shown in Fig. 9. The interpretation of this curve is as follows: if point P_1 is placed at the origin, then the value of $\gamma_{12}(0)$ is read off by placing the point P_2 at the appropriate point along the ordinate. The degree of coherence between points P_1 and P_2 is 0·5; between P_1 and P_4 is −0·1. The negative value of $\gamma_{12}(0)$ has to be interpreted as a phase change, i.e., the interference fringes formed will have a central minimum not a central maximum of intensity. Figure 10(a) shows the resultant fringes when $\gamma_{12}(0) = 0·7$ and Fig. 10(b) when $\gamma_{12}(0) = -0·13$. Note that the fringes in Fig. 10(b) have a central minimum. The fringes have equal spacing in both Figs. 10(a) and (b) and to produce these fringes the separation of the two points was constant and the size of the incoherent circular source changed. The appropriate plots of $\gamma_{12}(0)$ are shown as 10(c) and (d). The source diameters were in the ratio 1:1·71 (Thompson, 1958; Taylor and Thompson, 1957).

IV. Measurement of Degree of Coherence

A. Fringe Visibility Method

The earliest measurements on the coherence of optical fields involved the determination of the visibility (Zernike, 1938) by using a Young's two-beam interference experiment. The field of interest is sampled by two pinholes in an otherwise opaque screen and the fringes viewed at some distance behind the screen. (The arrangement is similar to that shown in Fig. 6). The intensities of the maximum and the adjacent minimum are measured for a given separation of the two pinholes. These measurements are then used to give the modulus of the degree of coherence.

$$\mathcal{V} = |\gamma_{12}(0)| = \frac{I_{max} - I_{min}}{I_{max} - I_{min}}. \tag{2.48}$$

The field is then explored by moving the two pinholes about in the field and varying their separation to give a two-dimensional plot of the modulus of the degree of coherence.

The Michelson stellar interferometer also relies on a fringe-visibility measurement for its operation. The method is not for measuring the coherence but uses the visibility to determine the diameter of the star. However, it could be used as a method of determining the modulus of the degree of coherence. Figure 11 shows the principle of the Michelson stellar interferometer; the light from a distant star is sampled by the two aperture P_1 and P_2 via the two sets of mirrors M_1, M_3 and M_2, M_4. The fringes are viewed by the telescope and are formed in the focal plane of the objective. Mirrors M_1 and M_2 are adjustable and the visibility of the fringes changes as the separation between these mirrors is varied. (In determining the stellar diameters the telescope was a reflector and hence the objective lens of Fig. 11 is really a mirror.) The fringes are viewed and the separation of the mirrors M_1 and M_2 increased until the fringes disappear and hence $|\gamma_{12}(0)| = 0$. If the star has a uniform intensity distribution the angular diameter is determined by solving Eq. (2.47) with the right-hand side set equal to zero. It must be stressed that in Michelson's original concept the analysis and explanation of the interferometer depended upon an integration over the source. Here we have used the concept of degree of coherence and the van Cittert–Zernike theorem to achieve the same end.

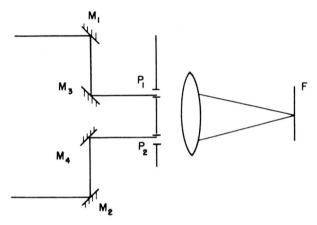

Fig. 11. Schematic diagram illustrating the principle of the Michelson Stellar interferometer.

The theory of the Michelson stellar interferometer was first discussed in terms of the theory of partial coherence by van Cittert, 1934 and later by Zernike, 1948.

B. Two-Beam Interference Techniques

The more usual method of measuring the degree of coherence is a variation on the fringe-visibility methods just discussed, and is of particular importance in work connected with optical transforms. It is interesting to note that many of the original experimental verifications of the theoretical predictions of the theory of partially coherent light were carried out using an optical diffracto-meter (Thompson and Wolf, 1957; Thompson, 1958) and the results immediately applied to improving the operation of that instrument (Taylor and Thompson, 1957). Other experimental results have been presented by Baker (1953) and by Arnulf et al. (1953).

Figure 12 shows the diffractometer used in the experimental measurement of coherence. The source of light S_0 is imaged by a lens L_0 on to a pinhole S_1 and the light that emerges from the pinhole is rendered parallel by a lens L_1. A second lens L_2 strictly similar to L_1, brings the interfering beams to a focus in the focal plane F of the lens L_2. An optical flat silvered on the front side (mirror M in the figure) is used to reduce the overall length of the instrument. The Fraunhofer pattern of any object placed on A is seen at F and viewed through the microscope.

As an example, as well as to illustrate the method, we will use the specific experimental results of Thompson and Wolf. The focal lengths of the three lenses L_0, L_1 and L_2 were $f_0 = 200$ mm, $f_1 = f_2 = 1\cdot52$ m. The diameter of L_0 was 50 mm, the distance from L_0 to S_1 was 400 mm, and the separation of L_1 and L_2 was 140 mm. The mirror M was at a distance of 850 mm from

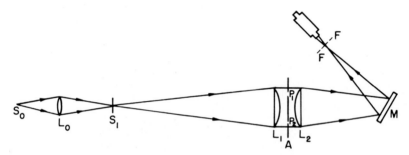

Fig. 12. Schematic diagram of an optical diffractometer.

L_2. The source S_0 was a mercury-vapour compact-source lamp and a filter selected the mercury yellow doublet, with mean wavelength $\bar{\lambda} = 5790$ Å. Over this range of wavelengths, the output of the compact-source lamp shows the mercury yellow doublet and no other spectral lines; a certain amount of background is, of course, passed by the filter. The secondary source S_1 had the form of a circular pinhole of diameter $2r_1 = 0\cdot09$ mm 10^{-2} cm. Now the area of coherence at an image of a light source is of the order of the effective area of a diffraction pattern of a single source point, and its diameter, in the case of the pattern formed by L_0, is of the order of $5\cdot7 \times 10^{-3}$ mm, i.e., about one-twentieth of the diameter of the pinhole. Hence, under these circumstances, the secondary source acted effectively as an incoherent primary source.

A number of experiments are carried out with series of opaque screens with pairs of pinholes of diameter $2a = 1\cdot4$ mm placed in them. The separation $2h$ of the two apertures was varied in steps of 1 mm from 6 mm to 30 mm. Two beams of light emerged from the two apertures and the degree of coherence $\gamma_{12}(0)$ between the two beams depends on the separation of the apertures. In the focal plane F a Fraunhofer diffraction pattern of a circular aperture is observed whose bright disc is about 1 mm in diameter, and this pattern is crossed by interference fringes. The system is clearly analogous to the Michelson stellar interferometer since lens L_1 collimates the light from the source.

The intensity distribution in the plane F is given by

$$I(\alpha, h) = 2\left(\frac{2J_1(v)}{v}\right) \left\{1 + \left|\frac{2J_1(u)}{u}\right| \cos\left[\beta_{12}(u) - \delta\right]\right\}$$

$$v = \frac{ka\alpha}{f_2}, \qquad u = kr_1\frac{2h}{f_1}, \qquad \delta = \frac{2h\alpha}{f_2}, \qquad (2.49)$$

where α is the coordinate in the plane F and $\beta_{12}(u)$ is the phase of the degree of coherence. The first Bessel function term $(2J_1(v)/v)$ is the diffraction pattern produced by each of the apertures individually; the second Bessel function term $|2J_1(u)/u|$ is the modulus of the degree of coherence produced by the incoherent circular source. The values of $I_{\max}$ and $I_{\min}$ are given by

$$I_{\max}(\alpha, h) = 2\left(\frac{2J_1(v)}{v}\right)^2 \left\{1 + \left|\frac{2J_1(u)}{u}\right|\right\}$$

$$I_{\min}(\alpha, h) = 2\left(\frac{2J_1(v)}{v}\right)^2 \left\{1 - \left|\frac{2J_1(u)}{u}\right|\right\}$$

$$(2.50)$$

Figure 1 actually represents a series of photographs of the resulting fringe pattern in such an experiment. Thompson and Wolf compared a number of their photographic results with the equivalent calculated curves. Figure 13 shows a typical result of this experiment in which the photographic record is compared with the calculated curves from Eq. (2.50). In this particular result $2h = 25$ mm and $|\gamma_{12}(0)| = 0.062$. Photoelectric measurements of the fringe patterns were also made using this same experimental arrangement to obtain more detailed quantitative data.

The fringes shown in Figs. 10(a) and (b) are from a related experiment in which the value of r_1 was changed to produce the coherence functions shown in Figs. 10(c) and (d). It is clear from this result that measurements of this type can reveal more than just the modulus of the degree of coherence. The presence of a central minimum in the fringe pattern of Fig. 10(b) shows that $\gamma_{12}(0)$ is negative. Again distinct comparisons have been made between the experimental results and the theoretical curves with excellent agreement being found (Thompson, 1958).

A detailed understanding of this experiment clearly allows the diffracto-meter (or any other diffraction apparatus) to be correctly designed to achieve adequate coherence in the plane of the diffracting object. The competing factor is how much light gets through the system; making the aperture $2r_1$ too small will give good coherence but cause loss of light. How much coherence is needed? The answer depends upon the particular application. For qualitative studies of diffraction a value of $\gamma_{12}(0) = 0.5$ between the extremities of the

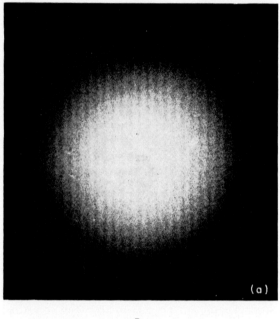

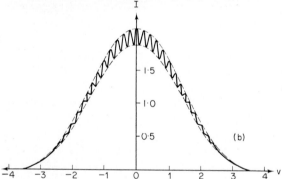

Fig. 13. Comparison between the photographic record of two-beam interference fringes with $2h = 2.5$ cm, $\gamma_{12}(0) = 0.062$ and the equivalent calculated curve. The upper and lower broken lines are I_{max} and I_{min} respectively.

diffracting object is often sufficient. For quantitative studies considerably better conditions need to be arranged (see Sections II.E and II.F). We will return to the question further in Section II.G.

C. Intensity-Correlation Technique

Brown and Twiss (1956, 1957a, b) have used a different type of interfero-meter for the measurement of star diameters. The radiation from the star is received by two mirrors that focus the light onto two photomultipliers. The two mirror-photomultiplier combinations are the detectors and their separations can be varied. The currents produced by the photomultipliers are amplified and then multiplied. The linear multiple is then fed into a correlator that measures the correlation in the *fluctuations* of the two currents. The quantity that is being measured is the time-averaged quantity $\langle \Delta I_1(t) \Delta I_2(t) \rangle$ where

$$\Delta I(t) = I(t) - \langle I(t) \rangle. \tag{2.51}$$

$I(t)$ is the instantaneous intensity at the photomultiplier which produces a current proportional to $I(t)$.† It can be shown that

$$\langle \Delta I_1(t) \Delta I_2(t) \rangle \propto |\gamma_{12}(0)|^2 \tag{2.52}$$

i.e., the square of the modulus of the degree of coherence is measured by this interferometer. Again, as in the Michelson interferometer, this result is then used to compute the star diameter.

D. Polarization Methods

A number of polarization interferometers with a fixed shear have been designed to measure the degree of coherence by the application of harmonic analysis. An excellent review of these techniques has recently been written by Françon and Mallik (1967). In particular the Savart polariscope and the Wollaston prism have been used in a configuration similar to that shown in Fig. 14(a). Light from the source S is collimated by lens L_1 and the field is sampled by the two apertures at P_1 and P_2. The polariscope—the modified Savart shown in Fig. 14(b)—is placed between two polarizers; the second lens L_2 images the source onto the detector. If the eye is placed behind the polariscope then fringes are observed on the image of the source. One method of using the device is to rotate the polariscope about an axis parallel to the direction of the fringes; the maximum and minimum current readings, C_{max} and C_{min}, given by the photo detector are used to compute the modulus of the degree of coherence

$$|\gamma_{12}(0)| = \frac{C_{max} - C_{min}}{C_{max} + C_{min}}. \tag{2.53}$$

† A correct analysis of this system requires a quantum-mechanical treatment which is not appropriate here.

The reader is referred to the review article by Françon and Mallik (1967) for proof of Eq. (2.53) and for details of the methods of using the technique for coherence measurements.

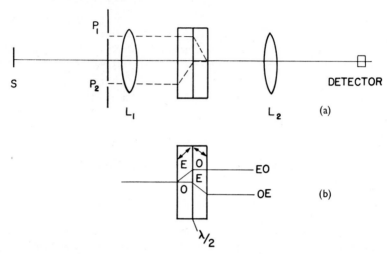

Fig. 14. Schematic diagram of a polarization method of measuring the degree of coherence.

V. Diffraction with Partially Coherent Light

A number of studies have been made of diffraction with partially coherent light. We will follow quite closely the paper by Shore *et al.* (1966) in this section.

The need for an understanding of the effects of departures from purely coherent light in diffraction experiments is vital whenever measurements have to be made on the resulting pattern. The initial theoretical results were obtained by Parrent and Skinner (1961) for the Fraunhofer diffraction pattern of a slit aperture illuminated with partially coherent quasi-monochromatic light.† They used an exponential form for the mutual-intensity function across the diffracting aperture. Subsequently Shore (1962) used the same form of the mutual intensity to discuss the Fraunhofer pattern of a circular aperture; Cathey (1964) has also used this same mutual-intensity function and a slit in which one half is out of phase with respect to the other half.

Although the exponential correlation does not correspond to an easily realizable experimental situation, the van Cittert–Zernike theorem does make possible the design of experiments with a great deal of control over the form of the mutual-coherence function. In particular, any function that can be

---

† At about the same time Bakos and Kantor (1961) examined theoretically and experimentally the diffraction pattern of a slit illuminated by an incoherent slit source.

expressed as the Fourier transform of a non-negative function is an easily realized mutual-coherence function. It is obtained by creating an incoherent source described by the appropriate non-negative function. The resulting distant field is partially coherent and has the desired mutual-coherence function.

Preliminary experimental studies (Thompson, 1964a, b), using both uniformly illuminated slit and circular apertures as primary incoherent sources, produced results that were qualitatively as well as quantitatively different from those that had been predicted by the exponential model; a more detailed study was, therefore, necessary in which a direct comparison between theory and experiment would be possible. Essentially, the problem consists of determining the Fraunhofer intensity distribution arising from an aperture illuminated with quasi-monochromatic, partially coherent light. For the purpose of the evaluation we make use of the Fourier-transform relationship derived by Schell

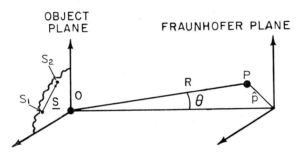

Fig. 15. The coordinate system for diffraction with partially coherent light.

(1961) for partially coherent illumination (see Shore, 1962), which states that the intensity distribution in the Fraunhofer diffraction pattern of a plane, quasi-monochromatic spatially stationary aperture distribution is proportional to the Fourier transform of the product of the aperture autocorrelation function and the normalized aperture mutual-intensity function:

$$I(P) = A \int \gamma_{12}(0) C(\mathbf{S}) \exp(ik \sin \theta \hat{\mathbf{p}} . \mathbf{S}) d\mathbf{S}, \qquad (2.54)$$

where A is a constant, λ is the mean wavelength, $k = 2\pi/\lambda$, θ, $\hat{\mathbf{p}}$ (a unit vector) and S are defined in Fig. 15, $\gamma_{12}(0)$ is the complex degree of coherence, and $C(\mathbf{S})$ is the autocorrelation function of the aperture amplitude.

Shore (1962) has evaluated Eq. (2.54), assuming $\gamma_{12}(0)$ to be an exponential function, $\exp(-x/L)$, where L is the correlation interval. Shore et al. (1966) have used an IBM 7094 computer to evaluate several other examples which are specifically related to the following physically realizable situations: (1) a circular aperture illuminated by a circular incoherent source, (2) a slit aperture

3

illuminated by a circular incoherent source, and (3) a slit aperture illuminated by an incoherent slit source. In the first two examples $\gamma_{12}(0)$ is of the form $2J_1(x/L)/(x/L)$ and in the third example $\gamma_{12}(0)$ is the form $\sin(x/L)/(x/L)$. This method is the most elegant way for evaluation of the result using a computer. However, conceptually it is perhaps easier to understand the result as the convolution of the characteristic intensity distribution of the Fraunhofer diffraction pattern of the aperture produced by a point source and the image of the actual source.

A. Diffraction by a Slit Aperture

The problem of diffraction by a slit can be solved quite readily in closed form and hence it will be instructive to carry through the analysis to illustrate the statements made above. Assume that the diffracting aperture, of width $2a$, is illuminated with a "collimated" beam of quasi-monochromatic light from an incoherent slit source of width $2r$. The complex amplitude $\Psi(x)$ in the resulting diffraction field from a typical point in the source is

$$\Psi(x) = 2a \operatorname{sinc}\left[ka\left(\frac{x}{f}+\frac{\alpha}{f_c}\right)\right], \tag{2.55}$$

where f_c is the focal length of the collimating lens, α is the coordinate in the source plane and f is the focal length of the lens producing the diffraction pattern. The resultant intensity $I_R(x)$ for all such points in the incoherent source is, therefore,

$$I_R(x) = \int_{-r}^{r} 4a^2 \operatorname{sinc}^2 ka\left(\frac{x}{f}+\frac{\alpha}{f_c}\right) d\alpha.$$

The equation may be rewritten using the rectangular function notation for the aperture, $R(\alpha,r)$, and

$$I_R(x) = \int_{-\infty}^{\infty} R(\alpha,r) \operatorname{sinc}^2 ka\left(\frac{x}{f}+\frac{\alpha}{f_c}\right) d\alpha. \tag{2.56}$$

Equation (2.56) is a convolution integral and represents a convolution of the source function with the intensity distribution in the Fraunhofer pattern of the aperture. This convolution may be re-expressed in terms of the Fourier transform of the two functions; hence

$$I_R(x) = \int \left[\left(\int R(\alpha,r)\exp\left(\frac{ik\alpha\xi}{f}\right)d\alpha\right) \times \left(\int \operatorname{sinc}^2 ka\left(\frac{x}{f}+\frac{\alpha}{f_c}\right)\exp\left(\frac{ik\alpha\xi}{f}\right)d\alpha\right)\right]\exp\left(\frac{ikx\xi}{f}\right)d\xi. \tag{2.57}$$

The integral in the first bracket is the Fourier transform of the uniform source which by the van Cittert–Zernike theorem is the mutual-intensity function of the illumination in the aperture plane. The other integral in the second bracket is the autocorrelation function of the aperture amplitude which in the example is a triangular function. Hence Eq. (2.57) is essentially the result quoted in Eq. (2.54). Equation (2.57) can be evaluated to give

$$I(U, V) = \frac{1}{V}\left\{Si(V + U) + Si(V - U) - \frac{1 - \cos(V - U)}{(V - U)} - \frac{1 - \cos(V + U)}{(V + U)}\right\}.$$

$$(2.58)$$

Here $Si(x)$ is the sine integral

$$\left(\int_0^n \frac{\sin x}{x}\,dx\right),$$

$u = 2ka\theta$, $v = 2a/L$. In the coherent limit, $v = 0$, this expression reduces to the familiar result

$$I(u, 0) = [\sin(u/2)/u/2]^2,$$

while in the incoherent limit, $v = \infty$, the intensity distribution normalized to its value at $\theta = 0$ reduces to the constant value of unity.

Theoretical results are shown in Fig. 16 for various values of the parameter $c = 2\pi ar/\lambda f$; $c = 0$ is the coherent limit and the incoherent limit is reached as $c \to \infty$. In the curves of Fig. 16 the width of the diffracting aperture is left constant and the size of the source changed to produce the appropriate change in c. The main effects to be seen are the gradual loss of contrast of the maxima and minima and a shift in the position of these features—sufficient in some cases to produce a reversal of the maxima and minima ($c = 4 \cdot 0$). This result means that measurements based on diffraction-pattern techniques can be in error if correct coherence conditions are not used. Values of c up to about unity appear to be acceptable for the one-dimensional example.

B. Diffraction by a Circular Aperture

Equation (2.54) has also been used to calculate the diffraction pattern of a circular aperture illuminated by a circular incoherent source. This is more realistic than the one-dimensional example and has direct application in the design of diffraction experiments and measurement techniques. The theoretical curves show very similar trends to those of Fig. 16.

Experimental measurements have been made of this pattern for comparison with the theoretical results (Shore et al., 1966). In the experimental arrangement the primary source was a mercury arc filtered for the green line at 5,461 Å. The arc was imaged onto a small pinhole that acts as the incoherent source for the system. This pinhole of diameter $2r$ was in the back focal plane of a collimating lens. The "collimated" beam illuminated the diffracting aperture and the diffraction pattern was recorded in the transform plane, which is the focal plane of a second lens. The degree of coherence across the aperture was varied by changing the value of $2r$. The same effect could be achieved by varying the diameter of the diffracting aperture, but the former procedure is preferred since it keeps the major physical dimension of the system constant. The diffraction patterns were recorded on Pan-X film, and, by exposing a step wedge

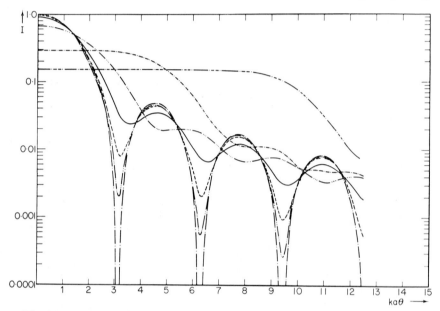

Fig. 16. Theoretical Fraunhofer intensity distribution for a slit aperture illuminated by a slit source for various values of c; $c = 0$ (—·—), $c = 0.5$ (—··—), $c = 1.0$(————), $c = 2.0$(————), $c = 4.0$ (—···—), $c = 10.0$ (——·——), $c = 20.0$ (—·——·—). (After Shore *et al.*, 1966.)

on the same film, the relative intensity distribution across the pattern was obtained from microdensitometer traces. To facilitate the recording of both the high intensity at the centre of the pattern and the relatively low intensity of the outer rings, a number of exposures were made of each diffraction pattern. To obviate the difficulties that arise from halation and scattering a small central stop was used in the longer exposures to remove the high-intensity peak at the centre.

In the actual experiments $2r$ was varied from 52 to 270 μm, the focal length of the collimator was 250 mm and the focal length of the transform lens 1·20 m; the diffracting aperture had a diameter of 2·2 mm. Two typical experimental results are shown in Fig. 17 together with the corresponding theoretical curves.

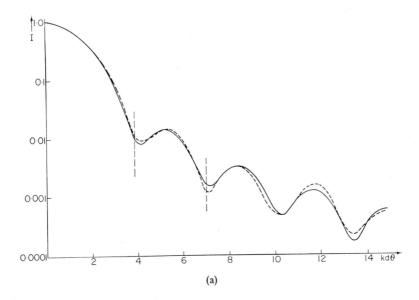

(a)

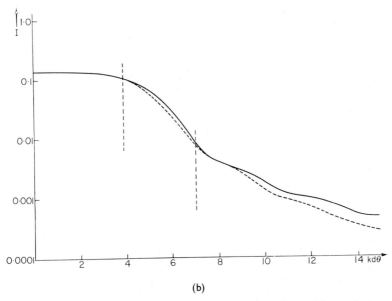

(b)

Fig. 17. Comparison between theoretical (———) and experimental (-----) curves for a circular aperture illuminated by a circular source. (a) $c = 1 \cdot 0$ (b) $c = 4 \cdot 62$ (experimental) $c = 5 \cdot 0$ (theoretical). (After Shore *et al.*, 1966.)

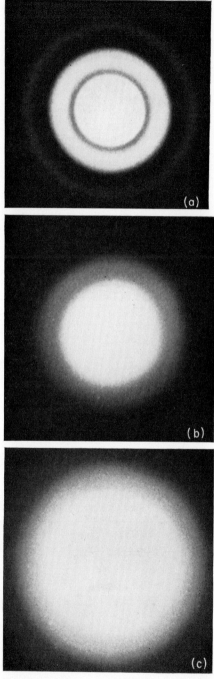

Fig. 18. Photographs of the diffraction patterns of a circular aperture with partially coherent light. (a) $c = 0.2$ (b) $c = 1.5$ and (c) $c = 4.6$. (After Parrent and Thompson, 1969.)

The experimental results support the theoretical predictions. Photographs corresponding to some of these experimental results are shown in Fig. 18; (a) is the result for $c = 0.2$ and a very slight filling in of the first dark ring is noticed; in (b) $c = 1.5$ and in (c) $c = 4.6$ corresponding to the data in Fig. 17(b).

The theoretical and experimental studies of the two-dimensional circularly symmetric problem again clearly point out the shift in the position of maxima and minima. This leads to the conclusion that the precise form of the mutual-intensity function is important not only for quantitative descriptions but also for qualitative understanding. The shift that occurs is particularly important in measuring instruments based on diffraction effects, since the results are in error proportional to the shift.

VI. INTERFERENCE WITH PARTIALLY COHERENT LIGHT

The subject of two-beam interference by division of wavefront has been discussed in earlier sections and the general law is stated in Eq. (2.23). It should be mentioned that two-beam interference by division of amplitude has also been discussed for partially coherent light (Hariharan and Sen, 1961) and a generalized theory of interferometers has been attempted by Steel (1965, 1967).

Multiple-beam interference by division of wavefront is a fundamental to the process of optical transformation hence coherence effects can play an important role. In this section we will develop some analysis of multiple-beam interference with partially coherent light and use the approach and results given by Thompson (1966).

A. Multiple-Beam Interference—Regular Arrays

1. One-dimensional Regular Array

For the general case of N apertures let the intensities of the individual beams be $I_1, I_2 \ldots I_N$. The resultant intensity for quasi-monochromatic illumination is then

$$
\begin{aligned}
I_R = & I_1 + I_2 + I_3 + \ldots + I_N \\
& + 2(I_1 I_2)^{1/2}|\gamma_{12}(0)| \cos[\beta_{12}(0) + \delta_{12}] \\
& + 2(I_1 I_3)^{1/2}|\gamma_{13}(0)| \cos[\beta_{13}(0) + \delta_{13}] \\
& + \ldots \\
& + 2(I_n I_m)^{1/2}|\gamma_{nm}(0)| \cos[\beta_{nm}(0) + \delta_{nm}] \quad m > n \\
& + \ldots \\
& + 2(I_{N-1} I_N)^{1/2}|\gamma_{N-1,\,N}(0)| \cos(\beta_{(N-1)N} + \delta_{(N-1)N}). \qquad (2.59)
\end{aligned}
$$

Here we have used the quasi-monochromatic approximation; $|\gamma_{nm}(0)|$ represents the modulus of the complex degree of coherence between I_n and I_m. Equation (2.59) may be collected into a general form

$$I_R = \sum_{n=1}^{N} \sum_{m=1}^{N} (I_n I_m)^{1/2} |\gamma_{nm}(0)| \cos [\beta_{nm}(0) + \delta_{nm}]. \tag{2.60}$$

It is noted that

$$\left. \begin{array}{l} |\gamma_{nm}(0)| = 1 \text{ when } n = m \\ |\gamma_{nm}(0)| = |\gamma_{nm}(0)|. \end{array} \right\} \tag{2.61}$$

a. Slit incoherent source. Let us now consider that the array described above is illuminated by an incoherent slit source. By application of the van Cittert–Zernike theorem, the modulus of the complex degree of coherence produced in the plane containing the array of apertures is a $|\sin x/x|$ function, where x depends upon the source dimensions (width $2r$) and 2θ is the angle subtended by the source at the plane containing the apertures, i.e., $x = (2\pi/\lambda)2r\sin\theta$. Under the condition, then, that the van Cittert–Zernike theorem is valid, $\sin\theta \simeq \theta$. An important property of the function $\mathrm{sinc}(x) = \sin x/x$ is that the zero values of the function occur at equally spaced intervals. Hence, it is possible to illuminate a one-dimensional regular array such that the illumination is coherent over a single aperture but incoherent with respect not only to its immediate neighbour but with respect to every other aperture in the array. Hence, the intensities of the individual diffraction patterns are added, i.e.,

$$\begin{array}{l} |\gamma_{nm}(0)| = 0 \text{ for } n \neq m, \\ |\gamma_{nm}(0)| = 1 \text{ for } n = m, \end{array} \tag{2.62}$$

$$\therefore \quad I_R = \sum_{n=1}^{N} I_n. \tag{2.63}$$

This is a difficult result to achieve if the spacing of the array is the spacing determined by the first zero value of $\sin x/x$ (here called the coherence interval). This difficulty occurs since $|\gamma_{nm}(0)|$ is the correlation existing between two points whereas we actually select by two finite-size apertures. However, if the spacing of the array is that determined by the second or third zero of the coherence function, then the condition expressed in Eq. (2.64) is more easily realized.

b. Circular incoherence source. For a circular incoherent source, the van Cittert–Zernike theorem gives the modulus of the complex degree of coherence as $|2J_1(x)/x|$ where x is as previously defined. Equation (2.59) is still applicable but the $|\gamma_{nm}(0)|$ values are now different. Setting the fundamental separation of the apertures in the array so that they are just incoherent with respect to each other (i.e., when $2J_1(x)/x = 0$ for the first time) does not make all the

$|\gamma_{nm}(0)| = 0$ for $n \neq m$ since the $2J_1(x)/x$ function does not have equally spaced zeros. Hence, if we make the closest separation of the apertures in the array equal to the coherence interval for zero correlation, then

$$|\gamma_{n,\,n+1}(0)| = |\gamma_{n+1,\,n}(0)| = 0,$$
$$|\gamma_{n,\,n+2}(0)| = |\gamma_{n+2,\,n}(0)| = 0{\cdot}045, \qquad (2.64)$$
$$|\gamma_{n,\,n+3}(0)| = |\gamma_{n+3,\,n}(0)| = 0{\cdot}040.$$

Only an approximately incoherent addition of the individual diffraction patterns is obtained.

2. Two-dimensional Array

Equation (2.59) is still valid and the $|\gamma_{nm}(0)|$ have to be taken between every pair of apertures in the array. The actual $|\gamma_{nm}(0)|$ again depends on the geometry of the source. We consider two sources here which are of practical interest.

a. *Rectangular incoherent sources.* For any rectangular two-dimensional array the coherence between all possible pairs of apertures in the array can be arranged to be zero at the same time by suitable choice of the source dimensions.

b. *Circular incoherent sources.* The same arguments apply here as applied to the one-dimensional array: only the nearest neighbours along array directions can be arranged to be incoherent with respect to each other. Fig. 19(a) shows a square array of 49 circular apertures of 1 mm diameter of basic separation $S = 10$ mm. Figure 19(b) shows the Fraunhofer diffraction pattern of such an array when the illumination is coherent over the whole array. In Fig. 19(c) the illumination provided by a high pressure mercury arc source filtered for the green line at 5,461 Å has been arranged so that the distance S corresponds to the coherence interval of the illumination. The addition is not entirely incoherent and the pattern shows some structure determined mainly by the distance S'. The area shown in these illustrations is limited to the central maximum in the diffraction pattern of one of the individual circular apertures that form the elements of the array. Figure 20 shows a more interesting illustration; (a) shows an array of groups of apertures, each group actually representing a projection of a molecule of hexamethyl-benzene. The centre-to-centre spacing is 12 mm. The illumination was arranged so that the coherence interval was 12 mm. Hence, while each group was illuminated coherently, the groups were illuminated incoherently with respect to each other. The resultant diffraction pattern is shown in Fig. 20(b). This represents an incoherent addition of the diffraction pattern formed by each group, resulting in an intense display of the diffraction pattern. For comparison, Fig. 20(c) shows the diffraction pattern of a single group. The intensity in Fig.

20(b) is 49 times greater than in Fig. 20(c). This type of arrangement allows sufficient intensity to be obtained in a diffraction pattern for display purposes, without the necessity of using a laser.

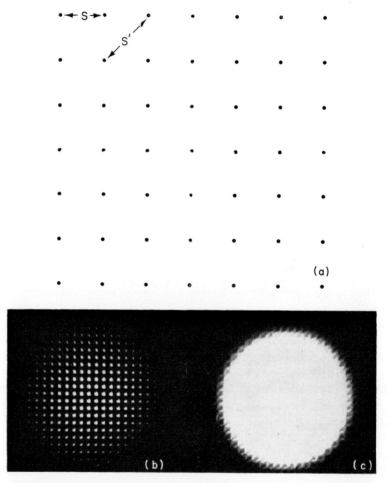

Fig. 19. Multiple-beam interference with partially coherent light of the array (a); with (b) coherent illumination and (c) coherence interval = S. (After Thompson, 1966.)

In Section V the idea was expressed that in a diffraction problem with partially coherent light the resultant intensity distribution could be thought of as the convolution of the source function with the characteristic diffraction pattern of the aperture. This statement is true for multiple apertures or single apertures and hence can be applied to the present discussion.

For a large array of circular holes the coherent Fraunhofer diffraction pattern is another array in which each element is the diffraction pattern produced by the aperture which limits the object array. This is always true

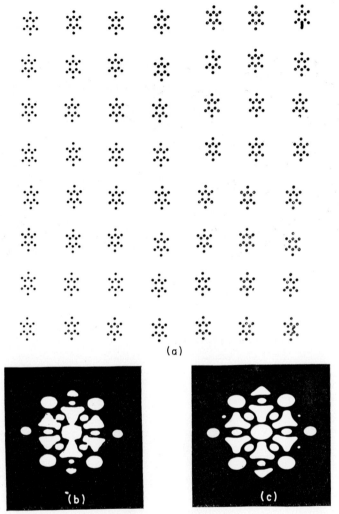

(a)

(b) (c)

Fig. 20. Multiple-beam interference with partially coherent light of the array (a); with (b) coherence interval equal to centre-to-centre spacing of groups; (c) diffraction pattern of one group alone. (After Thompson, 1966.)

for coherent illumination since the image of the illuminating light source in the diffraction plane is smaller than the diffraction pattern of the aperture bounding the object array. Very often, however, the light is partially coherent and this last condition does not hold. For example, in Fig. 21(a) an array of circular aperture was illuminated from an incoherent circular source such that the source image was larger than the diffraction pattern of the boundary of the object array; hence the actual diffraction pattern consists of an array of images of the source. In Fig. 21(b) a similar experiment was conducted with a rectangular incoherent source somewhat larger than the circular

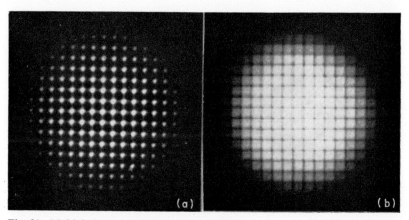

Fig. 21. Multiple-beam interference with partially coherent light of a two-dimensional regular array. An illustration of the effective replication of the source: (a) incoherent circular source; (b) incoherent rectangular source.

incoherent source used previously; the images of the rectangular source are seen clearly in the diffraction pattern.

B. Multiple-Beam Interference—Irregular Arrays

Multiple-beam interference from irregular arrays is, of course, an old subject forming essentially the basis of Young's eriometer (see, e.g., Ditchburn, 1951). In the eriometer, the irregular array is incoherently illuminated but with sufficient coherence to be coherent over an individual blood cell or other particle; hence, an intensity addition of the individual diffraction patterns results. It would be expected that if the same array had been coherently illuminated a similar result would be achieved since the interference fringes would essentially cancel out. Some care has to be taken in making this statement.

Figure 22(a) shows the Fraunhofer diffraction pattern of an irregular array of many thousands of lycopodium powder particles formed when illuminated with light with a coherence interval equal to a few thousand particle diameters and Fig. 22(b) with a coherence interval only a few particle diameters. These photographs are seen to look very similar. However, Fig. 22(a) looks more

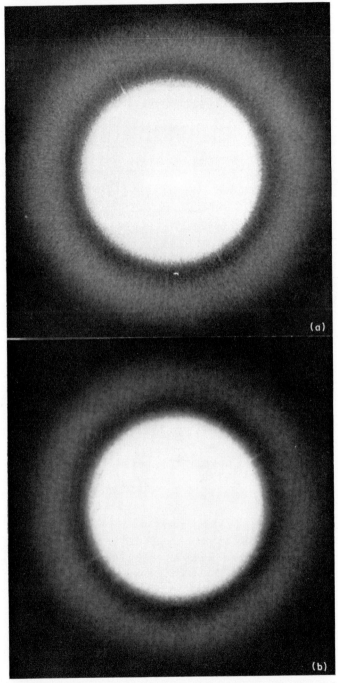

Fig. 22. Multiple-beam interference by an irregular array of many thousand particles;
(a) illumination coherent over a few thousand particle diameters; (b) illumination coherent
over a few particle diameters.

grainy than Fig. 22(b); the grainy appearance is produced by residual inter-ference effects. The other difference that is not apparent in the photographs is a bright spot at the centre of the pattern in Fig. 22(a).

VII. DESIGN OF OPTICAL DIFFRACTION SYSTEMS

The clear purpose of the foregoing discussions is to provide some guide to the design of optical diffraction systems from the point of view of the coherence requirements. Hence in a sense this section is the summary of the earlier discussions. The design problem will be considered in two parts determined by the type of light source used.

A. Incoherent Primary Source

The spatial coherence can be controlled by changing the size of the incoherent source. Usually this is achieved by using an arc lamp and imaging the arc onto a small aperture. The image of the arc can be considered incoherent provided that the width of impulse response of the imaging system is very much smaller than the aperture that it illuminates. The required spatial coherence can then be predicted by application of the van Cittert–Zernike theorem. In fact any described mutual intensity function can be created providing its Fourier transform is a non-negative function. Whilst adequate coherence can be obtained by this method complete coherence cannot. However, control is available for special purposes such as the one discussed in Section II.D and illustrated in Fig. 20 or for designing systems that will be used for replication of a particular distribution (Fig. 21). This technique has been suggested for replicating circuit layouts for use in the microelectronics industry (Parrent and Thompson, 1964; Lowenthal et al., 1968).

Another advantage of the incoherent-source approach is the excellent control over the uniformity of the intensity distribution across the diffracting object. The disadvantages are lack of energy and lack of temporal coherence. The temporal coherence of the optical field is determined either by the natural line shape of the particular spectral line chosen or the spectral characteristic of the filter used. A usual arrangement is to image the arc onto the small aperture (discussed earlier when spatial coherence was considered) by means of a pair of microscope objectives so that the filter may be placed in the collimated region between the objectives. Figure 23 illustrates the lack of temporal coherence in a typical situation. Two-beam interference fringes are formed in the usual two-pinhole experiments. Figure 23(a) shows the high-quality fringes formed by the field produced by a high-pressure mercury arc using the natural pressure-broadened line width of the line at 5,461 Å. Clearly the field has excellent spatial coherence with good fringe visibility noted over the entire fringe field. Hence if a plane diffracting object is used that has dimensions commensurate with the separation of the two pinholes then adequate spatial and temporal coherence are available. However, if an extra optical path is introduced into one of the two-interfering beams of Fig. 23(a) then the coherence length can easily be exceeded and Fig. 23(b) results—an

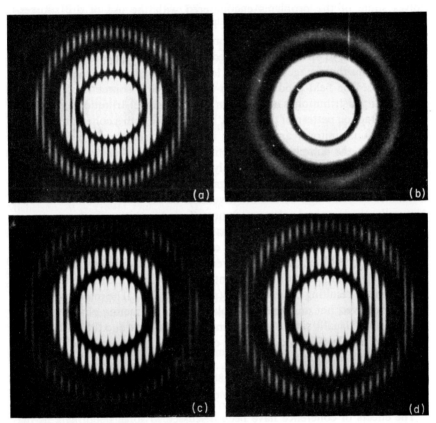

Fig. 23. Effect of temporal coherence of the light source (a) two-beam interference fringes formed with light from a mercury-arc source (b) fringes are absent when a small extra optical path is introduced into one of the interfering beams, (c) two-beam interference fringes formed with a laser light source (d) fringes still present when extra optical path is added.

intensity addition of the two beams. The extra optical path added here was that given by a microscope cover slip. In setting up diffraction experiments the maximum path difference that will be encountered has to be determined to verify the adequacy of the temporal coherence. Some control is available by choosing the characteristics of the spectral filter; the more coherence length

CHAPTER 3

Determination of Crystal Structure

B. Chaudhuri

*Department of Physics, University of Gauhati,
Gauhati, Assam, India*

I. Introduction

It is well known that in all methods of structure determination by analysis of diffraction by matter, the problem resolves into one of designing suitable strategies for determination of phases of the diffraction spectra. There are now two general approaches to the solution of this "phase problem" in crystal structure analysis from X-ray diffraction data. In one, commonly known as the direct method the entire work is presented in the form of a purely mathematical problem utilizing the services of the modern computer. In the other, the subject is treated as a physical problem of recognizing the structure of an object directly from its X-ray Fraunhofer diffraction pattern. An approximate answer to the problem is immediately sought from a composite strategy in which due considerations are given to crystal symmetry, stereo-chemistry and the physical relationship that exists between an object and its diffraction pattern. The second approach is akin to the historical trial-and-error method but with its elegance and objectivity increased manyfold by intro-duction of new experimental techniques for testing trial structures (Hanson *et al.*, 1953) and gradual realization of the importance of the theoretical relationships that exist between an object and its diffraction pattern in terms of Fourier transform and convolution theories. The mathematical formalisms of these concepts have been fully treated by Wrinch (1946), Titchmarsh (1948), Lipson and Taylor (1958) and Hosemann and Bagchi (1962).

II. FOURIER TRANSFORMS

A. Introduction

Fourier transformation and convolution theories were first introduced into problems of structure determination by Hettich (1935) and Ewald (1940). Knott (1940) first showed how to represent the Fourier transform of a planar molecule as a contour map and illustrated its scope of application by taking the case of naphthalene (Banerjee, 1930), a known structure. The method has since been successfully applied to unknown structures of varying complexities by Waser and Lu (1944), Klug (1950b), Stadler (1953), Sim *et al.* (1955), Bailey (1958), Hall and Nobbs (1966) and others.

B. Mathematical Formulation

As we have seen, the Fraunhofer diffraction of any radiation produced by any object is equivalent to the mathematical process of Fourier transformation. If a crystal is irradiated by a beam of parallel monochromatic X-rays of wavelength λ, in the resulting X-ray Fraunhofer diffraction the amplitude of the diffracted beam in a particular direction is related to the electron density distribution $\rho(\mathbf{r})$ in the unit cell containing N atoms, by the Fourier-transform relation

$$G(\mathbf{S}) = \int \rho(\mathbf{r}) \exp(2\pi i \mathbf{r}_j \mathbf{S}) \, dv_r. \tag{3.1}$$

Equation (3.1) may be written in terms of atoms with scattering factor f_j at location r_j ($j = 1$ to N), when it takes the form

$$G(\mathbf{S}) = \sum f_j \exp(2\pi i \mathbf{r}_j \mathbf{S}) \tag{3.2}$$

where

$$\mathbf{r}_j = x_j \mathbf{a} + y_j \mathbf{b} + z_j \mathbf{c} \tag{3.3}$$

is the positional vector of the jth atom with fractional parameters x_j, y_j, z_j on a lattice defined by repeat distances $\mathbf{a}$, $\mathbf{b}$, $\mathbf{c}$. The vector $\mathbf{S}$ is the difference between the wave-vectors of the scattered and incident radiation (Fig. 1). Thus

$$\mathbf{S} = \mathbf{s} - \mathbf{s}_0$$

where

$$|\mathbf{s}| = |\mathbf{s}_0| = \frac{1}{\lambda}.$$

The vector $\mathbf{S}$ is in the reciprocal space (Chapter 1, Section I.G) in which the Fourier transform of an object is described;

$$\mathbf{S} = h\mathbf{a}^* + k\mathbf{b}^* + l\mathbf{c}^* \tag{3.4}$$

where $\mathbf{a}^*$, $\mathbf{b}^*$, $\mathbf{c}^*$ are the primitive translations of the reciprocal lattice. The function $G(\mathbf{S})$, representing the Fourier transform of the unit-cell contents may be calculated for integral and non-integral values of h, k, l. The structure

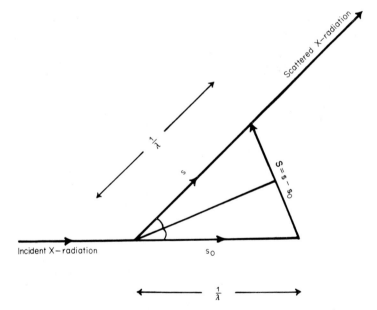

Fig. 1. Vectorial representation of the process of X-ray scattering.

factors $F(hkl)$ are the values corresponding to the integral values of h, k, l—that is, the Fourier transform calculated at the reciprocal-lattice points. Equation (3.2) may be written in the form

$$G(S) = \sum f_j \cos 2\pi(r_j S) + i \sum f_j \sin 2\pi(r_j S) \qquad (3.5)$$
$$= A(S) + iB(S). \qquad (3.6)$$

For a centrosymmetric structure the transform is real $[B(S) = 0]$ and transform regions have only two possible phases 0 and π separated by nodal lines of zero amplitude. For non-centrosymmetric distribution the transform is complex. In such cases the real and imaginary parts are to be calculated and plotted in separate contour diagrams (Klug, 1950b). Alternatively, the transform is presented in the form of separate contour maps to represent the distribution of amplitudes $\sqrt{A^2(S) + B^2(S)}$ and phase angles $\phi = \tan^{-1}[B(S)/A(S)]$ (Lipson and Taylor, 1958). To obtain a three-dimensional representation of a Fourier transform, contours at different levels have to be drawn giving a separate diagram at each level. For a planar configuration of atoms, the transform is particularly simple; if variation in the atomic scattering factor is ignored—that is, if the atoms are regarded as point atoms—the transform has identical sections perpendicular to the atomic plane, a property that has been found very helpful in solving such structural problems.

C. Uses and Limitations

From the point of view of the Fourier-transform approach to structure analysis, the problem resolves into finding a relative orientation of the reciprocal lattice and the calculated Fourier transform of the contents of the unit cell, so that modulus of the transform at each reciprocal-lattice point corresponds to the structure amplitudes of the X-ray reflexions. The process has obvious advantages (Chapter 1, Section I.G) over the conventional trial-and-error method in which observed intensities of the various diffraction spectra are compared with the calculated structure factors, i.e., with the Fourier transforms of the trial structure calculated only at the reciprocal-lattice points. The complete Fourier transform, being a continuous function, demonstrates clearly the nature of gradients in the neighbourhood of the reciprocal-lattice points and thus may indicate how a particular postulation differs from the correct one or may suggest what changes are necessary. An investigation of this kind is generally promising if there is only one planar centrosymmetric molecule in the unit cell or one projected centrosymmetric molecule in the effective unit cell of projection. It is also essential that the shape of the molecule be approximately known. The problem becomes difficult if there are several molecules in the unit cell related by symmetry. For a non-centrosymmetric molecule the process becomes complicated as the correctness or otherwise of any postulated orientation has to be tested by vectorial addition of the real and imaginary parts of the transform. When there are several such molecules related by symmetry operations, application of the method may not be profitable at all, unless favoured by special circumstances giving symmetry of high order in the transform, as in the case of triphenylene (Klug, 1950b).

A serious drawback of this method is obviously the severity in the computational labour involved. The Fourier transform (Eq. 3.2) has to be calculated for all possible values of S and not merely at the reciprocal-lattice points defined by integral values of h, k, l. If the molecule is not planar and its approximate shape is not known, the computational labour and presentation of the transforms as contour diagrams is formidable enough even for laboratories equipped with a high-speed computer. Thus the method, in spite of its theoretical appeal, has so far been applied to only a few favourable cases. However, methods evolved for the rapid production of Fourier transforms optically (Chapter 1, Section I.B; Hanson et al., 1953) have considerably widened the scope of applicability of Fourier-transform theory in the field of crystal-structure analysis.

III. OPTICAL TRANSFORMS

A. Historical Introduction

The well-known Fourier-transform relation (Eq. 3.1) applies to all diffraction phenomena involving light rays, X-rays, or beams of electrons or neutrons. For the diffraction of X-rays by atoms, $\rho(\mathbf{r})$ is the density of electrons, the scattering matter in the electronic shells of the atoms. For electron diffraction it is the electrostatic potential due to the atomic nuclei and their electron cloud, and for neutron diffraction it is the delta-function potential of the nuclear forces. In dealing with diffraction of light rays, $G(\mathbf{S})$ may be taken to represent the amplitude distribution of light coherently scattered by an object having transparency distribution $\rho(\mathbf{r})$. As we have seen in Chapter 1, the idea of exploiting such a situation to demonstrate the optical analogue of the X-ray diffraction effects with simple optical experiments, by simulating a beam of X-rays by a beam of light rays was first suggested by Bragg (1939). The possibility of reproducing images of crystal structures by optical Fourier synthesis was first demonstrated successfully for (010) projection of diopside (Bragg, 1939). Buerger (1950) and Hanson et al. (1951) generalized the method, but a more fruitful utilization of the idea was made by Taylor and Lipson (1951) who showed that the Fourier transform of any planar configuration of atoms could be easily simulated optically by representing the atomic arrangement by a set of holes in a mask and photographing its Fraunhofer diffraction pattern.

B. *Simulation of Scattering Factor Curve*

A specially designed optical instrument, suitable for producing small-angle Fraunhofer diffraction pattern of large-scale objects, was first described by Bragg and Lipson (1943). The instrument has since undergone various modifications and details of designs of a precision model of the instrument which is now known as the optical diffractometer (Chapter 1, Section I.D) are given by Taylor *et al.* (1951), Hughes and Taylor (1953) and Taylor and Thompson (1957). Simpler versions of such equipment are also described by Buerger (1950), Berry (1950), Clastre and Gay (1950), Hosemann (1962), Wyckoff *et al.* (1957), Aravindakshan (1957) and others.

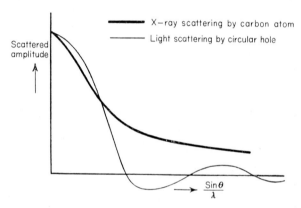

Fig. 2. Comparison of amplitude distribution curve for light scattered by a circular aperture and X-ray scattering factor curve for carbon. Reproduced from Taylor and Lipson (1964) with permission.

The simulation of the atomic scattering of X-rays by the scattering of light by circular holes may appear theoretically somewhat inconsistent. The variation with $\sin\theta/\lambda$ of X-ray scattering factor f of an atom is related to the radial distribution of electron density $U(r) = 4\pi r^2 \rho(r)$ of a spherically symmetric atom by the Fourier-transform relationship. The resulting f curve which is the Fourier transform of the atom has a maximum value equal to the atomic number at $\theta = 0$ and falls off asymptotically towards zero with increasing scattering angle. On the other hand, the amplitude distribution curve of light scattered by a circular aperture undulates through positive and negative values (Chapter 1, Fig. 10) exhibiting alternate maxima and minima. Nevertheless, it has been shown by Taylor and Lipson (1964) that the two scattering curves are significantly similar (Fig. 2) in the low-angle region to justify the simulation.

C. Theoretical Aspects of the Optical-Transform Approach in Two Dimensions

A photographic record of an optical diffraction pattern is however a record of intensity only and thus does not strictly represent a Fourier transform which is generally presented in contour maps in two parts as amplitude and phase diagrams (Section II.B). There is however close resemblance between the photographically recorded optical diffraction patterns and the amplitude diagrams of calculated Fourier transforms and the optically derived Fourier transforms are called optical transforms (Chapter 1, Section I.C).

Though all diffraction processes are three-dimensional in character, optical transforms are always recorded in two dimensions and are therefore most profitably used in studying crystal structures in projection. Two circumstances have combined to make this procedure particularly appropriate. Firstly, in contrast to the problem of scattering of X-rays by crystalline material where the interatomic separations are comparable with wavelengths of X-rays, in all optical-transform work with visible light the wavelength λ used is about 10^4 times smaller than the average separation of the scattering units represented by the holes in the mask. The sphere of reflexion (Ewald, 1921) having radius $1/\lambda$ can therefore be regarded as almost planar. Secondly, the Fourier transform of the transparency distribution of a set of holes in a mask is in general a three-dimensional function but has similar cross-sections parallel to the plane of the mask. Thus the plane central section of the reciprocal solid (Lipson and Taylor, 1958), cut by the sphere of reflexion parallel to the plane of the mask and normal to the incident radiation, gives the observed diffraction pattern which is the complete representation of the transform. It is also well known that a zone of X-ray diffraction data arising from intersection of the sphere of reflexion with any central layer of the reciprocal lattice is the Fourier transform of the structure projected along the direction normal to the layer. These are the essential theoretical backgrounds with which optical transforms of two-dimensional objects are used in crystal-structure analysis.

D. Extension to Three Dimensions

A method for producing optical transforms that can be compared with X-ray diffraction data corresponding to the intersection of the sphere of reflexion with non-central layer of the reciprocal lattice has been described by Harburn and Taylor (1961). The Fourier-transform equation representing optical diffraction by a hypothetical array of holes in three dimensions may be written as

$$G(\mathbf{S}) = \sum f_j \exp 2\pi i(\mathbf{r}_j \mathbf{S}) \qquad (3.7)$$

where f_j is the scattering function of the jth hole, $\mathbf{r}_j$ its vector distance defining its position from an arbitrary origin and $\mathbf{S}$ is the vector in the reciprocal space.

If the holes simulate an atomic arrangement in a unit cell with translation vectors **a**, **b**, **c** and reciprocal translation vectors **a***, **b***, **c***, then

$$G(\mathbf{S}) = \sum f_j \exp 2\pi i (hx_j + ky_j + lz_j) \tag{3.8}$$

where x_j, y_j, z_j are the components of $\mathbf{r}_j$ parallel to the crystallographic axes, expressed as fraction of **a**, **b**, **c**, and h, k, l are the components of **S** along the reciprocal axes (may be integral or non-integral multiples of **a***, **b***, **c***). Equation (3.8) may be re-written as

$$G(\mathbf{S}) = \sum f_j \exp i\phi_j \exp 2\pi i (hx_j + ky_j) \tag{3.9}$$

where $i\phi_j = 2\pi l z_j$.

The relation (3.9) is basically an equation for Fourier transformation representing diffraction by holes in a two-dimensional mask, where the scattering function of the jth hole is modulated by a phase operator $\exp i\phi_j$. If only the values of h, k, l that are integral multiples of **a***, **b***, **c*** be considered, then

$$G(\mathbf{S}) = F(hkl) = f_j \exp i\phi_j \exp 2\pi i (hx_j + ky_j) \tag{3.10}$$

where $F(hkl)$ is the structure factor of the reflexion hkl. Thus, to produce an optical transform that will correspond to a non-zero-layer X-ray diffraction photograph, one has only to control appropriately the phase of the light transmitted by the holes in a mask representing a projection of the structure along any crystallographic axis. If the phase operator is calculated taking all possible values of l (integral and non-integral) the complete three-dimensional optical transform may be derived. To control the phase of the transmitted light Harburn and Taylor (1961) used methods that will be described in Chapter 6.

E. Properties of Optical Transforms

1. Translation

It can be shown that if $G_1(\mathbf{S})$ be the Fourier transform of a function $\rho(\mathbf{r})$ with respect to a particular origin the Fourier transform of $\rho(r - a) = \rho(\xi)$, i.e., the function $\rho(\mathbf{r})$ with origin shifted by **a** is given by

$$G_2(\mathbf{S}) = G_1(\mathbf{S}) \exp\{-2\pi i(\mathbf{a}.\mathbf{S})\}. \tag{3.11}$$

For

$$G_2(\mathbf{S}) = \int \rho(\mathbf{r} - \mathbf{a}) \exp\{-2\pi i(\mathbf{r}.\mathbf{S})\} \, dv_r$$

$$= \int \rho(r - a) \exp\{-2\pi i(\mathbf{r} - \mathbf{a}).\mathbf{S}\} \, dv_r \exp\{-2\pi i(\mathbf{a}.\mathbf{S})\}$$

$$= \exp\{-2\pi i(\mathbf{a}.\mathbf{S})\} \int \rho(\xi) \exp\{-2\pi i(\boldsymbol{\xi}.\mathbf{S})\} \, dv_\xi$$

$$= G_1(\mathbf{S}) \exp\{-2\pi i(\mathbf{a}.\mathbf{S})\}.$$

This shows that a shift of origin produces a phase shift in transform space while the modulus remains unaffected. In optical transforms, which are only records of the square of the modulus, the phase shift produced by the shift of origin, that is by translation of an object parallel to itself, is not observed. The effect of parallel translation of an object is thus not detectable in optical transforms.

2. *Rotation*

Rotation of an object in real space, about any axis, produces rotation of the transform about a parallel axis, with the same angular velocity. Any change of location of the axis of rotation in real space may be regarded as a combination of the operations of rotation and translation. Since translation in real space has no effect on an optical transform, change of location of the rotation axis in the real space does not produce any corresponding change in the rotation of the transform which always rotates about a parallel axis passing through the centre of the transform.

3. *Addition*

If two structures are added in real space the transform of the combination is the vector sum of the transforms of the individual structures referred to a common origin. Thus, the transform of any complicated structural unit may be regarded as the sum of the transforms of a few simple units into which it can be broken up. Figure 3(a) shows the transform of a mask representing the (010) projection of benzo[1,2:4,5]dicyclobutene (Lawrence and Mac-Donald, 1969). The mask may be regarded as the addition of two configurations of holes—an outer group of six holes arranged in the form of a distorted hexagon and an inner group of four holes distributed at the corners of a parallelogram. The transforms of the two units are shown in Figs. 3(b) and (c) with indication of the phases of the various transform regions (see Section VI.A). The vector addition of these two transforms gives the Fig. 3(a).

For two similar structures with parallel orientations but separated by a vector distance **a** the sum of the transforms $G(\mathbf{S})$ and $G'(\mathbf{S})$ is given by

$$G(\mathbf{S}) + G'(\mathbf{S}) = G(\mathbf{S}) + G(\mathbf{S})\exp\left[-2\pi i(\mathbf{a}.\mathbf{S})\right]$$
$$= G(\mathbf{S})\left[1 + \exp\{-2\pi i(\mathbf{a}.\mathbf{S})\}\right]. \qquad (3.12)$$

Thus the combined transform is just the transform of only one structure modulated by straight fringes as represented by the function $[1 + \exp\{-2\pi i(\mathbf{a}.\mathbf{S})\}]$. This is illustrated in Fig. 4. Such fringes sometimes lead to molecular location in the process of structure determination (see Section IV.D). If the two structural units do not have parallel orientation but are related by centre of symmetry, as for a non-centrosymmetrical projection of molecules arranged in a centrosymmetrical plane group, the fringes will be wavy in nature (Taylor and Lipson, 1964), as shown in Fig. 5.

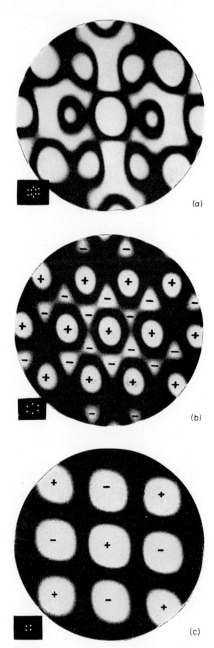

Fig. 3. Illustration of addition property of transforms. (a) Optical transform of a molecule of benzo[1,2:4,5]dicyclobutene in the (010) projection. (b) Optical transform of the outer ring of six atoms of the molecule with indication of phases of the transform regions. (c) Optical transform of the inner ring of four atoms of the molecule with indication of phases of the transform regions.

4. Multiplication and Convolution

According to the convolution theorem (Chapter 10, Section II.F), if $G_1(S)$ and $G_2(S)$ are the Fourier transforms of functions $\rho_1(r)$ and $\rho_2(r)$ then the Fourier transforms of the product $G_1(S)G_2(S)$ is a function $\rho_{12}(r)$ with distribution given by the convolution product of $\rho_1(r)$ and $\rho_2(r)$ as defined by the convolution integral (Ewald, 1940)

$$\rho_1\rho_2 = \int \rho_2(r)\rho_1(r-r')\,dv_{r'} \qquad (3.13)$$

where r and r' are vectors in real space independent of each other. Conversely, it can be shown that Fourier transform of the convolution is the product of the two transforms. That is,

$$G\rho_1\rho_2(S) = G_1(S)G_2(S). \qquad (3.14)$$

The concept of convolution product of functions is of great value in problems of structure analysis. The convolution may be regarded as an operation by which position vectors of the two functions to be convoluted are vectorially added while the corresponding scattering functions are multiplied. It is thus a mathematical device by which a function may be reproduced at various points defined by another function (Lipson and Taylor, 1958). From the standpoint of convolution theory, a crystal structure is regarded as the convolution of the function representing the distribution of scattering matter in the unit cell and the lattice point function. The X-ray diffraction pattern is therefore the product of the diffraction pattern of the unit cell contents and the diffraction pattern of the lattice point function. The latter is the reciprocal lattice (Chapter 1, Section I.E). The transform of the structure is thus the transform of the unit-cell contents sampled at the reciprocal-lattice point, in much the same way as the diffraction pattern of a diffraction grating is the single-slit pattern sampled at the positions of the various spectra of the grating (Chapter 1, Fig. 8). In optical-transform work the convolution-multiplication property plays a very useful role as it enables structural investigations to proceed quite effectively using the diffraction pattern of a mask representing the contents of only one unit cell of the structure (Chapter 1, Section I.H). Such a possibility of dispensing with the necessity of preparation of large mask to represent an extended structure with many unit cells is a great advantage in practice.

IV. OPTICAL TRANSFORMS IN PRACTICE

A. Optical Transform as an Aid to the Trial-and-Error Method

The testing of the correctness of a trial structure is the most obvious use of optical transforms in structure analysis. In the trial-and-error method an approximate model of the structure is postulated from chemical and other evidence and its correctness or otherwise is tested by comparing a set of experimentally derived structure amplitudes with those calculated from the atomic parameters deduced from the postulated model. The process is repeated till one can produce a reasonably correct model suitable for refinement by routine procedures. The burden of computational labour slows down the process of such an approach even for a simple structure and with increasing complexity the work may soon become overwhelming unless a modern computer is available. Besides the burden of computation, one inherent disadvantage of this hit-or-miss method is its inability to provide any guide-line to changes in the atomic parameters that may be necessary to approach the correct structure. The use of optical transforms (Taylor and Lipson, 1951; Hanson et al., 1953) provides a rapid method for testing trial structure. The technique has added a new vigour to the old trial-and-error method by providing an elegant experimental equivalent that dispenses with much tedious calculation at the initial stages of structure determination. Further the method can indicate how far a particular trial model departs from the correct one and what possible changes in atomic positions may be necessary to improve the model (Chapter 1, Section I.G).

B. *Adoption of Standard Scales and Hole Sizes*

The first step in the procedure for testing trial structures by the optical-transform method is to punch holes in an opaque mask to represent the pattern of atomic arrangement of the unit cell, projected on to an appropriate plane. For an optical diffractometer having $4\frac{1}{2}$-in. diameter lens system, masks with holes representing atomic arrangement on a scale of $\frac{1}{6}$ cm $= 1$ Å has been found convenient. This is achieved after reducing a drawing on a scale of 2 cm to 1 Å with a pantograph punch (Hughes and Taylor, 1953) having a reduction ratio 12:1. For a non-planar molecule, a mask can be punched directly from a three-dimensional wire model to obtain projection of any desired orientation (Hughes et al., 1949). The wire model is mounted above the pantograph table and location of each atomic site in projection is found by looking down at its image on a small mirror fitted on the free end of the movable pantograph arm which is moved till alignment with the object atom is obtained at the centre of a cross engraved on the mirror. The small-angle Fraunhofer diffraction patterns of these masks, produced by the optical diffractometer, are the optical transforms which are enlarged to a suitable scale for comparison with the X-ray diffraction data (see Section IV.C).

Hole sizes normally used range from $\frac{1}{2}$ mm to $1\frac{1}{2}$ mm in diameter, but for problems with atoms of nearly same scattering factor $1\frac{1}{2}$ mm or 1 mm is most convenient. When representation of atoms of different scattering factors becomes necessary the simplest way is to vary the hole sizes so that areas of the holes are roughly proportional to the scattering factors of the atoms they are supposed to represent. Such a procedure may however fail to keep the proper adjustment in the amplitude ratio of the scattered beam at all angles as it is well known that the larger the hole size the sharper is the fall of the intensity of the central maximum in the diffraction pattern of a circular aperture (Chapter 1, Section II.C). A solution to this practical problem has been suggested by Harburn (1961) who has described a photo-etching technique for the preparation of gauzes of different transmission factors to adjust the intensity of transmitted light from holes of the same size.

C. *Drawing of Weighted Reciprocal Lattice*

An undistorted visual representation of X-ray diffraction data, suitable for direct comparison with optical transforms, is best obtained by drawing the corresponding weighted reciprocal-lattice section. For drawing such a section, a scale of 5 cm to 1 Å^{-1} is convenient for routine work. The reciprocal points are weighted by drawing them as black disks with diameters proportional to the structure amplitudes of the various reflexions. Usually the structure amplitudes are divided into the following six groups—$0·1\,M$ to $0·3\,M$, $0·3\,M$ to $0·5\,M$, $0·5\,M$ to $0·7\,M$, $0·7\,M$ to $0·9\,M$ and $0·9\,M$ to M, where M is the maximum value amongst the observed structure amplitudes. The first group is generally omitted. It is worthwhile mentioning that, although the optical transform is a record of intensity and not of amplitude, because of the response of most photographic materials the transform simulates amplitude diagrams and reciprocal-lattice sections with weighting proportional to the structure amplitudes have been found quite satisfactory. The best presentation of weighted reciprocal lattices is, however, obtained by using unitary structure factors (Harker and Kasper, 1948) as this emphasizes structure-sensitive high-angle reflexions that otherwise remain insignificant due to a general fall-off of scattering-factor curves and attenuation by temperature. Such high-angle reflexions are important in the optical-transform approach as they very often provide precise structural information. A procedure for deriving unitary structure factors directly from the observed intensity data is described by Woolfson (1961). Figure 6(a) shows the *h0l* section of the weighted reciprocal lattice of the red form of 5-methoxy-2-nitrosophenol (Crowder *et al.*, 1959), and the optical transform of the unit-cell contents projected on to (010) plane is shown in Fig. 6(b) for comparison.

Though most optical-transform work is two-dimensional, on special occasions presentation of a complete three-dimensional weighted reciprocal lattice may be of interest. This can be easily achieved by drawing the weighted reciprocal-lattice sections corresponding to different layers separately on clear glass plates and mounting them on a frame with appropriate spacings (Kenyon and Taylor, 1953). Such three-dimensional reciprocal lattices when viewed from different angles may often unfold interesting aspects of structural features (Taylor and Lipson, 1964).

D. Testing of Approximate Structure

In order to test a trial structure by the optical-transform method, a comparison has to be made between the optical transform of the structure projected on an appropriate plane and the corresponding weighted reciprocal-lattice section so as to seek an agreement between the strong and weak regions of the transform with the strong and weak reflexions of the weighted reciprocal-lattice section. Normally a suitable transparency of the weighted reciprocal-

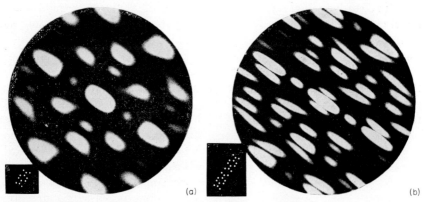

Fig. 4. (a) Optical transform of a single molecule of p-nitrobenzoic acid projected on to (010). (b) Optical transform of two molecules of *p*-nitrobenzoic acid projected on to (010) showing straight fringes.

lattice section is superposed upon the transform in appropriate orientation to observe the match. It is pointed out in Section III.E that according to the convolution-multiplication property of transforms a crystal structure is the convolution of the unit-cell contents with the lattice point function, and hence the transform of the structure is the product of the transform of the unit-cell contents and the transform of the lattice. In other words the transform of the structure is essentially the transform of the unit-cell contents observed only at the reciprocal-lattice point. Therefore for testing the correctness of a structure it is sufficient to take the transform of only one unit cell (Chapter 1, Section I.H). Further, if two molecules in the unit cell of any structure have parallel orientations in a projection on to a plane, consideration of the transform of only one molecule is necessary (Section III.E). For example, a situation of this kind arises in all structures having centrosymmetric molecules (or centrosymmetric projections of molecules) with plane group $p2$ of projection. In such cases a correspondence between strong regions of the transform of one molecule with the strong reflexions may be regarded as the necessary criterion for the correct shape of the molecule. Weak or zero X-ray intensities may appear in strong regions of the transform of a single molecule due to occurrence of nodal lines in the composite diffraction pattern (see Fig. 4) of the centro-symmetrically related molecules in the effective unit cell of the projection. The orientation and separation of these nodal lines, once recognized in the weighted-reciprocal lattice section, may give the separation of the molecular centre and may locate the position of the molecule immediately (Section V.B). Even when molecules have non-centrosymmetric projections, the transform of only one molecule need be considered while investigating the molecular

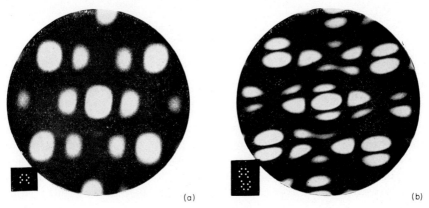

(a)   (b)

Fig. 5. (a) Optical transform of a single molecule of catechol projected on to (010). (b) Optical transform of two molecules of catechol projected on to (010) showing wavy fringes.

shape. The nodal lines in the interference pattern of the centrosymmetrically related pair of molecules which are themselves non-centrosymmetric are however wavy in nature (see Fig. 5) and may not help direct evaluation of molecular position. However, if the molecules have some centrosymmetric groups in themselves and these groups have their own characteristic transform peaks, one may look for more-or-less straight nodal lines in the corresponding regions of the weighted reciprocal-lattice section to obtain any possible clue for molecular location (Section V.C). More objective methods for molecular location, utilizing such fringe functions in transform space, are described by Taylor (1954a) and Taylor and Morley (1959).

If a plane group involves a unit cell with two or more molecules related by a mirror or glide-plane, the optical-transform test may still be carried out quite effectively using the transform of one molecule. It is well known that reciprocal space has the same symmetry as the point-group symmetry of the plane group in real space. In addition reciprocal space is centrosymmetric.

4

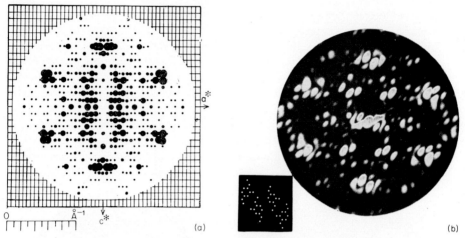

Fig. 6. Visual comparison of weighted reciprocal lattice section with optical transform. (a) *h0l* section of the weighted reciprocal lattice of the red form of 5-methoxy-2-nitrosophenol. (b) Optical transform of the contents of a unit cell in (010) projection of 5-methoxy-2-nitrosophenol (plane group *pgg*).

A given X-ray reflexion in these cases thus occurs four times, distributed in the four quadrants. For adjudging a correct shape all that is necessary is to ensure that strong reflexions in the weighted reciprocal lattice lie on the strong parts of the transform *or* its mirror image. In other words, if the *hk*0 reflexion is strong then either the *hk*0 or the *h$\bar{k}$*0 reflexion should lie on the strong regions

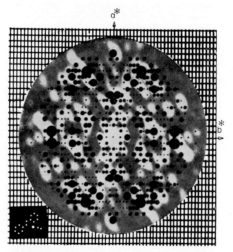

Fig. 7. Optical transform of single molecule of azobenzene-2-sulphenyl cyanide in (001) projection with the weighted reciprocal-lattice section superimposed upon it.

of the transform. This is illustrated in Fig. 7 for the correct shape of a single molecule of azobenzene-2-sulphenyl cyanide (Kakati and Chaudhuri, 1968) for which the plane group of the (001) projection is *pgg*. In extending the optical transform test to the complete unit-cell contents, it has to be remembered that for plane groups involving a glide plane, the optical transform does not exhibit the exact symmetry of the corresponding X-ray diffraction pattern which displays the full point-group symmetry of the plane group. That is, the transform regions of the two adjacent quadrants are not exact mirror images of each other (see Fig. 6b) so far as their shapes are concerned, although the intensity at the reciprocal-lattice points are the same (Taylor and Lipson, 1964).

E. Extension to Testing Positions of Hydrogen Atoms

A quick optical-transform procedure for testing the correctness of any postulated configuration of the hydrogen atoms in a structure is suggested by Pinnock and Lipson (1954). A mask is made to represent only the hydrogen atoms, the positions of which may be based upon standard bond lengths and angles. Such speculation becomes easy for hydrogen atoms in aromatic ring compounds where the hydrogen atoms are co-planar with the ring. Superposing a transparency of the reciprocal lattice net on the resulting transform, one can make a quick visual estimate of the relative magnitudes of the contributions of the hydrogen-atom configuration to the various reflexions at low angles where the X-ray scattering by hydrogen atoms has significant values. For a centrosymmetric projection it is possible to derive quickly the signs of the hydrogen-atom contributions optically (see Section VI). This may immediately suggest whether the inclusion of the hydrogen atoms in the structure-factor calculation produces any improvement in the agreement. It is true that difference synthesis (Cochran, 1951) provides a more objective method for location of hydrogen atoms, but the optical method provides a quick guide and may even suggest whether the data are accurate enough for a difference synthesis to be useful. Moreover, while the difference synthesis is confined to centrosymmetric structures the optical method may be extended to non-centrosymmetric projections as well.

V. STRUCTURE DETERMINATION

A. Basic Strategy

Structure determination by optical-transform methods may be described as a composite strategy involving two basic operations. Firstly, a careful examination of the weighted reciprocal lattice is made to obtain the highlights of information about the whole structure or any part of it. Secondly, an attempt is made to co-ordinate these items of information with whatever knowledge of chemical constitution, stereochemistry and crystal symmetry is available, such that the optical transform fits well with the weighted reciprocal-lattice section. A solution is thus sought from an overall assessment of the merits of the diffraction pattern taken as a whole. The process of structure determination starts with simple experiments immediately after collection of intensity data without any involvement in numerical work needing the service of computer which is often considered almost indispensable in other methods of structure analysis right from the beginning.

B. Evidence from Weighted Reciprocal Lattice

1. Introduction

Though, in principle, the optical-transform approach to the problem of structure analysis consists in fitting the transform of the unit-cell contents of the structure with the weighted reciprocal-lattice section, usually the process of structural investigation proceeds in the reverse direction. That is, opportunity for giving a correct start to the investigation often emerges from a careful study of the weighted reciprocal lattice. This furnishes a visual perspective of the X-ray diffraction pattern as a whole and one may recognize in it significant structural features to form the basis of the structure determination. In other words, in optical-transform work the weighted reciprocal lattice may be used as an equivalent of the conventional Patterson function map. The only difference is that computation of the Patterson function converts intensity data into real space in the form of information about inter-atomic vectors from which the crystal structure has to be built up, while in using the reciprocal lattice one has to work out the pattern in real space directly from the pattern in reciprocal space.

2. Recognition of Molecular Features

An approximate idea about the general feature of the molecular shape and size may often be obtained from a survey of the general characteristics of the transform. The symmetry in a transform conforms to the point-group symmetry (crystallographic or non-crystallographic) of the object. One may therefore always look for evidence of symmetry in the weighted reciprocal lattice to derive any possible information about symmetries in molecular shape. Figure 8(a) shows the $h0l$ section of the weighted reciprocal lattice of coronene (Robertson and White, 1945). The general hexagonal symmetry that is observed in it conforms to the hexagonal symmetry of the coronene molecule (Fig. 8b). On account of the reciprocal relationship between an object and its transform, a molecule with elongated structure has transform peaks that are narrow in the direction parallel to the length of the molecule (see Fig. 14b).

In the reciprocal-lattice section these are manifest in the form of clusters of reflexions that are narrow along the length of the molecule and elongated perpendicular to it (see Section V.C). In general, the direct evidence regarding the molecular shape is borne out by the origin peak (see Figs. 12f and 13f) which arises due to superposition of fine fringes produced by atoms having the largest possible interatomic separations that define the shape and size of the molecule. Further, the smaller the molecule, the larger are the transform peaks, so that several reciprocal points may lie on a transform peak; with

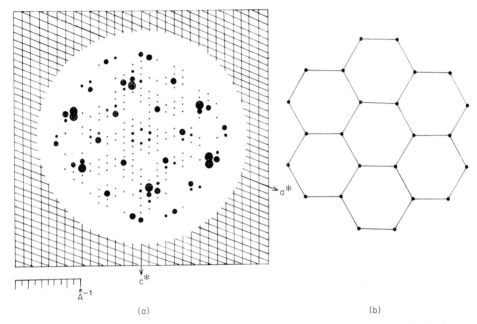

Fig. 8. (a) $h0l$ section of the weighted reciprocal lattice of coronene. (b) The idealized coronene molecule.

large molecules and hence with large unit cells, the general texture is finer, and reciprocal-lattice points are distributed more or less over adjacent peaks.

Structural information derived from such considerations is only qualitative, but one cannot ignore the heuristic value of the approach and the ease with which the information is derived.

3. Recognition of Groups; the Benzene Transform

A usual feature in many organic compounds is the occurrence of structural groups in the form of ring configurations. Such groupings of atoms in a molecule give rise to their own characteristic traces in transforms and their identification in the weighted reciprocal lattice often furnishes the necessary clue to the solution of a structure. One of the most commonly occurring ring structures of atoms in organic compounds is the planar hexagonal aromatic benzene nucleus. Figure 9(a) shows the optical transform of an idealized benzene ring. The transform has six prominent peaks which are called benzene peaks (Taylor and Lipson, 1964) and they form a regular hexagon. If the side of the hexagonal ring is taken as $1\cdot4$ Å, the centres of the peaks lie upon a circle of radius $0\cdot82$ Å^{-1}, called the benzene circle (Taylor, 1952b). Each of these peaks arises due to spacing of magnitude $(1\cdot4\sqrt{3}/2)$ Å of planes in the object. The calculated Fourier transform of a benzene nucleus is shown in Fig. 9(b) where the corresponding peaks at $0\cdot82$ Å^{-1} from the origin are regions of positive amplitude in the transform. It has been pointed out earlier that the three-dimensional Fourier transform of a planar molecule has identical sections perpendicular to the molecular plane. Figures 9(a) and 9(b) therefore represent only right central sections of the complete three-dimensional benzene transform. In an oblique section of the transform the benzene peaks lie on an ellipse representing the transform of the projection of a tilted benzene ring. The contraction of the ring in the projection results in expansion of the transform in the direction along which the ring contracts. Consequently some of the benzene peaks may be pushed well beyond the benzene circle. The optical transform of a tilted benzene ring in projection is shown in Fig. 9(c). The six benzene peaks which lie on an ellipse can be readily recognized. The geometrical guide-line which indicates their location is that the vector distance between the centres of any two adjacent peaks must be equal to the vector distance of the centre of the third adjoining peak from the origin. The following outlines of construction may be used to derive the benzene ring in projection from the benzene peaks. In Fig. 9(d) which depicts the important features of Fig. 9(c), the centres of the centrosymmetrically equivalent benzene peaks such as A_1, A_4, etc. are joined by straight lines passing through the origin. If the scale in the transform space is known, the distances OA_1, OA_2, OA_3, etc. give the distances of this benzene peaks in Å^{-1}. The spacings of the planes giving rise to the corresponding benzene peaks are obtained in Å from the reciprocals of these distances. For a scale of 2 cm = 1 Å, which is normally used for drawing figures in real space (Section IV.B), points B_1, B_2, B_3, etc., are marked on the lines OA_1, OA_2, OA_3, etc., respectively, such that $OB_1 = 2/OA_1$ cm, $OB_2 = 2/OA_2$ cm and so on. The normals drawn to OA_1, OA_2, OA_3, etc., respectively at points B_1, B_2, B_3, etc., outline the shape of the projected benzene ring. In X-ray diffraction patterns of compounds containing a number of benzene rings in parallel orientation or of any condensed-ring compounds involving hexagonal ring configurations, the characteristic peaks of the benzene transforms may be very often recognized in the weighted reciprocal-lattice section in the form of six different clusters

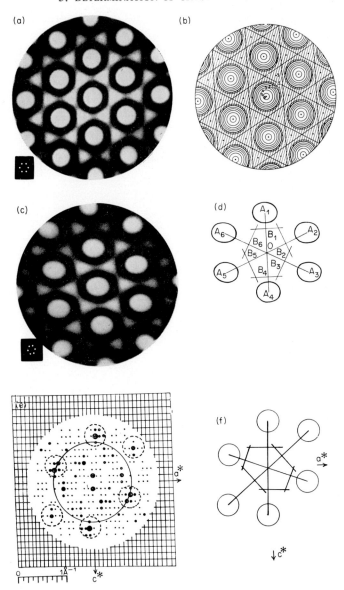

Fig. 9. Benzene transform. (a) Optical transform of an idealized benzene ring. (b) Calculated Fourier transform of an idealized benzene ring. (c) Optical transform of a tilted benzene ring in projection. (d) Construction of the benzene ring in projection from the benzene peaks of (c). (e) $h0l$ section of the weighted reciprocal lattice of bishydroxyduryl-methane (Chaudhuri and Hargreaves, 1956) showing the benzene peaks and benzene circle. (f) Construction showing the derivation of benzene ring in projection for the (010) projection of bishydroxydurylmethane.

of strong reflexions, unless scattering from heavy atoms, or large number of other atoms, eclipses them. Thus, if the benzene peaks are correctly identified in the weighted reciprocal-lattice section of any structure, the orientation of the ring in the projection can be immediately derived as shown in Figs. 9(e) and 9(f). Such considerations also apply to compounds with alicyclic rings having chair or boat conformations which may have hexagonal projections; the X-ray diffraction pattern of the corresponding zones will then display the characteristic benzene-transform patterns.

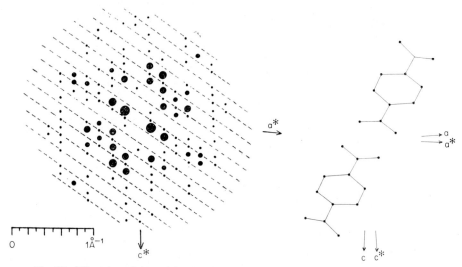

Fig. 10. $h0l$ section of the weighted reciprocal lattice of p-nitrobenzoic acid (with weights proportional to the structure amplitudes) showing straight nodal lines. The two interfering molecules in (010) projection are shown inset.

Similar possibilities are there for identifying characteristic peaks in the transform of other commonly occurring groups in organic compounds, such as five-membered heterocyclic furan, pyrrol, oxazole rings. These peaks in the X-ray diffraction pattern, if correctly recognized, may immediately provide useful structural information relating to these groups.

4. *Fringes*

The presence of two or more molecular units in parallel orientation should produce equidistant interference fringes which may help in molecular location (Section IV.D) in real space unless the interfering units are so far apart that the fringes are too fine to be recognized. The $h0l$ section of the weighted reciprocal lattice (Fig. 10) of p-nitrobenzoic acid shows the production of equidistant straight nodal lines in the interference pattern of the two centrosymmetrically related molecules *cf.* Fig. 4— which are themselves structurally centrosymmetrical and thus have parallel orientation in the projection. The fringe separation of about 0.15 Å^{-1} corresponds to a molecular centre separation of 6.7 Å which agrees well with value arrived at by Sakore and Pant (1966).

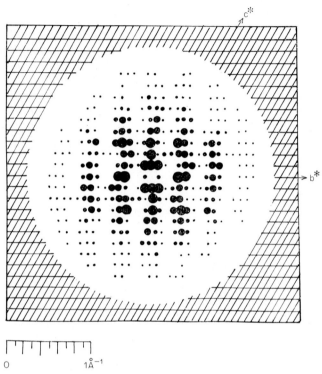

Fig. 11. $0kl$ section of the weighted reciprocal lattice of 9-p-carbethoxyphenyl-9-stibia-fluorene (Chaudhuri, 1955) showing the fringes produced by two centrosymmetrically related heavy antimony atoms.

A molecule having a long regular chain structure of similar atoms gives rise to prominent fringes in reciprocal space owing to the cumulative effect of vector addition of interference fringes produced by numerous pairs of atoms having the same vector separation. Such fringes immediately give the chain direction and the interatomic separation in the chain (see Section V.C).

Sharp fringes are generally observed in the weighted reciprocal-lattice section (Fig. 11) if the unit cell of the projection contains a pair of heavy atoms. Such fringes can be compared with the Young's fringes in optics originating from interference of two coherent sources. The vector separation between the heavy atoms can be deduced from the fringe separation or if the pair originates from a centre of symmetry, the position of the heavy atom is directly located.

Thus, in the optical-transform procedures the weighted reciprocal lattice plays a role somewhat analogous to that of the Patterson function in conventional methods of structural investigation. The optical-transform methods, however, have some obvious advantages. Important structural evidences are almost immediately obtained without any serious calculations. This is decidedly an advantage not only to researchers who may not have easy access to computer facility but also to those who consider it worthwhile to cut down computing hours as far as possible. Moreover, structural evidence is often displayed more distinctly in the weighted reciprocal lattice than in the Patterson function map. For, while parallel orientation of molecules or presence of chain molecules give rise to prominent fringes in the weighted reciprocal lattice, in the Patterson map evidence of such structural feature appears in the form of a lone peak which may not often be easily sorted out from the rest. The Patterson peaks representing interatomic vectors between adjacent atoms of a benzene ring may be weak as, according to the theory of Patterson function (Patterson, 1935), such peaks are produced by high-angle reflexions which are normally weak. On the other hand use of the unitary structure factors enables prominent display of these weak reflexions in the weighted reciprocal lattice.

C. Examples

1. Introduction

In applying the principles of optical-transform methods in the solution of structural problems, the approach may differ according to the different natures of the problems. This is best explained by examples. The most suitable problems are obviously those which involve planar molecules of approximately known forms and have atoms of nearly the same scattering factor. Organic compounds generally provide such examples. The methods have also been applied successfully to solve, in projection, numerous organic structures with non-planar molecules having structural grouping of atoms in the form of ring structures. A few examples of these categories are described below. In addition, for demonstrating the scope of the method, the applications of the optical-transform principles are also described in relation to the solutions of a few structures that have been solved by other methods. It should be emphasized that the main objective in the strategy of the optical-transform method is to provide the approximate structure that may be subsequently refined by the routine procedures.

2. Diphenylene Naphthacene

Bennett and Hanson (1953) used optical-transform methods to obtain the trial structure of diphenylene naphthacene, $C_{30}H_{16}$, which has an approximately planar molecule of the form as shown in Fig. 12(a). The transform

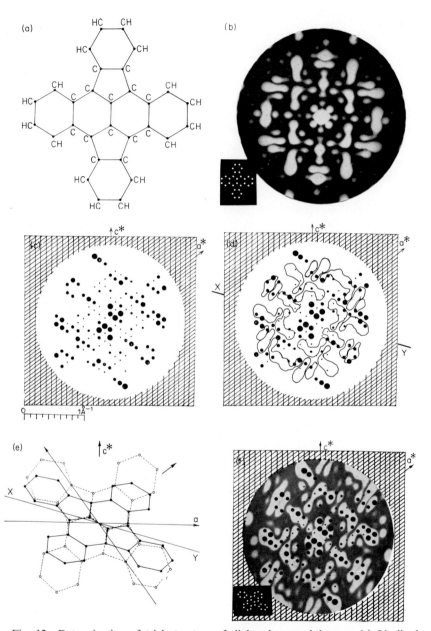

Fig. 12. Determination of trial structure of diphenylene naphthacene. (a) Idealized molecule of diphenylene naphthacene. (b) Optical transform of the idealized molecule. (c) $h0l$ section of the weighted reciprocal lattice. (d) The correct relative orientation of (c) with the prominent features of (b). (e) Derivation of the tilted molecule in projection (bold line). (f) Optical transform of the tilted molecule in projection with the weighted reciprocal-lattice section superimposed upon it to demonstrate the approximate match for the trial structure.

of the molecule is shown in Fig. 12(b). The space group is $P2_1/c$ with unit-cell dimensions $a = 11{\cdot}0$, $b = 5{\cdot}15$, $c = 19{\cdot}6$ Å, $\beta = 126°$. The unit cell contains only two molecules, each lying on centres of symmetry at $(0,0,0)$ and $(\frac{1}{2},\frac{1}{2},0)$. The effective unit cell of projection on (010) contains one molecule and the plane group is $p2$. The structural investigation may therefore be made in relation to the transform of only one molecule (Section IV.D). Figure 12(c) shows the $h0l$ section of the weighted reciprocal lattice. From Fig. 12(d), which depicts the correct relative orientation of the features of Figs. 12(b) and (c), it is clear that to obtain a match between the two, the transform is to be expanded perpendicular to the line XY shown inset. This is equivalent to tilting the molecule about this line out of the plane (010) by an appropriate angle so that molecule contracts in projection in a direction perpendicular to the tilt axis. The necessary factor of expansion of the transform gives the secant of the angle of tilt. Conversely the cosine of the angle of tilt gives the factor of contraction of the projected molecule. Figure 12(c) shows the derivation of the tilted molecule in projection. The required angle of tilt was found to be about $51{\cdot}5°$. The agreement between the transform of the tilted molecule in projection and the weighted reciprocal-lattice section is shown in Fig. 12(f).

3. Tetrabenzmonazaporphin

Tetrabenzmonazaporphin, $C_{35}H_{21}N_5$, has planar molecules of approximately known shape and thus affords an ideal problem for application of optical-transform principles. The structure has been solved completely from two-dimensional X-ray diffraction data by Das (1967) who has used optical transforms freely to obtain the trial structure. The crystal of tetrabenzmonazaporphin belongs to the monoclinic space group $P2_1/a$ with $a = 17{\cdot}6$, $b = 6{\cdot}1$ and $c = 12{\cdot}5$ Å, $\beta = 122{\cdot}7°$ (Woodward, 1940) and has two molecules in the unit cell. Space-group considerations require the molecule to be centrosymmetric. This is however not satisfied by the chemical requirement of the molecule of tetrabenzmonazaporphin which is similar to that of phthalocyanine except that three out of four centrosymmetrically arranged outer nitrogen atoms are replaced by CH groups (Figs. 13a, b). The deviation from the requirement of an exactly centrosymmetrical molecule is however small as the substitution of N by CH produces little change in the effective scattering of X-rays.

Fig. 13. Determination of trial structure of tetrabenzmonazaporphin. (a) Idealized molecule of tetrabenzmonazaporphin. (b) Idealized molecule of phthalocyanine. (c) $h0l$ section of the weighted reciprocal lattice of tetrabenzmonazaporphin. (d) Optical transform of the idealized molecule of tetrabenzmonazaporphin. (e) The derivation of the tilted molecule in projection (bold line). (f) Optical transform of the tilted molecule in projection with the weighted reciprocal lattice-section superimposed upon it to demonstrate the approximate match for the trial structure.

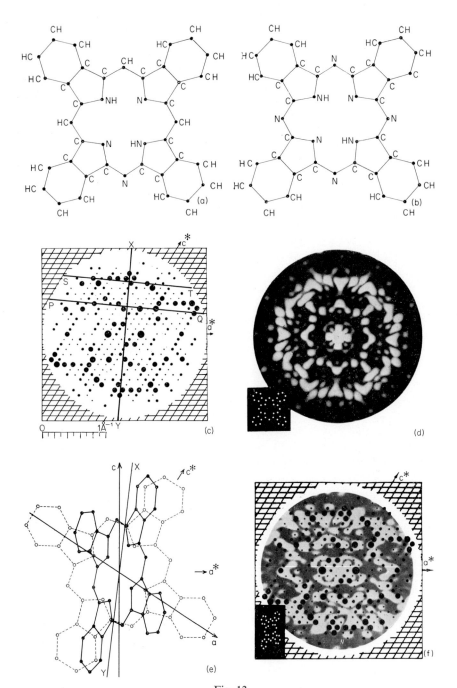

Fig. 13.

The main features of the $h0l$ section of the weighted reciprocal lattice (Fig. 13c) of tetrabenzmonazaporphin are, firstly the group of strong reflexions around the origin, secondly the group of comparatively strong reflexions lying about the line PQ at a distance of 0.4 Å^{-1} from the origin, and lastly a second group of reflexions lying about the line ST at a distance of about 0.84 Å^{-1} from the origin. The effective unit cell of projection along (010) contains only one molecule. Thus for optical-transform considerations it is sufficient to compare the outstanding features of the reciprocal-lattice section with those of the transform (Fig. 13d) of one idealized molecule; these features are the prominent central region, the square-shaped rings of strong peaks at distances of 0.4 Å^{-1} and the outermost prominent ring of peaks at a distance of about 0.84 Å^{-1} from the origin. Obviously, the last two features in the transform account for the two lines of strong reflexions in the weighted reciprocal-lattice section. The stretching of these reflexions along these lines suggests that the molecule is steeply tilted about the tilt axis XY (Fig. 13c) out of the plane of projection. From the point of view of structure determination all that was necessary was to find the direction XY in the idealized molecule and to estimate the magnitude of the tilt angle. A trial rotation of the weighted reciprocal-lattice section with respect to the transform of the idealized molecule immediately revealed the direction of the tilt axis in the molecule and its orientation relative to the crystallographic axis. The estimated value of the tilt angle of the molecule for producing the necessary expansion of the transform, so that it fitted with the weighted reciprocal-lattice section, was found to be $60°$. The projection of the tilted molecule on to (010) plane is shown in Fig. 13(e) where the direction of the tilt axis XY is also shown inset on the idealized molecule. In Fig. 13(f) the approximate match between its transform and the weighted reciprocal-lattice section is demonstrated. A quicker but more expensive procedure for deducing orientation of such planar molecule have been described by Menarry and Lipson (1957) by direct presentation of the transform on television monitor and tilting the mask in a gimbal arrangement. Such information about the orientation of molecule of known shape, in the unit cell, would also follow from packing consideration (Woodward, 1940), but a more objective approach was afforded by the optical-transform methods.

4. 1,5-Dihydroxyanthraquinone

The crystal structure of 1,5-dihydroxyanthraquinone, $C_{14}H_8O_4$, has been solved by Hall and Nobbs (1966) using the calculated Fourier transform of an idealized planar molecule of the compound with standard bond lengths and bond angles. The form of the molecule is shown in Fig. 14(a). There are two molecules in the unit cell having dimensions $a = 15.755$, $b = 5.308$, $c = 6.003$ Å, $\beta = 93° 37'$. The space group is $P2_1/a$. The two molecules therefore lie on special positions and the effective unit cell of the projection on (010)

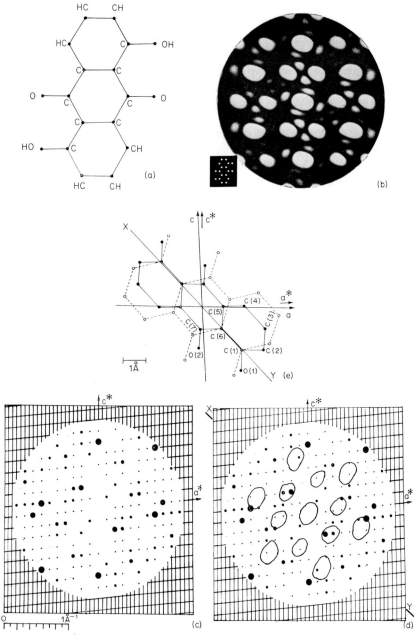

Fig. 14. Determination of trial structure of 1,5-dihydroxyanthraquinone. (a) Idealized molecule of 1,5-dihydroxyanthraquinone. (b) Transform of the idealized molecule. (c) $h0l$ section of the weighted reciprocal lattice. (d) Correct relative orientation of (c) with the prominent feature of (b). (e) Derivation of the tilted molecule in projection (bold line).

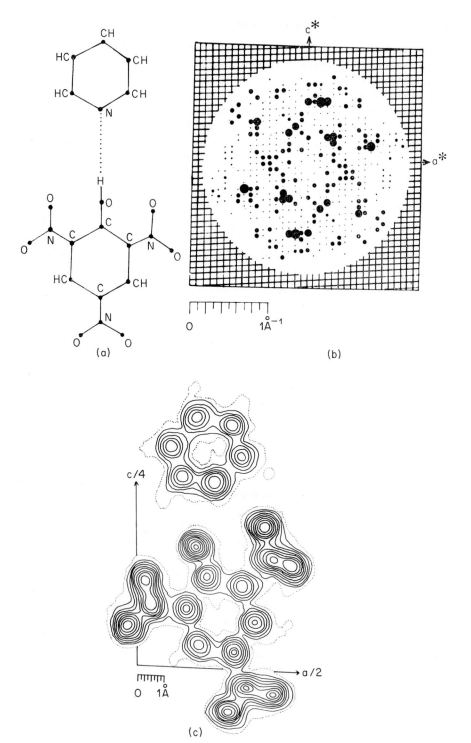

Fig. 17. (a) Idealized molecule of pyridine picrate. (b) $h0l$ section of the weighted reciprocal lattice. (c) (010) electron-density projection.

8. *Catechol*

Brown (1966) made use of a few outstanding reflexions having structure amplitudes near the maximum possible values, to derive the trial structure of catechol, $C_6H_6O_2$. The atomic co-ordinates were adjusted by trial to make the calculated structure amplitudes large for these reflexions. The space group is $P2_1/a$ and there are four molecules in general positions placed in a unit cell with dimensions $a = 10\cdot94$, $b = 5\cdot51$, $c = 10\cdot07$ Å and $\beta = 119°$. A simpler approach perhaps follows from optical-transform consideration. The $h0l$

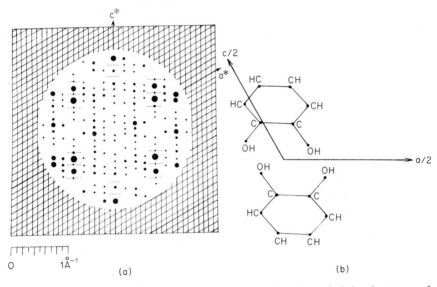

Fig. 18. (a) $h0l$ section of the weighted reciprocal lattice of catechol showing traces of fringes in the regions of the benzene peaks. (b) Two centrosymmetrically related molecules of catechol in (010) projection.

weighted reciprocal lattice section is shown in Fig. 18(a) which displays the benzene-transform pattern. Assuming the normal C—O bond distance and utilizing the benzene peaks to derive the orientation of the phenyl ring of the catechol molecule in projection, a correct molecular model was quickly obtained from the optical-transform test. The two centrosymmetrically related molecules (Fig. 18b) in the effective unit cell of projection on (010) give rise to wavy nodal lines (*cf.* Fig. 5b) which are however difficult to trace out in the weighted reciprocal-lattice section. Nevertheless, in the six benzene peak regions, traces of straight fringes produced by the two centrosymmetrical phenyl rings of the two molecules are easily recognized (Section IV.D). The separation of the nodal lines directly suggests the location of the molecule in real space and gives $x = 0\cdot95$ Å, $Z = 2\cdot75$ Å as the co-ordinates of the centre of the phenyl ring. These compare well with the values $x = 1.30$ Å, $Z = 2.75$ Å found in the final analysis. However, for a direct approach to the problem of molecular location in such a situation, the method suggested by Taylor (1954a) and Taylor and Morley (1959) may be useful.

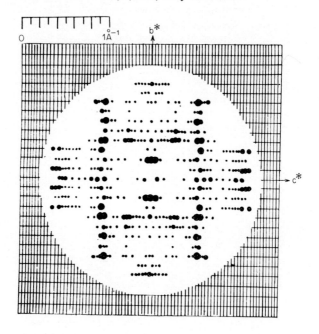

Fig. 19. $0kl$ section of the weighted reciprocal lattice of α-brassylic acid with the molecule shown inset.

9. α-Brassylic Acid

The $0kl$ weighted reciprocal-lattice section (Fig. 19) of α-brassylic acid, $COOH(CH_2)_{11}COOH$ (Housty, 1968), involving a polymethylene chain, affords an interesting example in which sharp fringes occur due to addition of fringes produced by pairs of atoms with the same vector separation. The space group is $P2_1/c$ with four molecules in the unit cell of dimensions $a = 5.59$, $b = 9.5$, $c = 37.6$ Å, $\beta = 135°$. The structure was solved by Patterson projections. Useful information can however be immediately derived from the weighted reciprocal-lattice section. The fringes parallel to b^* axis indicate that in the projection the direction of the chain is perpendicular to the $b*$ axis. A measurement made directly on the weighted reciprocal-lattice section gives a fringe spacing of 0.56 Å^{-1} which suggests a value of 1.80 Å for the projected repeat distance of the atomic units along the chain direction. This obviously corresponds to the projected separation between alternate carbon atoms of the methylene groups in the long zig-zag polymethylene chain of the molecule. If the normal C—C bond distance and tetrahedral bond angle ($109\frac{1}{2}°$) of a saturated carbon atom are assumed, the repeat distance works out at 2.52 Å. The chain axis is therefore inclined to the plane of projection at an angle of about $45°$, indicating that the chain is almost parallel to the c axis.

VI. OPTICAL PHASE DETERMINATION

A. Centrosymmetric Projections

For centrosymmetric projections of structures containing atoms of very nearly the same scattering factor, the optical-transform approach provides a simple experimental method for quick determination of the phase angles of the structure amplitudes. The importance of this method lies in the fact that it completely eliminates computational work which otherwise becomes indispensable, as structure factors of the various reflexions have to be calculated to know their phases before refinement of the structure is attempted by Fourier methods. The method can be effectively used in the earlier stages of refinement and routine calculations may be started or the computer may take over only at a fairly advanced stage in the process of refinement of a structure. The cut-down in the volume of computational work is decidedly an advantage in laboratories where computer facilities are not readily available or are expensive to obtain.

It is well known that for centrosymmetric projections the phase angles of the structure amplitudes with respect to the origin at the centre of symmetry are either 0 (positive) or π (negative). Once a reasonable agreement between the weighted reciprocal-lattice section and the optical transform of the corresponding trial structure is obtained, the signs of the structure factors may be determined if the positive and negative regions in the transform are recognized. This is easily done optically by punching an extra hole at the centre of symmetry of the mask representing the trial structure and obtaining its transform. The addition of an extra hole at the origin gives a positive bias to the whole transform, resulting in depression of the intensities of the negative regions and enhancement of the positive ones (Lipson and Taylor, 1951) as shown in Fig. 20 for a hypothetical case of a square ring of four atoms. A visual comparison at once leads to the identification of the positive and negative regions in the transform. When the unit cell or the effective unit cell of projection contains

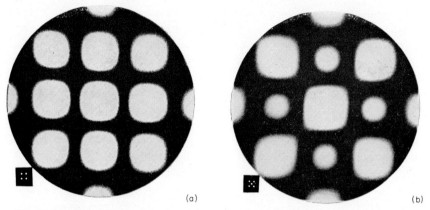

(a)
(b)

Fig. 20. Optical sign determination. (a) Optical transform of a hypothetical square ring of four holes. (b) Optical transform of the same square ring with an extra hole at the centre of symmetry.

a fairly large number of atoms, the scattering from one extra hole at the origin may not be adequate to produce any tangible effect on the transform. A larger hole at the origin may give desirable results in regions near the origin of the transform but may fail to produce any effect at its outer region due to the smaller area of the Airy disk in the diffraction pattern of the circular hole (Section IV.B). The best result in such circumstances is obtained by a hole of the same size and covering the rest of the mask with fine wire gauze that lowers down the transmission of the holes (Hanson *et al.*, 1953). Figure 21 demonstrates the process for the *b* axis projection of tetrabenzmonazaporphin

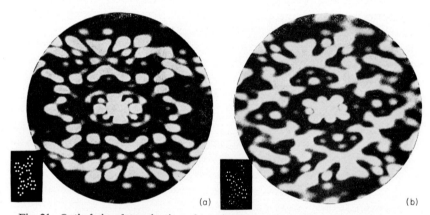

(a)
(b)

Fig. 21. Optical sign determination of tetrabenzmonazaporphin using gauze technique. (a) Optical transform of a mask representing the (010) projection of tetrabenzmonazaporphin (see Fig. 13). (b) Optical transform of the same mask as in (a) but all the holes, excepting an extra hole at the centre of symmetry, are covered with gauze to lower their transmission.

(Das, 1967). Once the signs of the transform peaks are identified, the signs of the reflexions may be read off by placing the weighted reciprocal-lattice section on the transform in appropriate position. When a transform is found to contain fine detail, as it will for a complicated structure, it is convenient to trace out the outline of various positive and negative peaks and to read off the signs of the structure factors by laying an unweighted reciprocal-lattice net on the tracing.

It is quite obvious that due to the presence of centre of symmetry in all diffraction patterns the process of sign determination for each reflexion may be carried in two opposite quadrants, thus giving independent checking of results. In a plane group of a projection involving the presence of a mirror or glide plane, the symmetry of the corresponding reciprocal space offers additional scope for independent checking of the correctness of sign assignments. Of course, in doing so, one has always to keep in view the sign relation that exists between equivalent reflexions occurring in the four quadrants (Pinnock and Taylor, 1955). Although the optical transforms of such plane groups may not display the full symmetry of the reciprocal-lattice section, (Section IV.D) the method has been successfully used by various workers (Taylor, 1952b; Pinnock et al., 1956; Crowder et al., 1959; Kakati and Chaudhuri, 1968).

B. Non-centrosymmetric Projections

For non-centrosymmetric projections the process of optical phase determination is however not straightforward. The real and imaginary parts of the transform have to be derived separately. The real part of the transform with respect to a given origin is obtained by taking the transform of a mask in which the structure and its centrosymmetrically related unit (centre of symmetry at the origin) are represented. To derive the imaginary part optically, the same mask is used, but the holes representing the centrosymmetrically related unit of the structure are covered with mica sheet of appropriate thickness to introduce a phase change of $180°$ to the light transmitted by these holes. The process is the physical equivalent of the mathematical procedure for separating the real and imaginary parts of the expression for Fourier transform by addition or subtraction of the complex conjugate term (Taylor and Lipson, 1964). Using such techniques, Hanson et al. (1953) derived optically the real and imaginary parts of the transform of the triphenylene molecule. These are compared with the calculated real and imaginary parts (Klug, 1950b) in Fig. 22. The phases in the real part are either 0 or π and in the imaginary part are either $\pi/2$ or $3\pi/2$. As usual, the identification of the phases of the various regions in both the transforms is made by addition of an extra hole at the origin. An approximate assessment of phases in the various regions in the transform may be made by giving appropriate weight to the contributions of the real and imaginary parts towards the overall intensity at any point in the transform and combining the phases of the two parts vectorially. Hanson (1953a) successfully applied this technique of phase

(a)

(b)

(c)

(d)

Fig. 22. (a) Real part of the optical transform of triphenylene. (b) Real part of the calculated Fourier transform of triphenylene. (c) Imaginary part of the optical transform of triphenylene. (d) Imaginary part of the calculated Fourier transform of triphenylene. Reproduced from Taylor and Lipson (1964) with permission.

determination in refining the structure of di-*p*-anisyl nitric oxide. The process is however laborious and ensures only a rough estimate of phase, but it amply demonstrates the scope of extension of the simple experimental approach of the optical methods to such situation.

VII. Conclusion

The foregoing discussions demonstrate the scope of optical methods in lessening the burden of computation in the process of crystal-structure determination from X-ray diffraction data. The apparent subjectivity and the lack of quantitative accuracy of the methods are often weighed against the objectivity and precision of a modern digital computer which is now so freely used in most of the direct methods of analysis. The optical-transform methods, however, have some distinct advantages. The approach is based on experiments of basic physics and requires no expensive equipment. The simple experimental techniques of preparation of optical transforms have speeded up procedures to a remarkable degree; although this cannot be compared with that of a modern computer, it may nevertheless be considered important by a researcher who does not have ready access to a computer laboratory. Still more relevant is the fact that results are immediate; although the computer is fast, there is always a lapse of time before its calculations are available.

The most noteworthy aspect of the process lies in the nature of the approach in which a physical harmony in the sequence of the procedures of structural investigation is maintained, right from the stage of collection of X-ray data to the final achievement of results, by exploiting the close relationship that exists between an object and its diffraction pattern. The method thus has an inherent appeal in the sense that results are visually presented by physical experiments, rather than achieved by mathematical calculations done by a machine.

The introduction of computer service in the field of crystal-structure analysis is a great step forward in extending the application of the process to more and more complicated problems. But excessive use of the computer is liable to turn the entire procedure of structure determination into a routine mechanical job and therefore may deprive the researcher of the satisfaction of appreciating the operation of the physical principles that are involved in the process. This may have the tendency to discourage a young investigator from cultivating the spirit of planning his own strategy to resolve any impasse arising in the course of an investigation. It is not improbable that over-zealous use of computer programmes, right from the beginning of a structural investigation, may slow down the progress by leading an observer astray to a pseudo-structure that may show a good measure of agreement, but the atomic positions of the achieved structure may not conform to the chemical requirement of the molecule. The optical-transform test on the other hand is always based on models that are chemically feasible and thus rules out such possibility. Thus a little consideration given to the transform approach at the initial stages of an analysis would not only be important from the educational point of view but also might be found helpful in the final achievement of results.

Optical methods are best suited for problems where the atoms are of about the same weight whereas the usual methods are better when a heavy atom is present. The heavy atom must, however, in some way distort the chemical arrangement and any result obtained should always be supplemented by work on the parent compound. The transform approach therefore may be regarded as a suitable supplement to conventional methods.

Finally, optical methods might be considered as a happy come-back of experimental physical optics in the subject of X-ray crystallography, which was discovered as a method for exploring the structure of matter on an atomic scale for the first time through simple experiments on X-ray diffraction phenomena. Brilliant handling of the subject by mathematically minded specialists (e.g. Woolfson, 1961) has since pushed the subject almost to the limit, but even then, from the point of view of physics, the transform approach may still deserve exploration.

References

Aravindakshan, C. (1957). *J. Sci. Inst.* **34**, 250.
Bailey, M. (1958). *Acta Cryst.* **11**, 103.
Banerjee, K. (1930). *Nature, Lond.* **125**, 456.
Bennett, A. and Hanson, A. W. (1953). *Acta Cryst.* **6**, 736.
Berry, C. R. (1950). *Am. J. Phys.* **18**, 269.
Bragg, W. L. (1939). *Nature, Lond.* **143**, 678.
Bragg, W. L. and Lipson, H. (1943). *J. Sci. Inst.* **20**, 110.
Brown, C. J. (1966). *Acta Cryst.* **21**, 170.
Buerger, M. J. (1950). *J. app. Phys.* **21**, 909.
Chaudhuri, B. (1955). Ph.D. Thesis. Manchester University.
Chaudhuri, B. and Hargreaves, A. (1956). *Acta Cryst.* **9**, 793.
Clastre, J. and Gay, R. (1950). *J. Phys. Radium* **12**, 75.
Cochran, W. (1951). *Acta Cryst.* **4**, 408.
Crowder, M. M., Morley, K. A. and Taylor, C. A. (1959). *Acta Cryst.* **12**, 108.
Das, I. M. (1967). D.Phil. Thesis, Gauhati University.
Ewald, P. P. (1921). *Z. Krist.* **56**, 129.
Ewald, P. P. (1940). *Proc. phys. Soc.* **52**, 167.
Hall, D. and Nobbs, C. L. (1966). *Acta Cryst.* **21**, 927.
Hanson, A. W. (1953a). *Acta Cryst.* **6**, 35.
Hanson, A. W. and Huml, K. (1969). *Acta Cryst.* **B25**, 1766.
Hanson, A. W., Taylor, C. A. and Lipson, H. (1951). *Nature, Lond.* **168**, 160.
Hanson, A. W., Lipson, H. and Taylor, C. A. (1953). *Proc. R. Soc. A.* **218**, 371.
Harburn, G. (1961). Ph.D. Thesis. Manchester University.
Harburn, G. and Taylor, C. A. (1961). *Proc. R. Soc. A.* **264**, 339.
Harker, D. and Kasper, J. S. (1948). *Acta Cryst.* **1**, 70.
Hettich, A. (1935). *Zeits. f. Krist.* **90**, 473.
Hosemann, R. (1962). *Polymer.* **3**, 349.
Hosemann, R. and Bagchi, S. N. (1962). "Direct Analysis of Diffraction by Matter," North-Holland Publishing Co., Amsterdam.
Housty, J. (1968). *Acta Cryst.* **B24**, 486.

Hughes, J. W., Phillips, D. C., Rogers, D. and Wilson, A. J. C. (1949). *Acta Cryst.* **2**, 420.

Hughes, W. and Taylor, C. A. (1953). *J. Sci. Inst.* **30**, 105.

Kakati, K. K. and Chaudhuri, B. (1968). *Acta Cryst.* **B24**, 1645.

Karle, J. and Karle, I. L. (1966). *Acta Cryst.* **21**, 849.

Kenyon, P. A. and Taylor, C. A. (1953). *Acta Cryst.* **6**, 745.

Klug, A. (1950b). *Acta Cryst.* **3**, 176.

Knott, G. (1940). *Proc. Phys. Soc.* **52**, 229.

Lawrence, J. L. and MacDonald, S. G. G. (1969). *Acta Cryst.* **B25**, 978.

Lipson, H. and Taylor, C. A. (1951). *Acta Cryst.* **4**, 458.

Lipson, H. and Taylor, C. A. (1958). "Fourier Transforms and X-ray Diffraction," Bell, London.

Menarry, A. and Lipson, H. (1957). *Acta Cryst.* **10**, 27.

Patterson, A. L. (1935). *Z. Krist.* **90**, 517.

Pinnock, P. R. and Lipson, H. (1954). *Acta Cryst.* **7**, 594.

Pinnock, P. R. and Taylor, C. A. (1955). *Acta Cryst.* **8**, 687.

Pinnock, P. R., Taylor, C. A. and Lipson, H. (1956). *Acta Cryst.* **9**, 173.

Robertson, J. M. and White, J. G. (1945). *J. chem. Soc.* 607.

Sakore, T. D. and Pant, L. M. (1966). *Acta Cryst.* **21**, 715.

Sim, G. A., Robertson, J. M. and Goodwin, T. H. (1955). *Acta Cryst.* **8**, 157.

Stadler, H. P. (1953). *Acta Cryst.* **6**, 540.

Talukdar, A. N. (1970). In preparation.

Taylor, C. A. (1952b). *Nature, Lond.* **169**, 1087.

Taylor, C. A. (1954a). *Acta Cryst.* **7**, 757.

Taylor, C. A. (1965). *J. Sci. Inst.* **42**, 533.

Taylor, C. A. and Lipson, H. (1951). *Nature, Lond.* **167**, 809.

Taylor, C. A. and Lipson, H. (1964). "Optical Transforms," Bell, London.

Taylor, C. A. and Morley, K. A. (1959). *Acta Cryst.* **12**, 101.

Taylor, C. A. and Thompson, B. J. (1957). *J. Sci. Inst.* **34**, 439.

Taylor, C. A., Hinde, R. M. and Lipson, H. (1951). *Acta Cryst.* **4**, 261.

Titchmarsh, E. C. (1948). "Introduction to the Theory of Fourier Integrals." Clarendon Press, Oxford.

Waser, J. and Lu, C. S. (1944). *J. Am. chem. Soc.* **66**, 2035.

Woodward, I. (1940). *J. chem. Soc.* 601.

Woolfson, M. M. (1961). "Direct Methods in Crystallography", Clarendon Press, Oxford.

Wrinch, D. (1946). A.S.X.R.E.D. Monograph No. 2.

Wyckoff, H. W., Bear, R. S., Morgan, R. S. and Carlstrom, D. (1957). *J. opt. Soc. Am.* **47**, 1061.

CHAPTER 4

Polymer and Fibre Diffraction

C. A. Taylor

Department of Physics, University College, Cathays Park,
Cardiff, South Wales

I. INTRODUCTION

A. Types of X-ray Pattern Obtained from Polymers

X-ray crystallographers, during the first 30 years or so after the discovery of X-ray diffraction, concerned themselves almost exclusively with the patterns produced by relatively perfect crystals. These patterns consist of sharp spots of varying intensity arranged in some regular geometrical pattern, and the appearance of diffuse spots, streaks, etc., was regarded at one time as a signal for the study of that particular specimen to be abandoned! During the second

115

30 years, interest in less-regularly crystallized material and the interpretation of the resulting diffuse X-ray patterns has gradually increased and the fact has emerged that a very high proportion of *all* X-ray diffraction photographs of so-called single crystals exhibit at least some element of diffuse scattering arising either from the presence of a proportion of less-well crystallized material, from disorder of some kind, or from thermal vibrations. In general, the smaller the molecular unit of structure, the more perfectly is it likely to crystallize, since, other things being equal, small molecules are more likely to be mechanically rigid and to pack regularly and systematically. Polymer molecules are necessarily long and hence unlikely to be rigid so that the chance of their packing in a regular way is greatly decreased. A mental comparison of the problem of packing steel balls into a box with that of packing long, flexible, helical springs will make the point. It is not surprising, therefore, that most polymeric materials give rise to more or less diffuse X-ray diffraction patterns.

The custom has arisen of classifying X-ray patterns from polymers into two distinct groups—low-angle scattering patterns and wide-angle scattering patterns. In some ways this classification is unfortunate—as are most "black and white" distinctions—since, of course, interesting information can be obtained from scattering at any angle. It would be out of place here to discuss in detail the methods of producing the various types of X-ray photograph used in polymer studies; a very useful survey of X-ray techniques is given by Alexander (1969).

Figure 1 shows six fairly typical examples of polymer scattering patterns of the kind with which we shall be concerned in this chapter. Figure 1(a) shows the simplest type of pattern, in which the scattering is confined to low angles and is little more than a diffuse ring; there is really very little information present. The specimen is an undrawn polyester fibre. Figure 1(b) (for a specimen of tricel—cellulose acetate) shows a little more information; the photograph resembles a powder photograph for poorly crystallized material except for the variation in eccentricity of the elliptical "rings". Figure 1(c) (for a specimen of acrilan) also resembles a powder photograph but with evidence of strong preferred orientation; this arises from the roughly parallel orientations of the polymer chains. For all these three the information is not only limited by the diffuseness of the rings and the general lack of detail but also by the angular extent of the scattering. The outer limit of detail in the diffraction patterns corresponds roughly to a spacing in the material of about 2·6 Å and hence one could not expect even statistical information about atomic separations; evidence even on a molecular scale would be only in very broad terms. Figure 1(d) (a photograph for a specimen of terylene) shows evidence of rather more order and there is detail out to angles corresponding to spacings of about 1·5 Å; the so-called "layer-lines" also begin to make their appearance. This

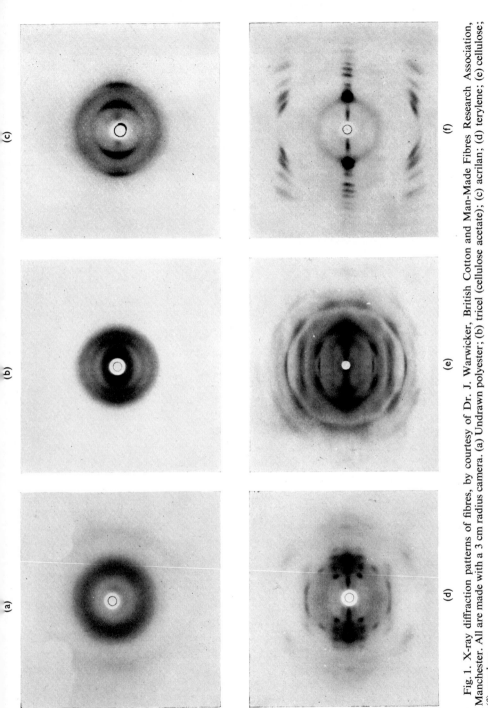

Fig. 1. X-ray diffraction patterns of fibres, by courtesy of Dr. J. Warwicker, British Cotton and Man-Made Fibres Research Association, Manchester. All are made with a 3 cm radius camera. (a) Undrawn polyester; (b) tricel (cellulose acetate); (c) acrilan; (d) terylene; (e) cellulose; (f) courlene.

5

name derives from the similarity with the corresponding lines of spots which occur in single-crystal rotation photographs, though of course in the fibre specimen there is no physical rotation; the fibres, fibrils, or individual chains occur in many different orientations within the same specimen, but always with their long axes approximately parallel to each other. Figure 1(e) shows another relatively well crystallized specimen, this time of a natural fibre— cellulose I; again the information extends out to about 1·5 Å and a fair amount of detail is present. In the last figure [1(f) for courlene] some quite sharp spots can be seen in addition to diffuse streaks and patches and the layer-line structure is well defined. One of the most interesting debates in the field of X-ray diffraction studies of polymers has been that centred round the interpretation of the significance of the ratio of the sharp spots to the diffuse areas. Various interpretations of the evidence have been given and so-called "degree of crystallinity" measurements often made from this kind of pattern will be discussed later.

B. The Questions to be Asked about Polymer Structures that may be Answered by X-ray Methods

The basic problem in conventional single-crystal structure determination is the phase problem; the correct phase must be associated with each "reflexion" before a complete image of the structure may be produced by Fourier synthesis and there is no completely objective method of determining the phases. Just as in focusing an optical image, some assumptions about the object must be made in order to achieve success; these assumptions might be that the atoms are discrete and spherical, that the electron density is everywhere "positive" (i.e., scatters in phase) or that the structure has sensible bond lengths, etc. One of the questions most frequently asked by students who begin to think seriously about the subject is "How, then, do we know that a particular solution put forward is unique?" The answer, of course, is that we do not know and no proof to my knowledge has ever been given that there can be only one set of phases that gives a reasonable solution. However, when one considers the relatively small number of parameters involved in a perfect crystal (the six lattice parameters and three position parameters for each atom in one unit cell) and the large number of elements of data (the intensities of individual spots from which may be derived the structure factors) in the three-dimensional X-ray pattern, the probability of more than one solution fitting the data is remote. On the other hand, in polymer structures—because there is no precise repetition of a given pattern unit—the number of parameters involved becomes fantastically large but the number of data elements does not increase in anything like the same way, even if one is prepared to make extremely accurate point-by-point measurements within the diffuse areas. The possibility of more than one solution for a given pattern must therefore be borne in mind.

One of the unfortunate facts in the history of X-ray diffraction studies of fibres has been that very frequently a researcher has formed a fairly rigid picture in his mind of the nature of a particular structure and has subsequently made an X-ray study by asking the question "Is this X-ray diffraction pattern consistent with my model?" Unless the comparisons are made very carefully indeed one can often reach the answer "Yes" when, in fact, the pattern would be equally consistent (or inconsistent) with other models. An alternative, though much more difficult, approach has been suggested by Taylor (1966, 1967). It is to ask the question "What kind of diffracting object produced this X-ray pattern?" It is more difficult because it suggests that no preconceived notions, no trial structures based on experience of related structures and no chemical or other data should be fed in. The problem should, initially at least, be treated purely as a physical diffraction problem. Although difficult, this approach has all the advantages of what has sometimes been described as "lateral thinking". It may enable one to shake off rigid preconceived ideas and to consider completely new models for a structure. It must be admitted immediately that the approach is only intended as a complementary technique. It will certainly not solve problems on its own but may be valuable in starting up new lines of thought.

The extent to which an answer to this new kind of question can be found depends, of course, on the complexity of the structure. For many structures it will be possible only to make statistical generalizations and there will be few X-ray diffraction patterns of fibre structures that will contain sufficient elements of information to permit the location of individual atoms in the way that is possible with highly crystallized material. Such statistical information, may, however, be extremely useful. An example would be the programme of investigation by Delf and Harburn (1971) in which statistical studies of the end-to-end distances in labelled polymer chains in dilute solution are being made. Statistical information on the distribution of the lengths derived from the X-ray studies will, it is hoped, give valuable insights into the process of coiling as the polymer chain length increases.

C. *The Difficulties of Interpretation of X-ray Scattering Patterns of Polymers and the Advantages of the Optical Transform Approach*

The primary problem of interpretation is that one has a very large number of parameters because of the absence of repetition and each molecule or polymer chain may be in a different orientation or position, or may even be of a different shape from its neighbours. The calculation of the intensity at a point in the X-ray diffraction pattern therefore involves summation of a far greater number of terms than is needed in single-crystal work and also, because the patterns are diffuse rather than discrete spots, intensities must be determined at a considerably greater number of points in reciprocal space. Calculation is therefore a major undertaking. Various new fast algorithms are now available (e.g., Cooley and Tukey, 1965) for computing Fourier transforms on this scale but even so considerable expenditures of computer time are needed and it is difficult to acquire a real familiarity with the relationships between real and reciprocal space unless very large numbers of corresponding objects and patterns have been studied. Nevertheless, such familiarity is an essential requirement in pursuing the approach suggested in the last section. Optical-transform methods provide a valuable aid. When these methods were first applied to the solution of single-crystal problems (e.g., Hanson *et al.*, 1953), it was found profitable to examine the transforms of very large numbers of different arrangements of holes representing real and hypothetical structures. Because the method is visual, instinctive notions of the relationships can be built up and because it is so rapid, even quite crazy ideas may be tried out with negligible loss of time; sometimes the crazy idea is the right one! The same principles are proving valuable in fibre studies and a systematic investigation of the various features of X-ray patterns and their relationship to various aspects of the fibre structure was begun by Mukhopadhyay (1968) and Mukhopadhyay and Taylor (1971) and more illustrations are given in the next section. These photographs have been prepared using the large laser diffractometer described in Section IV.B of this chapter.

The fundamental principle is exactly the same as that of earlier instruments —a two-dimensional mask of holes in thin opaque material is placed in a parallel beam of monochromatic light and the Fraunhofer diffraction pattern viewed in the back focal plane of a long-focus objective lens. With the laser system the coherence (Chapter 2) of the illumination is greatly improved and the total light available is so great that photographs are obtained, even after considerable direct enlargement of the pattern, with very brief exposures. The illustrations in the next section are reproduced just less than twice the actual size of the negative; the masks are reproduced almost actual size and most of the exposures were between 1/500 and 1/30 of a second.

II. Basic Concepts in Two Dimensions

A. The Characteristics of Patterns from Long-Chain Molecules and the Effects of the Shape of the Chain Unit

To illustrate the basic ideas of diffraction by long-chain molecules we have selected several basic chain types. They have not been chosen to represent any particular structures but rather to illustrate in a hypothetical way the

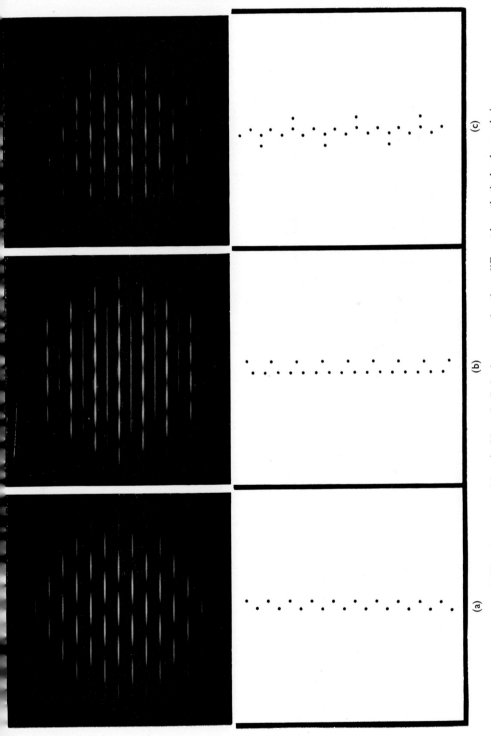

Fig. 2. Optical diffraction patterns (above) of 3 masks (below) representing three different hypothetical polymer chains.

(a)   (b)   (c)

optical effects of different features. This policy has been followed deliberately to be in line with the policy outlined in Section I.B, that is, always to start from the diffraction pattern rather than from a particular structural concept.

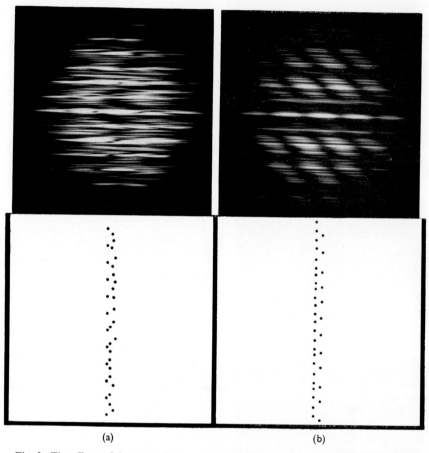

(a)   (b)

Fig. 3. The effects of departure from exact periodicity for the chain of Fig. 2b. (a) Two atoms out of three in each unit arranged randomly; (b) unit shape constant but periodicity along chain varying.

Figure 2 shows three basic chains together with their optical diffraction patterns. The important features to note are first, that the periodicity within the chain itself gives rise to the "layer line" effect (see also Fig. 6). This may seem a trivial and obvious point but it needs stressing because the existence of layer lines has occasionally been cited as evidence for crystallinity. The second point to note is that variations in intensity along the layer lines are

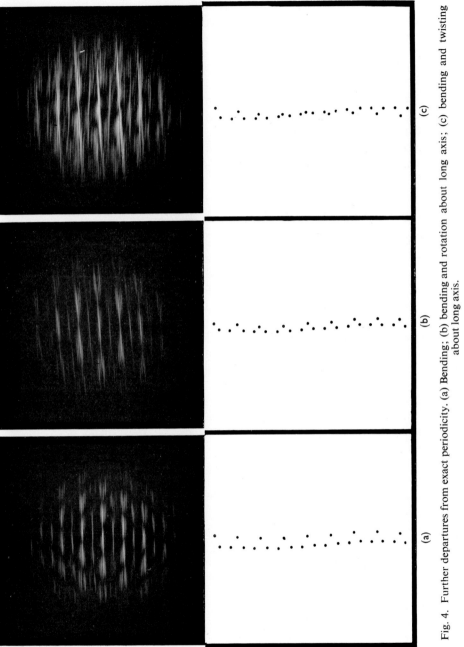

Fig. 4. Further departures from exact periodicity. (a) Bending; (b) bending and rotation about long axis; (c) bending and twisting about long axis.

strongly characteristic of the particular chain unit. Immediately, therefore, we have the possibility of separation of variables. Layer-line dimensions correspond to periodicities within the chain; layer-line distribution of intensity corresponds with the nature of the repeating unit. Figure 3 shows two examples of departure from precise periodicity within the chain of Fig. 2(b). In Fig. 3(a) one atom appears at regular intervals along the chain but the relative positions of the other two atoms in the unit vary; in Fig. 3(b) the unit shape remains the same but the periodicity varies. Figure 4 shows a further example of departure from strict periodicity; in this case the chain has been bent, rotated, or twisted. A study of Figs. 2, 3 and 4 forms a good basis on which to build experience of polymer diffraction phenomena.

B. The Effects of Lateral and Longitudinal Displacement of the Chains and of Variations in their Orientation

This section will consist mainly of a set of photographs, since by far the most effective way of transmitting information is by the study of photographs themselves and words are almost superfluous. The chain of Fig. 2(b) has been selected for this particular group of photographs and the sequences are as follows. In Fig. 5 we see first of all the build-up from a single chain to a group and it is important to notice the break-up of the layer lines, though the overall intensity distribution is still governed by that of the single chain. Figure 5(d) would represent a perfectly crystalline arrangement for this particular polymer. In Fig. 6(a) the effects of longitudinal displacement and distortion are shown; the lateral spacing remains as for Fig. 5(d). The projection of the new pattern on to the central horizontal line through the mask will be identical with that for the perfect crystal in Fig. 5(d) and consequently the horizontal row of spots in the diffraction pattern remains unaltered. Everywhere else in the pattern are, of course, effects of the random displacements; these result in the characteristic diffuse streaks. In Fig. 6(b) the vertical positions remain as for the perfect crystal and the lateral spacings are altered in a random way. The same argument can be applied and it is now seen that the central vertical row in the diffraction pattern remains unaltered while streaks appear everywhere else. The last effect to be considered in this section is that of the two kinds of rotation possible. Figure 7(a) shows the effects of rotation about the chain axes in a random way. The effect is somewhat similar to that of lateral displacement except that sharp spots remain in addition to the streaks and the intensity distribution along the streaks is changed. In Figs. 7(b) and (c) the chains cease to be parallel to each other and this introduces the spread in width of the layer lines, increasing outwards from the vertical centre line, so characteristic of many fibre photographs. Finally Figs. 8 and 9 show some combinations of the various kinds of lateral and rotational disorder that have

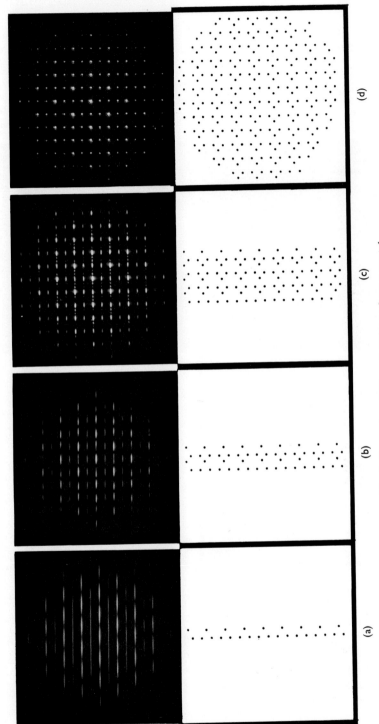

Fig. 5. Successive addition of chains to form a crystal.

been introduced and also comparisons for the three types of chains originally introduced in Fig. 2.

C. The Effects of Curvature and Twisting

All the departures from regularity so far introduced still involve identical rigid chains and it can be seen that the resultant diffusion in reciprocal space

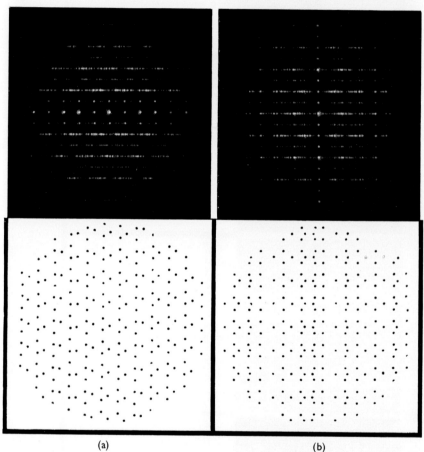

(a)   (b)

Fig. 6. The effects of displacement of the chains of Fig. 5(d). (a) Vertically; (b) horizontally.

is either confined to the layer lines or is in the form of fan-like spreading which increases in extent as one moves away from the central vertical axis. The effect of introducing bending of the chains is, however, significantly different. Figure 10 shows the result of bending the chains; a crystalline version is also given for comparison. The spreading of the spots is now quite different and the radial

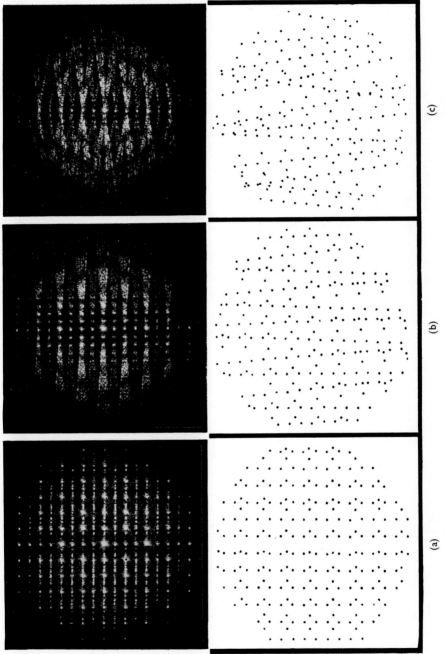

Fig. 7. The effect of rotation on the chains of Fig. 5(d). (a) Random rotation about long axes of chains; (b) random rotation about axes perpendicular to the plane of the mask (c) as (b) but larger rotation.

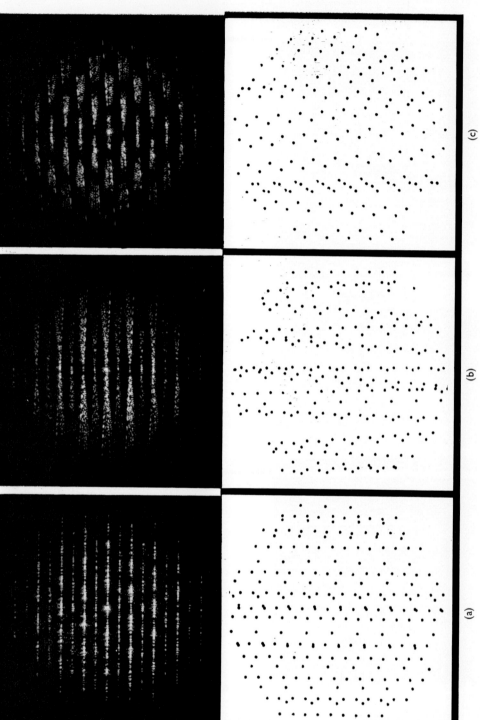

Fig. 8. (a) Chain of Fig. 2(b) with both vertical and horizontal random displacements; (b) chain of Fig. 2(b) with random rotations both about chain axes and axes perpendicular to the mask; (c) chain of Fig. 2(a) with non-parallel chains.

(a)　　　　　　　　(b)　　　　　　　　(c)

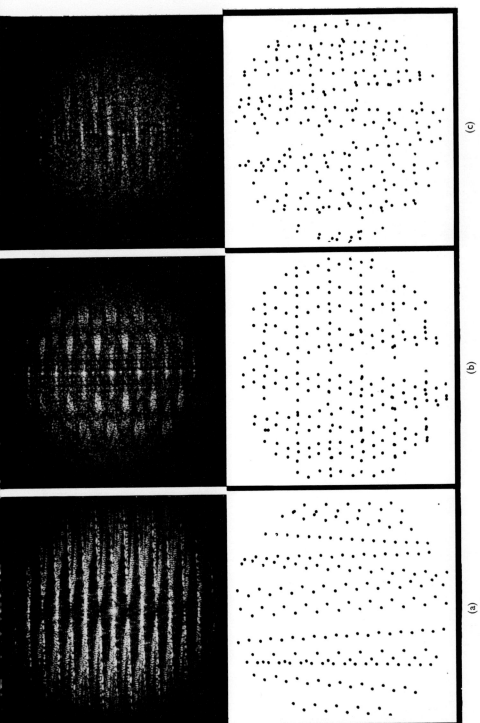

(a)

(b)

(c)

Fig. 9. (a) Chain of Fig. 2(a) with random rotations about both axes as for 8(b); (b) chain of Fig. 2(c) with non parallel chains; (c) chain of Fig. 2(c) with larger random rotations about both axes than for 9(a).

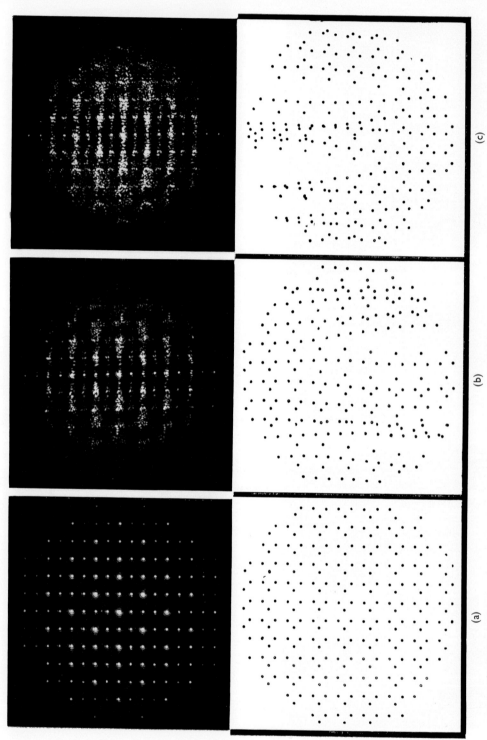

(a)

(b)

(c)

Fig. 10. The effect of bending on the chain of Fig. 2(b) with the crystalline version in Fig. 5(d) reproduced at (a) for comparison.

out many times (e.g., Taylor and Lipson, 1964) that the scattering properties of any object in real space may be summarized in terms of a three-dimensional figure in reciprocal space—often described as the "reciprocal solid". This structure is characteristic only of the object and not of the wavelength of the scattered radiation. It is simply the Fourier transform of the object and, to be complete, every point of the structure should have a phase and an amplitude associated with it. The most familiar example of such a structure is the reciprocal lattice, which is the reciprocal solid for the particular case of a three-dimensionally repeating object in real space. Non-regular objects produce more continuous structures and the opposite extreme would be a single spherical atom whose reciprocal solid would be a spherically symmetric distribution with a radial fall-off of amplitude corresponding to the scattering factor of the atom. When a beam of radiation falls on a particular real object, the scattering in various directions may be predicted by using the concept of the sphere of observation (crystallographers call it the sphere of reflexion). This sphere has a radius of $1/\lambda$ and hence the same reciprocal solid may be used to predict scattering with all kinds of radiations, at least as far as the geometrical aspects are concerned. For optical diffraction the relative scales of the significant distances in the masks and the wavelength (say 5 mm and 500 nm) are in the ratio of $10^4:1$ and hence the radius of the sphere is 10^4 greater than the significant distances in the reciprocal solid and the sphere may be considered effectively planar. Thus the optical patterns in the last section should really be considered as planar sections of three-dimensional solids, but, since the objects producing the patterns are infinitesimally thin, the reciprocal solids exhibit no variation in directions perpendicular to the planar intersection and hence all sections are identical. It may easily be shown that, if the mask represents a projection of a three-dimensional object on to a plane, then its optical transform or diffraction pattern is identical with the central section of the corresponding reciprocal solid—though non-central sections would be different.

Now let us consider X-ray scattering from polymers. The important dimensions, if we are concerned with individual atoms, will be of the same order as that of the wavelength ($\sim1\cdot5$ Å) and hence we should not expect the X-ray and optical patterns to be identical because of geometrical distortions arising from the substitution of the spherical section for the plane section. The distortions can be removed by using precession photographs (see Alexander, 1969, p. 87). Photographs taken with cylindrical cameras will avoid distortion for the equatorial region, which for some purposes is sufficient. If, however, one is concerned only with very-small-angle scattering—say up to angles corresponding to real space distances of 20 Å-or-so—the difference between the sphere and the plane is negligible and flat-plate photographs are quite acceptable. Thus one would expect the patterns of Fig. 2 to correspond very closely to the X-ray photograph taken on a precession camera for a single polymer chain—if that were physically possible. Similarly if the polymer chains were orientated differently in the X-ray beam a mask representing a different projection could be prepared and again the pattern would be correctly represented.

Let us now consider what happens in practice when we take X-ray photographs of fibrous polymers. The common feature is that the patterns remain unchanged when the fibre is rotated. In other words the reciprocal solids have cylindrical symmetry and any axial section is identical with any other. It is clear that this must arise from one of two causes; either (a) every separate

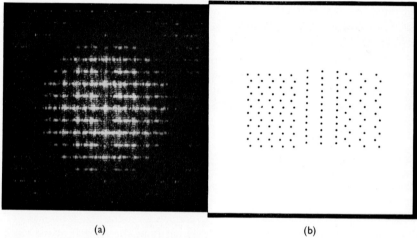

(a)   (b)

Fig. 14. A composite of three projections of a three-dimensional crystallite consisting of nine parallel chains.

polymer chain is independently orientated about its axis, and hence on the average the projection in any one direction perpendicular to the axis has the same statistical distribution of chain orientation as any other, or (b) groups of chains in sub-units—fibrils, micelles, fibres, etc.—may occur with some degree of order within each, but with the groups themselves randomly orientated. Thus in either case all projections are at least statistically equivalent as far as their distributions of orientations are concerned, even if they are not identical.

B. Relationships between Diffraction Patterns of Two-Dimensional Masks and Three-Dimensional Projections

The foregoing discussion permits us now to consider the relationships between optical patterns, such as those of Section II, and X-ray fibre photographs. The reciprocal solid for a complete fibre may be considered as the vector sum of all the solids for the individual fibres present. The result, of course, is a complicated interference pattern. If the polymer chains are regularly arranged with their axes parallel to form a perfect crystalline arrange-

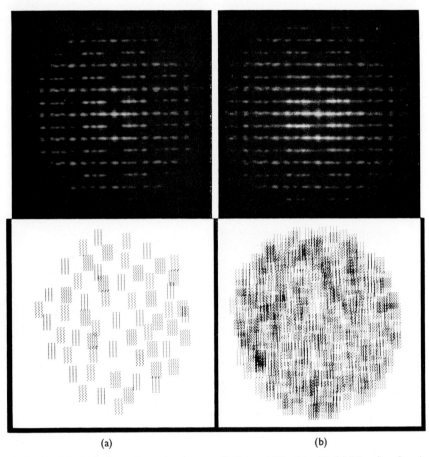

(a)   (b)

Fig. 15. Masks built up from the three projections at Fig. 14 with (a) 20 units of each and (b) 100 units of each. The "holes" are here relatively smaller than those of Fig. 14 and hence the extent of their Airy discs, and hence of the pattern, is greater than in Fig. 14.

ment, then a projection along one of the principal crystal axes perpendicular to the chains would be identical with, say, Fig. 5(d); hence that diffraction pattern would correspond exactly with an X-ray pattern produced by irradiating in the corresponding direction and no three-dimensional problems arise. To obtain the optical analogue of an X-ray photograph taken with the X-ray beam in any other direction it would simply be necessary to prepare a mask representing the corresponding projection. If the polymer concerned is thought to consist of groups of chains with a more-or-less crystalline arrangement within the group but with the groups randomly orientated about their long axes, it will be necessary to make a mask representing a composite of the various projections. Ideally the mask should represent a projection of the

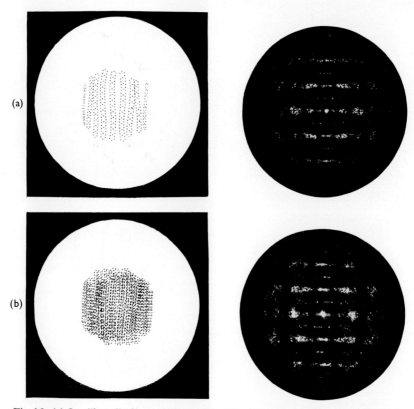

Fig. 16. (a) One "layer" of bent chains and its diffraction pattern; (b) a projection through a large number of similar though not precisely identical chains and its diffraction pattern. (By courtesy of Dr. U. Mukhopadhyay.)

whole fibre. In practice it turns out that a good indication of the general intensity distribution likely to occur may be obtained from a surprisingly limited number of units (Stokes, 1955, has given a theoretical approach to this kind of problem). There will, of course, be differences in the fine detail but not in the overall distribution. Figures 14 and 15 illustrate this. Figure 14 shows a composite of three different projections of a three-dimensional crystallite consisting of nine parallel simple chains arranged so that a section perpendicular to their axes would form a 3×3 square lattice. Figure 15 shows two masks built up from three different projections of this solid (a) with 20 examples of each randomly distributed in three dimensions and projected on to a plane and (b) with 100 examples of each randomly distributed in three dimensions and projected on to the same plane. The similarity of their diffraction patterns is quite remarkable. Figure 16 shows another comparison, this time between one layer of bent chain and a projection through a large number of layers with very similar though not identical distributions. Again the comparison is striking and provides evidence that the relatively small numbers of chain units used in the illustrations in Section II may still provide significant information in studying X-ray scattering from much more complicated three-dimensional distributions.

IV. TECHNIQUES

A. Mask-Making

The general problem of mask-making has been discussed in other chapters and also by Taylor and Lipson (1964); the special problems relating to applications in polymer work stem principally from the very large number of holes that need to be produced if realism is to be achieved. The usual techniques using the pantograph punch, photographic reproduction of drawings on glass plates used in a matching-oil bath, of etching, are all applicable to the final preparation of the mask; problems arise in the preparation of the original and the methods that have been suggested fall into several distinct categories.

1. Multi-Printing Techniques

Willis (1957) described a method of producing masks by a photograph printing technique in which the negative plate is moved by a double micrometer arrangement to different positions to permit repetitive patterns to be made. Mukhopadhyay (1968) has adapted this technique in two ways for producing masks representing polymer chains. The first involves a simple slide device which enables the photographic plate to be exposed in a series of steps so that a number of parallel representations of a given chain may be printed on the same plate. The second method involves a modification of the pantograph punch in which the punch itself is replaced by an optical projection device producing a reduced image of a larger mask; the image may then be printed down at many different places controlled by the pantograph arm. This technique has been developed into a more complicated system which will permit the images to be placed down in any orientation with a wide range of scales and also to permit the master mask to be changed very easily and rapidly

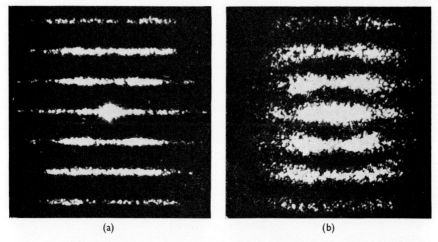

Fig. 17. Diffraction patterns of a random three-dimensional collection of about 30 helical springs, each about 0·25 in. in diameter and 6 in. long placed roughly parallel to each other. (a) All straight; (b) all bent in slightly different ways.

so that multiple complex masks may be produced (e.g., Figs. 15, 25 and 26). In all multi-printing systems the master units are holes in an opaque ground so that the outcome is a series of black dots on the plate. A final positive copy or an etching must be made before the mask can be used. In this way, however,

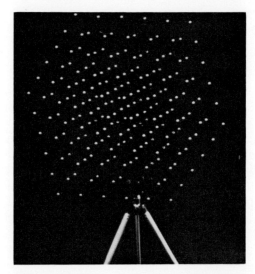

Fig. 18. Solid model of a single cubic lattice made of white polythene balls and black spokes which can be photographed directly to make a mask. (By courtesy of Dr. U. Mukhopadhyay.)

provided that the opaque areas are really dense, hundreds of exposures can be made on the same photographic plate in order to build up extremely complicated masks involving a great deal of overlapping.

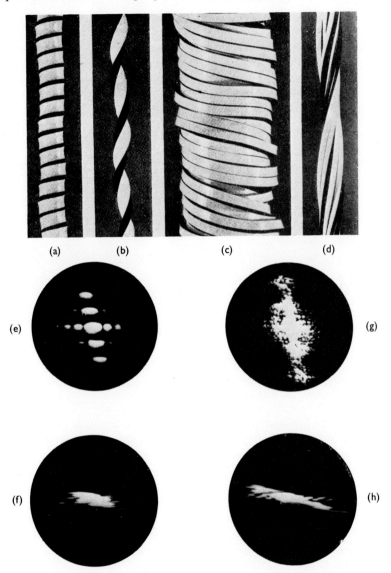

Fig. 19. (a)–(d) Photographs of models built by Morgan (1954) to represent single and multiple fibrillar crystals relaxed and stretched: (e)–(h) are the respective transforms of half-tone masks prepared directly from the photographs. Reproduced, with permission, from Taylor (1966).

2. *The Use of Three-Dimensional Models*

Various suggestions have been made for the use of three-dimensional models either directly in the diffractometer or as a means of making masks. Figure 17 shows the diffraction patterns of straight- and bent-wire helices placed directly in the diffractometer. In this case the helices were wound on a quarter-inch former. So far, however, this method has not proved to be of other than academic interest. A much more useful technique involves the photographing of three-dimensional models as a method of making masks. If the models are made of white polystyrene foam balls with black spokes and the photographs are taken against a black background the resulting photograph merely consists

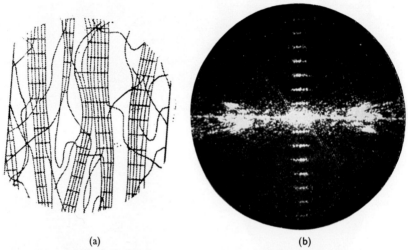

(a)   (b)

Fig. 20. (a) Drawing of a model proposed by Hearle (1965) and (b) transform of mask prepared directly from the drawing.

of black dots on the negative and a similar procedure to that described in the multi-printing technique may be used; that is, a series of photographs of the same model either in the same or different orientation can be placed side by side. Figure 18 shows a typical model standing on a tripod. A third way of using three-dimensional models is illustrated in Fig. 19; in this case photographs of some models devised by Morgan (1954) were available and the masks were prepared direct from photographs in the literature. The photographs already had a half-tone screen superimposed but this does not complicate matters and the accompanying diffraction patterns illustrate quite well the effects of stretching a polymer fibre. In passing, it could also be mentioned that many other types of drawn or photographed models available in the literature can be used directly to prepare a mask and Fig. 20 shows a further example from a line drawing due to Hearle (1965).

3. *The Use of Computers*

Various attempts have been made to use computers for generating masks, and there is great attraction in the notion of combining the speed and flexibility of a computer as a generator of distributions, orientations, etc., with the speed of the optical diffractometer in performing the actual Fourier transformation.

The simplest technique involves programming the computer to output a series of crosses on an automatic graph plotted from which masks may then be punched by hand using the pantograph. This is not such a tedious process as it sounds. Efforts are being made to adapt a graph-plotter output to produce dots which may be photographed directly and made into masks without intermediate punching. Figure 21 gives examples of the use of this technique. It begins with a set of 300 points which were generated by the computer using the following routine. Random numbers were generated to correspond to the three positional coordinates of a point in three dimensions; the result was then examined to ensure that the point lay within a sphere of a given radius and did not approach closer than a certain distance from any point already generated. If it passed this test it was retained and recorded as a two-dimensional projection on to a given plane on the graph plotter. As the process continues, of course, more and more points fail the test but it is not difficult to obtain the 300 points very quickly. The remaining figures are based on the first. Figure 21(b) shows the effect of placing two holes (in a dumb-bell like arrangement) with their centres of gravity at the points of Fig. 21(a) and with the dumb-bell axes all parallel to the same direction. In Fig. 21(c) each point of (a) is replaced by a small hexagon of six holes, all in parallel orientation. In Fig. 21(d) the hexagon of (c) are randomly orientated though with their planes still parallel to the plane of projection. Various proposals have been put forward for automated devices in which a computer would control a mask-making machine directly, and this remains an attractive possibility; so far no machine has, in fact, been produced. The principal difficulty has been expense, but recent improvements in both optical and electronic technology may permit advance in this direction in the not-too-distant future.

B. *The Requirements to be Satisfied by an Optical Diffractometer for Polymer Work*

In principle any optical diffraction equipment may be used for polymer work but it has been found in practice that the new laser-based diffractometer designed and build at Cardiff by Ranniko and Harburn (1971) has many advantages over earlier versions. The short exposures resulting from the high-power laser have already been mentioned. The extremely high spatial coherence of the illumination produces very clean, clear-cut patterns and the temporal

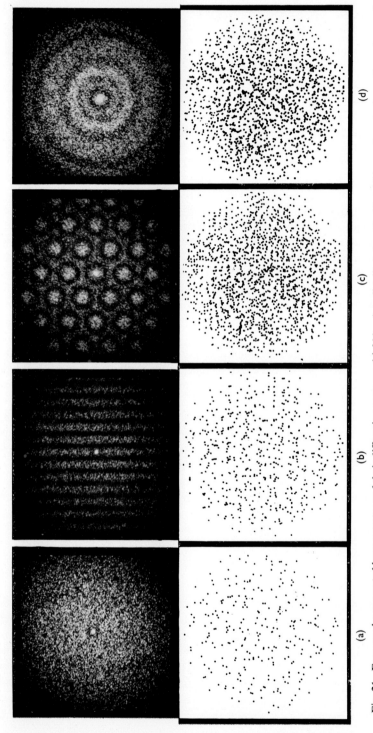

Fig. 21. Four masks generated by computer and their diffraction patterns. (a) 300 points distributed randomly within a sphere in three dimensions and projected on to a plane; (b) a "dumb-bell" pair of holes placed at each of the sites of (a) in parallel orientations; (c) a hexagon of six holes placed at the sites of (a) in parallel orientations; (d) a hexagon of six holes placed at the sites of (a) with planes parallel with the projection plane but in random orientations within the plane.

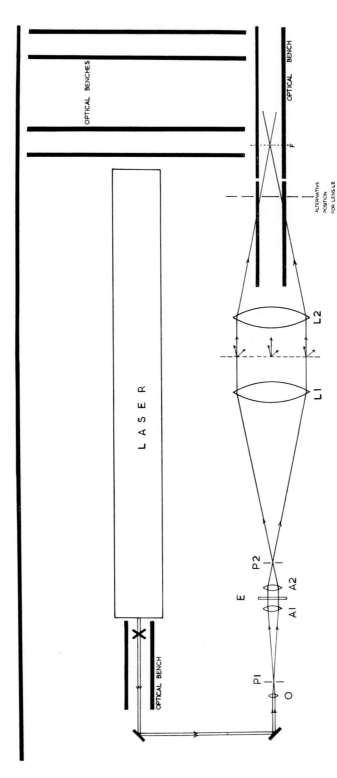

Fig. 22. Schematic diagram of the new diffractometer at Cardiff. (By courtesy of Dr. J. Ranniko.) A camera shutter is provided at point X; the system of lenses O, A1, A2 and pin holes P1, P2 provides beam expansion without loss of coherence and a plate E may provide compensation for the Gaussion distribution of amplitude across the primary base beam; Lenses L1 and L2 are the main collimator and objective; the diffraction pattern appears at F but auxiliary mirrors and lenses in various combinations may be mounted on the optical benches according to the demands of a particular experiment.

coherence (monochromaticity) removes many of the difficulties which were experienced with instruments in which high-pressure mercury arcs were used because of the high line breadth. The principal effect of this was that spots away from the centre of the field tended to be elongated radially. The diameter of the lenses should be as great as possible so that large masks can be used but, because masks used in polymer work are generally built up of very small holes the diffracting angles may be larger than for masks used in single-crystal structure determination; hence shorter focal length lenses may be used. Pre-magnification is usually incorporated so that the negatives do not need very much enlargement. With the older versions it was often necessary to enlarge up to 100 times; with the new diffractometer five times is sufficient.

There are considerable advantages in using the instrument horizontally particularly in relation to the use of the oil-immersion bath with masks made on photographic plates. It is obviously very much simpler for this to be inserted between the lenses if the instrument is horizontal. This configuration also has advantages in many other applications. A standard 35 mm camera body without a lens but incorporating a shutter can be usefully used for recording the photographs.

Figure 22 shows an outline of the Cardiff instrument which is built on three parallel horizontal 8 in. × 6 in steel girders on which are mounted optical-bench tracks which make it possible for the whole instrument to be used in a great many different modes. For example, it can be used for image recombination experiments (Chapter 6), for holography (Chapter 9), and for studying Fresnel diffraction patterns, etc.

V. TENTATIVE CONCLUSIONS AND IDEAS FOR THE FUTURE

The applications of optical-transform techniques to polymer problems are really just beginning to show results and a great deal remains to be done. It is hoped, however, that this chapter will have demonstrated the possibilities and potential of the method. It would at least seem that earlier doubts that have been expressed about the two-dimensional limitations of optical techniques do not really present any problem. Figures 14–17 among others make this point particularly well and a further illustration is given in Fig. 23. This figure consists of the optical diffraction patterns of two apparently random arrangements of holes. The diffraction patterns, however, reveal a totally different structure. One consists of a pile of a large number of square lattices in different orientations and the other consists of a large number of chains in parallel orientations. Both of these arrangements are essentially three dimensional but there is no problem in producing the diffraction patterns. This illustration also demonstrates the power of optical diffraction in determining distributions and periodicities in apparently random arrangements.

The second main point emerging from the chapter concerns the relative scales and it has been clearly shown that optical patterns may be used for elucidating small- and wide-angle patterns and that it is extremely important to study the scale of operations very carefully in practical work. The third important point emerging is that remarkably similar patterns may be derived from totally different arrangements. There are several figures associated with Section

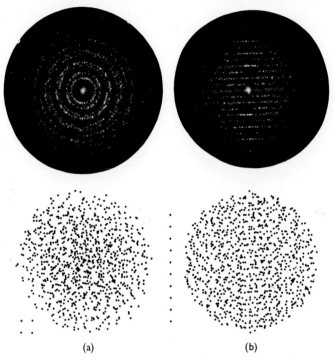

(a)   (b)

Fig. 23. (a) Projection through a superposition of a large number of square lattices in different orientations (one unit cell inset bottom left); (b) projection through a superposition of a large number of roughly similar chains in nearly parallel orientations (a typical chain is shown at left).

II which are superficially not dissimilar and yet the real-space objects are totally different. This supports the claim made at the beginning of the chapter that it is important to ask questions about the kind of diffracting object that might have produced a pattern rather than merely checking for consistency with a particular model. A further illustration of the possibility of almost identical patterns being derived from dissimilar objects is given in Fig. 24. In (a) a regular crystallite has a proportion of the atomic sites randomly vacant. In (b) a regular crystalline arrangement is surrounded by amorphous material, albeit in two dimensions; yet the diffraction patterns are very similar. In one case it could

be said that we have a crystalline portion and an amorphous portion whereas
in the other case there is obviously no division but merely a disordered crystal.
This particular illustration would seem to have important implications for the
methods of measuring the "degree of crystallinity" that have been proposed
at various times that are based on measurements of the sharp and diffuse
regions of an X-ray photograph. It would seem to be quite clear that measurements of this kind can have real significance only if they are related to a

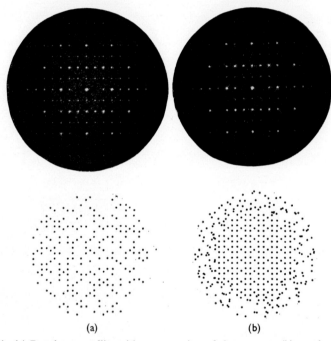

(a)   (b)

Fig. 24. (a) Regular crystallite with a proportion of sites vacant; (b) regular crystallite
surrounded by "amorphous" material.

particular model; two quite different models which would lead to similar
X-ray patterns might nevertheless lead to quite different estimates of the
degree of crystallinity or alternatively to a different meaning of the term
"degree of crystallinity". (See also Section II.C and Fig. 11.)

A great deal more work needs to be done particularly in relation to the
last point, i.e., the possibility that a number of different models may lead to
similar patterns; very accurate measurements with scintillation-counter techniques with very-high-density sampling may lead to the possibility of dis-
tinguishing between models. A really efficient system of mask-making preferably with on-line control from a computer would speed operations enor-
mously and would also make it possible to use very much larger numbers of

holes in the masks than has been possible up to now. The very much larger numbers of holes which would then be possible would greatly improve the resemblance between the optical patterns and X-ray photographs and would permit much more realistic statistical distributions to be used. In the last few

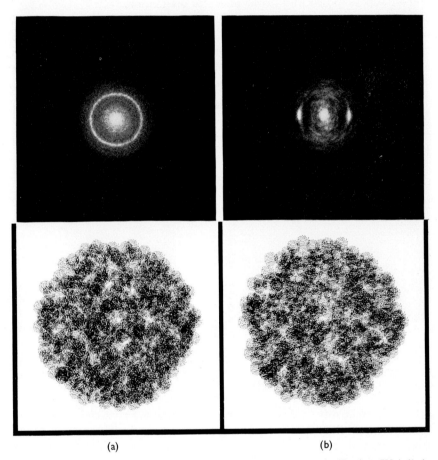

(a)   (b)

Fig. 25. Masks made with the optical-projection pantograph (Section IV.A.1) by superimposing 320 "crystallites" each of about 150 atoms arranged in very irregular chains roughly parallel to each other (a) in random orientation and (b) with strong preferred orientation. Compare Figs. 1(a) and (c) in general form.

months the development of the "optical-projection pantograph" mentioned in Section IV.A have made it possible to produce masks with up to about 50,000 holes in about 20 to 30 minutes. Figure 25 shows two masks, each made by superimposing 320 prints of a master mask of about 150 holes arranged in the form of roughly parallel but rather irregular chains. In (a) the 320

6

"crystallites" take up entirely random orientations whereas in (b) there is a strongly preferred orientation with the chains within ±5° of the vertical.

In Fig. 26 about 100 versions of each of three different arrangements of bent chains rather like those of Figs. 10(b) and (c) but with a different chain unit are superimposed with strong preferred orientation.

There is still, however, a considerable discrepancy between the number of holes in a mask and the number of atoms in a fibre and it is likely that the patterns will always remain somewhat "spotty". The spottiness makes measurement of the photographs by densitometer or isodensitracer somewhat difficult and it will be necessary to devise techniques for randomizing the

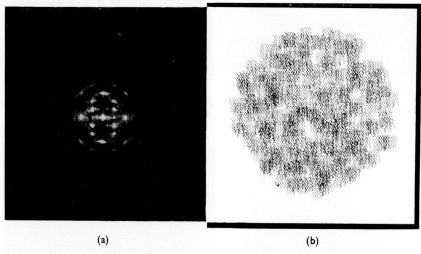

(a)   (b)

Fig. 26. Mask made as in Fig. 25 but with 100 crystallites of each of three arrangements of bent chains rather like those of Figs. 10(b) and (c) with strong preferred orientation. Compare Fig. 1(e) in general form only.

patterns in some way so that densitometer readings are possible. Suggestions have been made that the reduction in the coherence of the illumination would achieve this end. The problem then would be that the finer details of a pattern would also be lost and these are often significant. A more promising solution might be to produce a number of masks with similar though not identical distributions and to combine their optical diffraction patterns by straightforward overprinting, thereby eliminating interference effects, but leaving the overall intensity distribution unimpaired. Perhaps the most exciting possibility of all would be the development of a mask consisting of a sheet of some kind of "solid state" material which could be rendered transparent or opaque by the application of appropriate signals and which could be permanently connected to an analogue output from a computer, so that the configuration of a mask could be changed very rapidly while one was actually observing the pattern. Though this possibility is still very much in the future it seems much more probable that it will be achieved than perhaps might have been expected a few years ago.

REFERENCES

Alexander, L. E. (1969). "X-ray Diffraction Methods in Polymer Science," Wiley-Inter Science, New York.
Bonart, R. and Hosemann, R. (1962). *Kolloid-Z.* **186**, 16.
Cooley, J. W. and Tukey, J. W. (1965). *Math. Comput.* **19**, 90, 297.
Delf, B. W. and Harburn, G. (1971). Private Communication.
Hanson, A. W., Lipson, H. and Taylor, C. A. (1953). *Proc. R. Soc. A.* **218**, 371.
Hearle, J. W. S. (1965). *J. appl. Polym. Sci.* **7**, 1175.
Morgan, L. B. (1954). *J. appl. Chem.* **4**, 160.
Mukhopadhyay, U. (1968). Ph.D. thesis, University College, Cardiff.
Mukhopadhyay, U. and Taylor, C. A. (1971). *J. appl. Cryst.* **4**, 20.
Ranniko, J. R. and Harburn, G. (1971). (in preparation).
Stokes, A. R. (1955). *Acta Cryst.* **8**, 27.
Taylor, C. A. (1966). *Europ. Polym. J.* **2**, 279.
Taylor, C. A. (1967). *J. Polym. Sci. C.* No. 20, 19.
Taylor, C. A. (1969). *Pure Appl. Chem.* **18**, 533.
Taylor, C. A. and Lipson, H. (1964). "Optical Transform," Bell, London.
Willis, B. T. M. (1957). *Proc. R. Soc. A.* **239**, 184.

CHAPTER 5

Biological Studies

J. A. Lake

The Rockefeller University, New York, New York, U.S.A.

I. Introduction

Some of the most exciting recent developments in structural molecular biology have come about through application of the techniques of optical diffraction to electron micrographs, both with and without the combined use of X-ray data. In this chapter we will discuss and illustrate some of these developments involving optical diffraction and optical filtering of electron micrographs and consider, in particular, the reconstruction of three-dimensional structures from electron micrographs.

Klug and Berger (1964) were the first to analyse electron micrographs of biological molecules by optical diffraction using the micrograph as a diffraction mask. The method of Klug and Berger not only allowed the periodic nature of biological structures to be examined in an analytical manner, which had not previously been feasible, but also this approach to image analysis made apparent the advantage of extending the conceptual framework of X-ray crystallography to electron microscopy. A by-product of using optical diffraction to analyse electron micrographs of periodic structures was the realization that the "noise" in a micrograph could be removed by filtering. A "filtered" image is made by masking out noise at the diffraction plane of an optical diffractometer and by optically recombining only the diffracted rays containing the signal. Optical filtering also permits one to observe one side of a complex structure, such as a helix, when information from two sides, or layers, overlaps creating an image that is difficult to interpret. The ideas of optical filtering, particularly when applied to images of helices, prompted DeRosier and Klug (1968) to investigate how three-dimensional information is contained in electron micrographs, and to formulate an analytical method to extract this information.

II. OPTICAL DIFFRACTION

Klug and Berger (1964) demonstrated that the optical transform of an electron micrograph can be used to assess periodic structural details present in the micrograph. The benefits which followed as a result of analysing electron micrographs by optical diffraction are numerous. Unit-cell dimensions can be measured more accurately and conveniently from optical diffraction patterns than by direct measurement of electron micrographs. Symmetry elements, such as screw axes, which may not be obvious in an electron micrograph can be apparent in optical diffraction patterns of the micrograph. Optical diffraction patterns are useful in identifying moiré patterns. Finally, the optical diffraction pattern provides a quick estimate of the structural detail or resolution contained in an electron micrograph. If an electron micrograph of a periodic arrangement of molecules shows structural details that are consistently present within the repeating units of the array, then an optical diffraction pattern of the micrograph must contain diffraction spots, having intensities which can be observed above background noise, which appear at a resolution sufficient to delineate the observed structural details. Conversely, if diffraction maxima extend to a certain resolution and are discernible above background noise in the optical diffraction pattern of an electron micrograph, then one can conclude that structural details must be present in the micrograph at a resolution which is sufficient to account for all of the observed diffraction spots. Used in this manner, optical diffraction provides an analytical assessment of the preservation of detail in an electron micrograph.

Optical diffraction of electron micrographs has caused researchers to think consciously about relationships between the imperfect, noisy images seen in micrographs and optical transforms of the micrographs. Whereas X-ray diffraction patterns of proteins come from crystals containing typically 10^{15} molecules, electron micrographs of ordered arrays of molecules rarely contain more than 10^4 molecules. This naturally leads to more noise in optical diffraction patterns of electron microscope images than appears in equivalent

X-ray diffraction patterns. Indexing optical diffraction patterns of electron micrographs, i.e., the process of assigning a lattice which will accommodate all diffraction spots, is more difficult than indexing X-ray diffraction patterns because the high level of background noise present in optical diffraction patterns can obscure weak diffraction spots. Therefore one should make certain that any tentative indexing scheme adequately accounts for all spots.

In order to assure that optical diffraction patterns of micrographs have been properly indexed and to provide a check on the wide variation that is possible within an electron micrograph, it has been helpful to combine optical diffraction of micrographs of crystals with X-ray diffraction studies. The work discussed in the following two sections has utilized both methods in combination to obtain results which would not be easily found by using either method alone.

A. Bovine Liver Catalase

A study of bovine liver catalase by Longley (1967) demonstrates the power of optical diffraction and also affords a comparison of the relative merits of X-ray diffraction and of electron microscopy.

Bovine liver catalase is a heme-containing protein of molecular weight 250,000 which catalyzes the decomposition of hydrogen peroxide into water and oxygen. Catalase crystallizes into a trigonal form, space group $P3_1 21$. The crystals are large enough to be studied by X-ray diffraction and stable enough (after fixation in glutaraldehyde, dehydration with alcohol, embedding in plastic, and sectioning) to be studied by electron microscopy. The crystals may also be observed by negative contrasting with a heavy metal stain such as uranyl acetate, but we shall not discuss work done on these images.

Longley's results are summarized in Figs. 1 and 2. Figures 1(a), (b), and (c) are zero-layer precession photographs of catalase crystals taken with the X-ray beam perpendicular to the (10.0), (11.0), and (00.1) planes respectively. In Figs. 1(a) and (b) only every third diffraction spot is permitted along the c^* axis. (c^* is the Z direction of repeat in Fourier space (Section I.E, Chapter 1) and for this crystal is in the same direction as the real space z axis, c.) This implies that there is a three-fold screw axis along c. (A three-fold screw operating upon a molecule generates two other identical molecules spaced $c/3$ and $2 c/3$ above the molecule and rotated by 120° and 240° about the screw axis.) Figures 1(d), (e), and (f) are electron micrographs of sections of catalase crystals sectioned parallel to the (10.0), (11.0), and (00.1) planes respectively. Figures 2(a), (b), and (c) are optical diffraction patterns of Figs. 1(d), (e), and (f) respectively and approximately correspond, at low resolution, to X-ray diffraction patterns of Figs. 1(a), (b), and (c), respectively.

Although spots other than every third spot are present along the c^* axis in Figs. 2(a) and (b), these disallowed spots are present at reduced intensity.

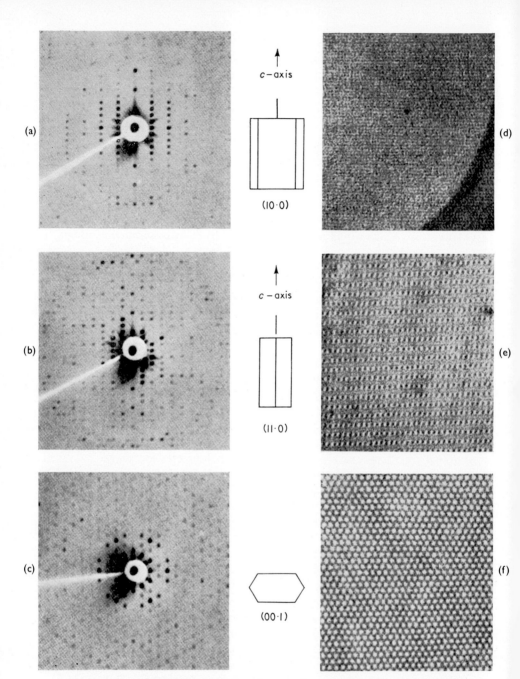

Fig. 1. (a, b, c), Zero-layer precession photographs of catalase crystals taken with the X-ray beam perpendicular to the (10.0), (11.0) and (00.1) planes respectively. No layer screen was used so that upper layers are present as well as the zero layer. The insert drawings show the orientations of the crystal in each case; (d, e, f), electron micrographs of crystals sectioned perpendicular to the (10.0), (11.0) and (00.1) directions. Protein is black. The insert drawings show the orientation of the crystal in each case. Magnification: 100,000 ×. Reproduced with permission from Longley (1967).

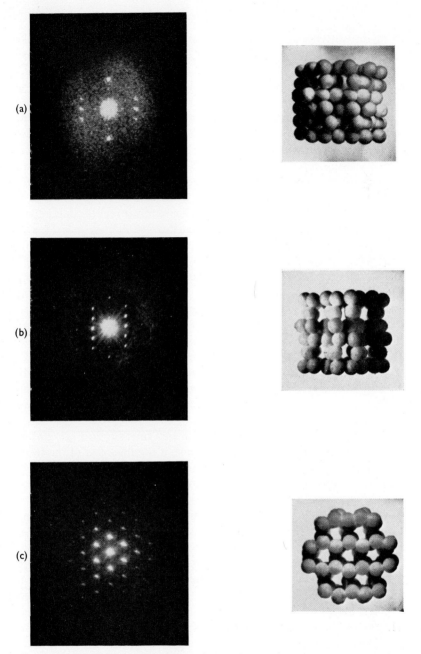

Fig. 2. (a, b, c), Optical transforms of the micrographs shown in Figs. 1 (d, e, f); (d, e, f), photographs of a model of the crystal viewed from the (10.0), (11.0) and (00.1) directions. Scale: 1 mm = 25 Å. Reproduced with permission from Longley (1967).

Therefore the existence of a three-fold screw axis might have been overlooked in the optical diffraction pattern if X-ray diffraction patterns of the crystal had not demonstrated it. The way in which electron micrographs (Figs. 1d, e) contain evidence for a three-fold screw axis is particularly interesting. We shall describe this in detail since it also provides a brief introduction to the Projection Theorem which we shall discuss later in this chapter in connection with three-dimensional reconstruction. Viewing Fig. 1(e) at a glancing angle perpendicular to the c axis, we note that the spacing in the c direction is one third the spacing seen when viewing Fig. 1(e) straight on. It is this "thirding" of the spacing in real space which causes a tripling of the spacing along the c^* axis in Fourier space. Viewed "edge on" the micrograph is, of course, a projection of the structure onto the c axis. The Projection Theorem is a relationship between projected structures in real space and sections through Fourier space. It states that the Fourier transform of a projected structure is equal to a central section through the Fourier transform of the unprojected structure (Section I.D, Chapter 1). This central section is perpendicular to the direction of projection. Therefore, if a repeat distance is reduced by one-third when a micrograph is viewed edge on then the spacing must be increased by three along the line of projection (the c^* axis), in the optical diffraction pattern, which is a Fourier transform. Thus a feature of the image, observed by viewing the image edge-on, is responsible for the missing spots along the c^* axis. Close inspection of Fig. 1(d) also reveals that the repeat distance along the c axis is "thirded" when viewed edge on perpendicular to the c^* axis, and the corresponding optical diffraction pattern (Fig. 2a) displays the appropriate tripling along the c^* axis.

The disallowed spots which appear at reduced intensity in Figs. 2(a) and (b) are artefacts and would presumably vanish if the thickness of the section of the catalase crystal were an exact multiple of the repeat distance, if the transmission of the micrograph were directly proportional to the projected protein density, and if dynamical electron scattering were not significant.

The structure which Longley inferred from his micrographs is shown in Figs. 2(d), (e), and (f). In the crystal, catalase molecules are arranged approximately along rows (Fig. 2f) in each of three layers (Fig. 2e). The three layers are identical and can be transformed into each other by rotations of 120°, i.e., by the 3_1 screw. In (11.0) views of the structure (Fig. 1e) one can see down the rows of molecules in every third layer. The other two layers appear more or less uniformly dense since one is not looking exactly down rows in the (11.0) view. The micrograph in Fig. 1(e) shows this feature of the structure. Longley's structure is based primarily upon this interpretation of the images and seems to explain the images to the resolution of the electron micrographs.

More systematic methods of determining three-dimensional structures from electron micrographs will be discussed shortly, but we note at this time that only three views were necessary to reconstruct the structure because additional views were generated by the screw symmetry of the catalase crystals. A more-detailed solution would require not just higher-resolution micrographs but more views as well.

B. Tropomyosin

The study of tropomyosin by Caspar *et al.* (1969) is an excellent example of the benefits obtained by using a combined X-ray diffraction and electron microscopy approach to a biological problem. Tropomyosin is a protein

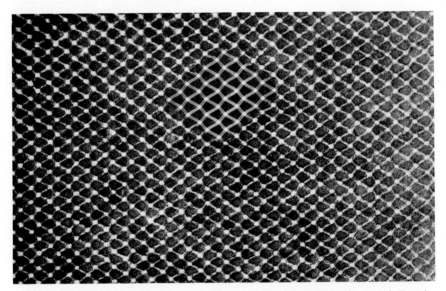

Fig. 3. View down the *a* axis of a negatively contrasted tropomyosin crystal. Protein is white. The inserted drawing located near the top centre shows the structure determined for tropomyosin. Reproduced with permission from Caspar *et al.* (1969).

present in all muscles and apparently fulfils both a regulatory and structural role. Tropomyosin has a molecular weight of about 70,000 and it is unique among fibrous proteins in forming true crystals. These crystals are suitable for X-ray diffraction and electron microscopy studies. Caspar *et al.* determined that the space group is $P2_1 2_1 2$ and that the orthorhombic unit cell has dimensions $a = 126$ Å, $b = 243$ Å, and $c = 295$ Å. Although they collected three-dimensional X-ray data, the data were most completely analysed in the planar projection down the *a*-axis.

Electron micrographs equivalent to projections along the *a*-axis were most easily obtained and interpreted. Figure 3 shows one such micrograph which contains an insert of the projected structure determined by Caspar *et al.* It

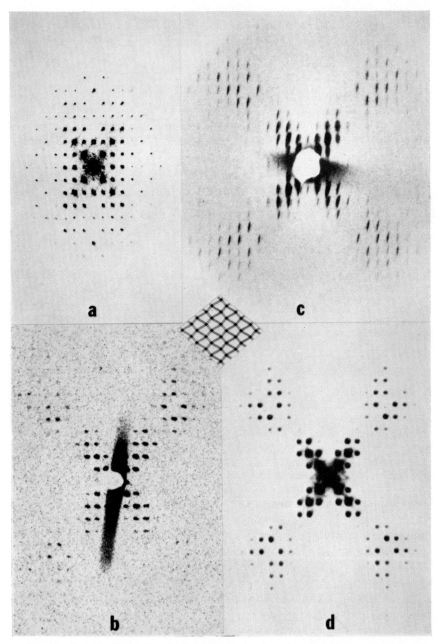

Fig. 4.

was hoped that the well-ordered electron micrograph in Fig. 3 could be directly compared with the X-ray diffraction pattern for the a-axis projection. This, however, was not the case. Figure 4(a) is the optical diffraction pattern of Fig. 3. The agreement with X-ray patterns (Figs. 4b, c) of equivalent projections is poor, even though the optical diffraction pattern has strong reflections down to 30 Å. The apparent reason for the discrepancy is that the dominant feature of the micrograph is the presence of nodes at intersections of the strands, while individual strands are less apparent. Because the electron micrograph is approximately an exponential function of the projected protein density, the contribution of the nodes is enhanced and the contribution of the strands is reduced.

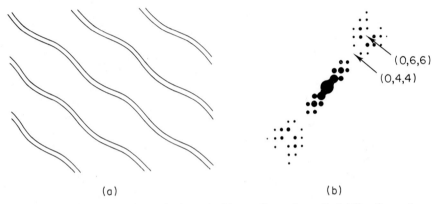

(a)   (b)

Fig. 5. (a, b), The set of strands shown in (a) contributes the optical diffraction pattern shown in (b). The two missing diffraction spots which were important in solving this structure are labelled.

Nevertheless, using the information contained in the micrograph Caspar *et al.* were able to phase† their X-ray patterns. The micrograph in Fig. 3 contains two features which played key roles in the phasing of their X-ray diffraction pattern. The first essential observation is that each of the strands forming the net appears to be divided into two filaments. The second point

† This is a shorthand notation, much used by crystallographers, to indicate the allotting of phase angles to the X-ray reflections.

Fig. 4. (a), Optical diffraction pattern of micrograph in Fig. 3. Note the absence of $0k0$ reflections with k odd. This indicates the presence of a two-fold screw axis. An examination of Fig. 3 viewed "edge on" perpendicular to the b (vertical) axis shows the expected "halving" of the repeat distance; (b) a still X-ray pattern viewed along the a axis of the crystal; (c), a precession photograph of the $0kl$ plane (a axis projection) of reciprocal space; (d), an optical diffraction pattern of the model structure shown in the insert. The model was constructed using phase information gathered from the micrograph in Fig. 3 and Fourier amplitudes from the X-ray patterns in this figure; thus the optical diffraction is in good agreement with the X-ray patterns. Reproduced with permission from Caspar *et al.* (1969).

is that the intersections of a strand with neighbouring strands are not evenly spaced but appear in a short-long-short sequence. Thus the region enclosed by four intersecting strands is kite-shaped, rather than a parallelogram. The bending of a strand must be periodic to make this type of intersection, but either localized or continuous bending of the filaments would be allowed.

Caspar *et al.* then constructed net models consisting of pairs of 20 Å diameter filaments with nodes of intersection at the spacings indicated in the electron micrograph shown in Fig. 3. The best fit between the calculated transforms and X-ray diffraction patterns was obtained using models with uniform bending of the filaments. It is thus possible to represent the structure analytically by a model built from sinusoidal filaments characterized by three

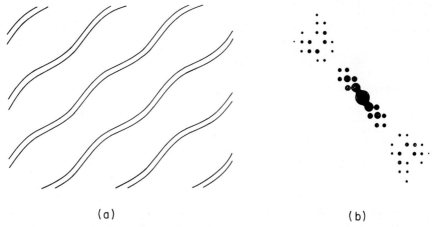

(a)   (b)

Fig. 6. (a, b), The set of strands shown in (a) produces the optical diffraction pattern in (b).

parameters. Furthermore, the diffraction pattern can be conveniently separated into diffraction from two sets of strands which can be considered separately. Fortunately, the diffraction patterns from both sets of strands do not overlap and they are shown in Figs. 5(a) and 6(a). The pattern shown in Fig. 5(b) is the optical diffraction pattern of the set of strands in Fig. 5(a), and the pattern in Fig. 6(b) is the optical diffraction pattern of Fig. 6(a). The transform shown in the X-ray diffraction patterns vanishes at 044, indicated by an arrow in Fig. 5(b). Caspar *et al.* demonstrated that this implies a separation of 25 Å, in projection, of the two filaments comprising each strand. The zero in the X-ray diffraction pattern at 066, also indicated by an arrow in Fig. 5(b), was shown to correspond to the first zero of the J_0 term of diffraction from the filament which is in projection a narrow, long-pitch helix, indicating that the amplitude of the sinusoidal wave of a filament is about 11·7 Å. Thus

Caspar *et al.* arrives at the model shown in the insert in Fig. 4. The optical diffraction pattern from the insert is shown in Fig. 4(d) and agrees with X-ray diffraction patterns of Figs. 4(b) and (c) in considerable detail.

This work with tropomyosin demonstrates the complementary nature of X-ray diffraction and optical diffraction of electron micrographs. The solution of this particular problem by X-ray means alone would have required phase information coming either from assumptions which might have been difficult to defend or from isomorphous derivatives which might have been difficult to prepare. Electron microscopy alone might have led to incorrect structures. By using phase information from electron micrographs together with the X-ray diffraction data, however, Caspar *et al.* were able to solve the structure in an unambiguous manner.

III. OPTICAL FILTERING

An electron micrograph is a two-dimensional structure from which it is possible to calculate a two-dimensional Fourier transform containing magnitudes and phases. A serious disadvantage of the straightforward application

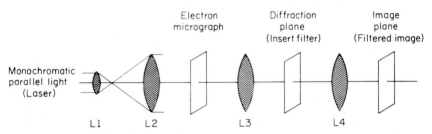

Fig. 7. An optical system typically used to diffract and filter electron micrographs.

of optical-transform methods to electron micrographs is that phase information which is contained within a micrograph is lost in photographically recording its optical diffraction pattern. A more desirable method of interpreting electron micrographs would be one which preserves phase information. One such method is optical filtering (see Section III.A). If an electron micrograph is used as a subject in an optical diffractometer and if only certain diffraction spots are allowed to pass through the diffraction plane (Fig. 7) and are then recombined by a lens (*L*4) it is possible to photograph the image produced by the allowed spots. This "filtered" image is also a two-dimensional structure and contains all of the phase information from the spots which were passed. This technique has been found to be very useful. Using the filtering technique it is possible to filter the transform of a periodic structure from the transform of another periodic structure (or from the transform of an aperiodic structure) even though the images of the two structures may have overlapped completely. This subject is considered in more detail in Section I.G, Chapter 11.

Optical filtering is a very useful technique but it can also be dangerous if the wrong diffracted rays are recombined to form an image. Therefore it is necessary to filter several images in order to determine the average features of a structure.

Electron micrographs of unstained specimens show very little contrast. In order to heighten the contrast observed in electron micrographs, either heavy atoms are attached at specific positions within the molecule (positive contrast), or the molecule is embedded in a layer of heavy atoms (negative contrast). In either case, electron micrographs primarily show the distribution of the heavy-metal contrasting stain. Structures may be partially or completely embedded in negative stain but negative contrast is formed only whenever the stain penetrates and fills vacant spaces within a structure. Thus the top of a particle which is embedded in a shallow layer of negative stain may not be surrounded by stain and will contribute very little to the image of the particle, while the bottom of the particle may be completely embedded and therefore provide the primary contribution to the image. These two contributions are spatially overlapped in an electron micrograph and cause a great deal of difficulty in interpreting images. This is particularly true if both surfaces are equally contrasted. If the two surfaces are periodic it is usually possible to separate the diffraction spots (or lines) into contributions arising from the front or back of the particle. The separation of the contributions from the front and the back of the structure is done by "indexing" the optical diffraction pattern. The pattern from a two-sided particle can be divided into two sets of lattice points. Both lattices may have slightly different lattice parameters since, for example, one side of a helix may contract in the drying negative stain to a greater degree than the other side, but the lattice parameters and the distribution of intensities at lattice points will be approximately the same. The decision as to which diffraction spots correspond to which lattice is frequently resolved by examining optical diffraction patterns of predominantly one-sided images, for in this instance one lattice of spots is considerably brighter than the other set and the choice of the basic lattice is simplified. Since the indexing procedure is a fundamental part of the structure-determining procedure, one must be sure that the indexing chosen includes all diffracted spots. If the two surfaces of the structure are curved (as in a helix) then diffraction spots from each side are broadened into layer lines and the lattices are correspondingly harder to index. The points concerning indexing will perhaps become more clear in the next section where they will be discussed further in connection with tubular variants of human-wart virus. Once the diffraction pattern has been indexed, either set of diffraction maxima may then be selected from the diffraction pattern by "filtering" and recombined into a filtered image. The filtered image can be observed or photographed and will be free of noise and of the confusion caused by the overlap of the image of the structure on the opposite side of the particle.

The procedure as outlined by Klug and DeRosier (1966) is demonstrated in the following figures taken from their paper. Figure 8(a) is a projection of a helical structure onto a plane in which both sides of the helix are represented. The parameters of the structure have been chosen so that they are the same as those of Tobacco Mosaic Virus at a radius of 100 Å. Figure

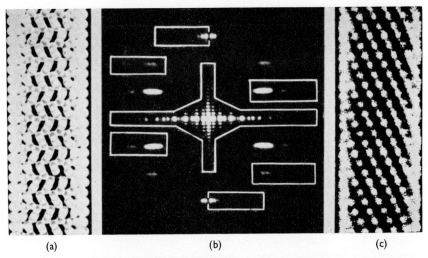

(a)   (b)   (c)

Fig. 8. (a), Positive replica of a photographic transparency representing the projection of a helical structure in the direction perpendicular to the helical axis; (b), optical diffraction pattern of (a); (c), filtered image of (a) obtained by passing only the diffraction rays enclosed by the boxes. Reproduced with permission from Klug and DeRosier (1966).

8(b) is the diffraction pattern of the transparency shown in Fig. 8(a). The parts of the optical diffraction pattern which were allowed through the filter are enclosed by the rectangles. Figure 8(c) is the filtered image of Fig. 8(a). Quite clearly, structure coming from one side of the helix has been removed. If the other set of spots had been allowed to pass through the filter the other side of the helix would have been imaged. Figure 9(a) shows the projection of only one side of the model of the helical structure used in Fig. 8(a). Figure 9(b) is the optical diffraction pattern of the "one sided" image in Fig. 9. This diffraction pattern, of course, shows spots corresponding only to the set which was allowed to pass to form Fig. 8(c). Figure 9(c) shows the image of Fig. 9(a) without any filtering. The loss of quality apparent in Figs. 8(c) and 9(c) arises from imperfections in the optical system but filtering of one side of the helix is convincingly demonstrated. The principal diffraction maxima of Fig. 9(b) lie at the approximate vertices of a lattice. The transform in Fig. 9(b) is similar to the diffraction pattern which would have been obtained from a planar sheet representing one side of a flattened helix. If one could simultaneously observe the front half of a helix and its diffraction pattern as the helix was being flattened, one would observe that the positions of layer lines would remain unchanged but that the intensity distribution on layer lines would gradually contract into discrete spots. Since it is much easier to index a lattice of discrete spots than it is to index a lattice of layer lines, it is frequently

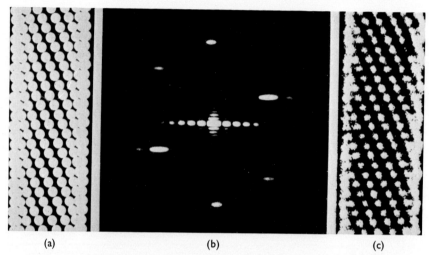

(a)   (b)   (c)

Fig. 9. (a), Projection of one side of the structure in Fig. 8(a); (b), optical diffraction pattern of (a); (c), the image of (a) without any filtering. The loss of quality in the image arises from imperfections in the optical system. Reproduced with permission from Klug and DeRosier (1966).

advantageous to examine images of flattened helices in the early stages of a structural investigation. We shall see in the following sections that since the curving of a planar array affords different views of the units composing the array, the broadened diffraction spots from helical arrays actually contain three-dimensional structural information from the image.

A. Tubular Variant of Human-Wart Virus

Human-wart virus infection is usually accompanied by the formation of hollow tubular structures of two kinds—wide tubes and narrow tubes. The structure of the tubular variants, presumably representing mistakes in virus assembly, is interesting because it provides an insight into the viral-assembly process. The wide tubes are composed of hexamers arranged on a hexagonal plane lattice rolled up into a tube. Such tubes are not particularly novel as they are of a type postulated by the theory of virus structure described by Caspar and Klug (1962). The narrow tubes, however, have a unique structure that was unanticipated by Caspar and Klug in their theory of virus construction, but which nevertheless is consistent with the postulate of bond specificity on which their theory is based.

The work of Kiselev and Klug (1969) in analysing the structure of narrow tube variants of human-wart virus demonstrates the power of the method of optical filtering and provides a clear insight into the principles involved. We will concern ourselves only with one of the two distinct subtypes of narrow tubes which were investigated, the "one-start" tubes.

An optical diffraction pattern of a one-start narrow tube is shown in Fig. 10(a). The diffraction pattern is composed of discrete spots rather than elongated layer lines which indicates that the helix is quite flattened. Nevertheless, the pattern in Fig. 10(a) is complex and there is a high level of background noise so that indexing is extremely difficult. The optical diffraction pattern is approximately symmetrical about the meridian, indicating that contrast has arisen evenly from both sides of the particle. This, of course, means that the particle is evenly surrounded by stain so that the image shows features coming from both sides of the helix. The pattern would be easier to index if primarily one side of the helix contributed to the image. Figure 10(d) shows a predominantly one-sided optical diffraction pattern. The reciprocal lattice used to index the "near" side of the tube has been drawn over the pattern. Even this one-sided pattern contains much noise and is difficult to index. The indexing of Fig. 10(d) accounts for the most intense spots while spots not included in the lattice arise from the weaker diffraction pattern of the "far" side of the particle. Both sides of the tube contributing to the diffraction pattern in Fig. 10(a) are indexed in Fig. 10(b) on the same lattice used in Fig. 10(d). The "near" side of the diffraction pattern filtered through the aperture shown in Fig. 10(c) appears in Fig. 11 (2057–10).

Surprisingly, the feature which emerges is that the tubes are made of pentamers. If a tube were cut along the axis and then opened with the inside facing up as in Fig. 12(a), one would see a structure similar to that shown in Fig. 12(b). This is, of course, a radial projection of the structure. Inspection of the filtered images of flattened helices shows that the plane group of the radial projection is $p2$. A unit cell and its two-fold axes have been included in Fig. 12(b). Each unit cell is seen to contain two molecules. Plane group $p2$ has four different types of two-fold axes which are located at the centre of the cell, at the corners of the cell, and one-half way along each of the two different types of cell edges (the two-fold axes are indicated by either closed or open ellipses). The structures seen in filtered images of the tubes have a novel property which is not required by the plane group symmetry: each pentamer is related in an identical manner through two-fold axes, to three equivalent pentamers. These special two-fold axes are indicated in Fig. 12(b) by closed ellipses. (Two two-fold axes which do not connect nearest neighbour pentamers are shown by open ellipses.) Thus we see that the conditions of constant bond length and bond angle which are required for the formation

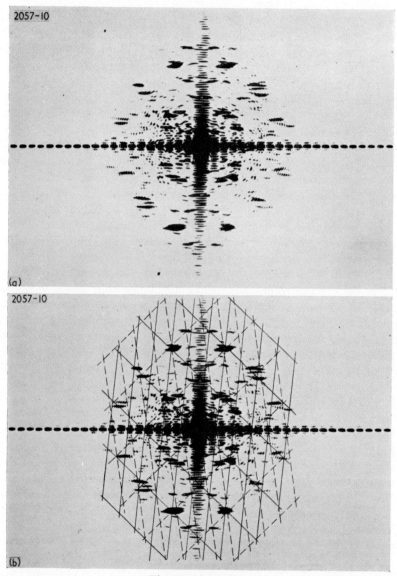

Fig. 10. (a) and (b).

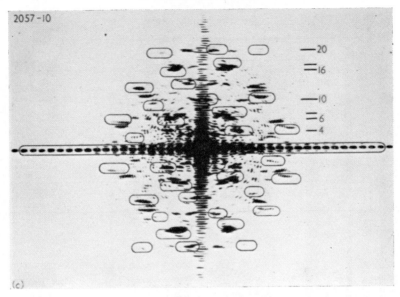

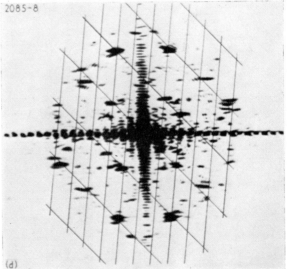

Fig. 10. (a), Optical diffraction pattern of a one-start narrow tube from human-wart virus. Contrast has arisen, more or less, evenly from both sides of the particle as judged by the symmetry of the diffraction pattern about the meridian; (b), the same pattern as in (a) indexed on two reciprocal nets corresponding to the near and the far sides of the tube; (c), the same pattern as in (a) indexed on a set of layer lines. The boxes show the apertures cut in the filter to produce the "near" filtered images in Fig. 11; (d), the optical diffraction pattern from another one-start narrow tube of human-wart virus. The reciprocal net is drawn for the side of the particle predominantly imaged in the micrograph. Reproduced with permission from Kiselev and Klug (1969).

One-start pentamer tubes

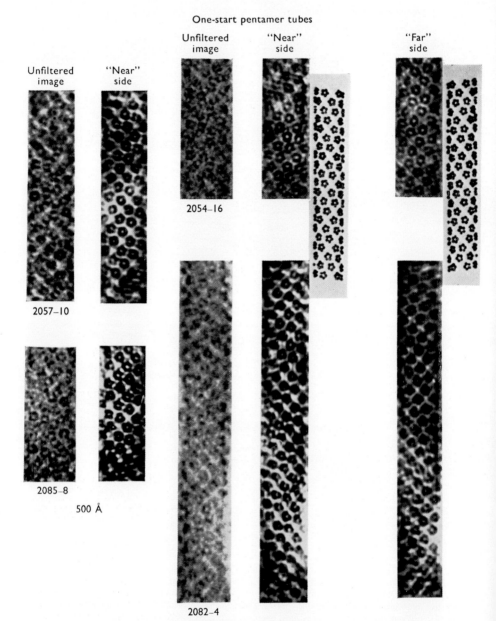

Fig. 11. Filtered images and electron micrographs of narrow, one-start, tubes of human-wart virus. Protein is black. Computer generated projections of a model structure are also shown for comparison. Reproduced with permission from Kiselev and Klug (1969).

of specific bonds between pentamers are fulfilled by each pentamer and its three nearest neighbours (i.e., the pentamers connected by the two-fold axes marked by closed ellipses). Caspar and Klug had postulated in their theory of virus assembly that bonds between subunits must occupy equivalent (or quasi-equivalent) positions with respect to neighbouring subunits. However this structure obeyed the requirements of Caspar–Klug quasiequivalence in a completely unexpected manner. Kiselev and Klug called this equivalently "bonded" arrangement of pentagons a pentagonal tessellation. Interestingly

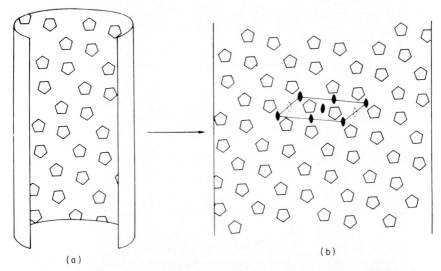

(a)   (b)

Fig. 12. The ideal pentamer lattice giving parameters chosen as close as possible to those observed for the one-start, narrow human-wart virus tubes; (a), the three-dimensional structure (illustrated as a sheet structure) opened and partially unrolled; (b), the completely unrolled structure, i.e., the radial projection of the structure. One unit cell and associated symmetry elements are shown.

in this arrangement each pentamer forms only three bonds with neighbouring pentamers. A pentamer is, of course, theoretically capable of forming five equivalent bonds, although not in a planar structure, so that the bonding potential of the pentamer subunit is clearly not maximized in this arrangement. A $T = 1$ icosohedron is a structure in which all five subunits of a pentamer could form equivalent bonds and structures of this size have been reported by others but not by Kiselev and Klug.

A well-preserved narrow one-start tube from human-wart virus and its optical diffraction pattern are shown in Fig. 13. The pattern shows a succession of diffraction maxima at approximate lattice positions instead of the lattice of sharper spots which is exhibited by flattened tubes. The spreading of the

J. A. LAKE

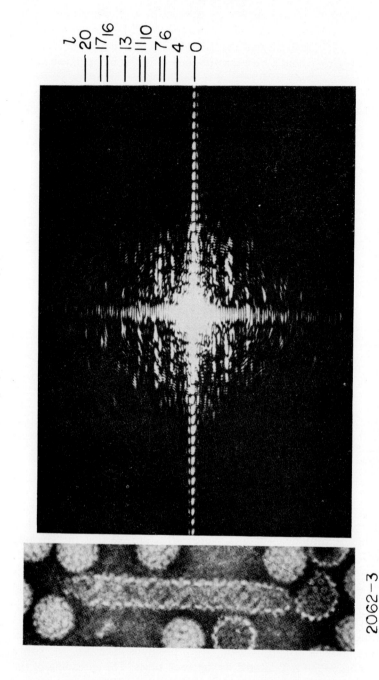

2062-3

Fig. 13. A well-preserved narrow, one-start tube from human-wart virus and its corresponding optical diffraction pattern. The pattern shows a succession of diffraction maxima on layer lines instead of the discrete lattice of spots seen in patterns from flattened tubes. Reproduced with permission from Kiselev and Klug (1969).

spots into lines indicates that the diffraction pattern contains three-dimensional information and brings us to our next topic, reconstruction of three-dimensional structures from electron micrographs.

IV. THREE-DIMENSIONAL RECONSTRUCTION

One normally forms a three-dimensional impression of an object by viewing it from several directions. Occasionally only two views are needed, one from each eye, but frequently one must move around in order to see hidden details. A direct analogy between visual processes and analysis of transmission electron-microscope images is, of course, not strictly valid, since we normally see objects by light scattered from their surfaces while in transmission electron microscopy we see images formed by transmitted electrons. Nevertheless, it is reasonable to ask why one does not use stereo microscopy to form three-dimensional images of structures. The answer is that in general only two views of a complex structure are not adequate to separate regions of overlapped structures. Parts of a complex molecule which are within the structure will almost certainly be hidden by other parts of the structure in both views. Stereo microscopy also has the disadvantage that structures are not presented in a tangible representation.

DeRosier and Klug (1968) formulated a process by which one could obtain a tangible "reconstruction" of a three-dimensional structure from transmission electron micrographs in an unambiguous manner provided that a sufficient number of views were available. They indicated that these views could be obtained by systematically tilting and photographing a specimen in the electron microscope or by photographing a few views of a structure containing rotation or screw axes. Since repeated exposure of a biological specimen to an electron beam destroys structural details, they emphasized that a small number of views (as few as one) could suffice provided the specimen has a sufficiently high rotational symmetry. The procedure demonstrated by DeRosier and Klug is illustrated in Fig. 14. Figure 14(a) shows a three-dimensional duck and Fig. 14(b) represents the three-dimensional Fourier transform of the duck. The Fourier transform has an inverse and one can mathematically calculate the three-dimensional transform† from the three-dimensional duck structure or vice versa. Figure 14(c) is a two-dimensional structure which represents a projection of the three-dimensional duck. By means of the two-dimensional Fourier transform one can calculate a two-dimensional Fourier transform (Fig. 14d) from the two-dimensional projected

---

† The Fourier transform is a complex function with two numbers associated with each point in space—a magnitude and a phase (Section I.A, Ch. 1). For simplicity, only the magnitudes of the Fourier transforms are shown in Fig. 14. It should be kept in mind, however, that both the magnitude and phase of the transform are used to calculate the corresponding three-dimensional (or two-dimensional) structure.

duck. Likewise one can perform the inverse operation from the Fourier transform to the projected duck structure. Figure 14(e) represents another "view" of the duck and Fig. 14(f) is the corresponding two-dimensional Fourier

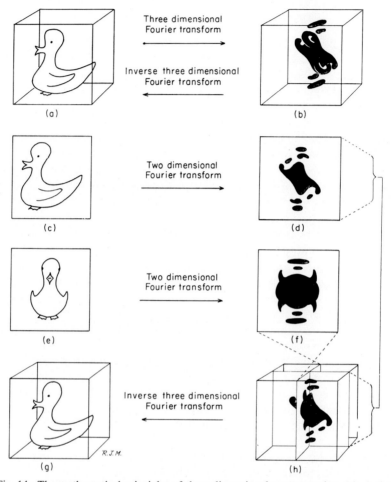

Fig. 14. The mathematical principles of three-dimensional reconstruction. (a), A three-dimensional duck and (b) its three-dimensional Fourier transform; (c), a projection of (a) and (d) the two-dimensional Fourier transform of (c); (e), another projection of (a) and (f) its two-dimensional Fourier transform; (g), the three-dimensional duck calculated from the three-dimensional Fourier transform (h) which was reconstructed by sampling three-dimensional Fourier space with the two-dimensional transforms (d) and (f).

transform. One can likewise transform from Fig. 14(f) to Fig. 14(e). The Projection Theorem (Section I.D, Chapter 1) states that the two-dimensional Fourier transform of a projected structure is the central section through the three-dimensional transform, which is perpendicular to the direction of view. Thus if one collects a sufficient number of two-dimensional Fourier transforms which correspond to different views of the structure and adequately samples three-dimensional Fourier space, one may determine the three-dimensional Fourier transform of the structure. This may then be inversely Fourier transformed to determine the three-dimensional structure. The number of views necessary to sample three-dimensional Fourier space depends, of course, upon the resolution desired in the three-dimensional reconstruction. Only two different views are shown for the purpose of illustration in Fig. 14, but this does not imply that these are sufficient to reconstruct the three-dimensional structure. From two views one could only reconstruct the duck structure to a very low resolution. In general, more sections must be included to obtain a reconstruction of high resolution.

A. Three-Dimensional Structure of the Tail of Bacteriophage T4

In their demonstration of the principles of reconstructing three-dimensional structures from electron micrographs, DeRosier and Klug (1968) reconstructed the structure of the tail of the bacteriophage T4. The tail of phage T4 is composed of a core surrounded by a contractile sheath. The core does not contract but the sheath contracts when the DNA, which is contained in the head of the phage, is injected into the host cell. By studying the structure of the uncontracted and contracted tail, DeRosier and Klug reasoned that it may be possible to discover details of the contraction and insertion processes.

Figure 15(a) is an electron micrograph of an uncontracted phage tail. Both sides of the image are equally contrasted as judged by the symmetry of its optical diffraction pattern (Fig. 15b) about the meridian. Figure 15(c) is the filtered image of the far side of the helix. As one examines Fig. 15(c) closely, it seems that there are two types of helices which spiral from the lower left of the filtered image to the upper right. One type of helix is near the outside of the tail, and the other has a radius about half of that of the tail. By carefully measuring the paths of these two helices one could presumably calculate the radii of the cylindrical surfaces on which they lie. However, the problem of determining the radius of the cylindrical sheet on which a helix lies is somewhat more complicated than this since the structural features observed are not precisely confined to a cylindrical sheet but, in fact, have finite thicknesses. The three-dimensional reconstruction procedure, while utilizing the same data, simultaneously considers structural details arising from all radii, and does it in a mathematically unique manner.

In order to sample Fourier space and perform a three-dimensional reconstruction, one must know both amplitudes and phases (Section I.D, Chapter 1). Fourier amplitudes can be obtained from photographs of the optical

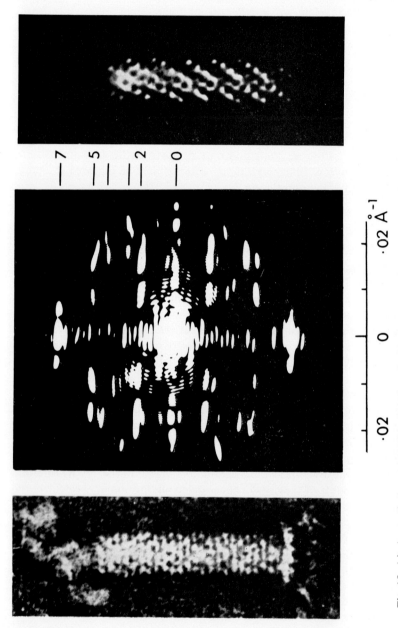

Fig. 15. (a), A negatively contrasted image of a tail of bacteriophage T4; (b), the optical diffraction pattern of the image in (a). The strong meridional spot on the seventh layer lines is contributed by the 38 Å spacing between annuli; (c) optically filtered image of the far side of the phage tail. Reproduced with permission from DeRosier and Klug (1968).

diffraction pattern, but Fourier phases must be measured by other methods. Phases can be measured directly from an optical diffraction pattern by interferometric means, i.e., by making a hologram (Chapter 9), or the phases (and the amplitudes) can be calculated by densitometering an electron micrograph at regularly spaced points on a grid and calculating the Fourier trans-

Fig. 16. Model of the three-dimensional structure of phage T4 tail as determined by DeRosier and Klug (1968). The fine black wires T follow the path of the helical tunnels located at an inner radius of about 60 Å. L and S refer to the large and small subunits described by Moody (1967).

form. The experimental difficulties in designing a good interferometric system led DeRosier and Klug to calculate their Fourier transforms on a computer.

The helical selection rule for the diffraction patterns of un-contracted T4 tails is $l = 2n + 7m$ (Moody, 1967), with the only permitted values of n being multiples of six. These selection rules imply that the structure can be visualized as six identical chains of subunits in register. Each chain contains seven subunits per repeat with each chain completing two turns per repeat. There are 42 (7×6) subunits per repeat. The 42 subunits per repeat provide 21 independent views of subunits (a view of a subunit is the same as a view of a subunit rotated 180° about the helical axis so that only 21 of the 42 views are independent). Twenty-one independent views were adequate to allow DeRosier and Klug to use data extending to 1/35 Å in Fourier space to calculate their density map of the phage tail.

The reconstruction of the tail is shown in Fig. 16. The model shows six helical tunnels at a radius of 65 Å, at approximately one-half the maximum radius. The tunnels are indicated by the T in Fig. 16. Protuberances extend to the outer edge of the tail and are indicated by the letters L and S in reference to the large and small subunits described by Moody (1967). Although the information presented in the paper by DeRosier and Klug could perhaps have been gleaned from a careful analysis of images of T4 tail or from filtered images of T4 tail, the three-dimensional reconstruction process treated the data in the micrographs in an objective manner and provided a tangible, unambiguous assessment of the data.

B. *Three-Dimensional Reconstruction of the Ribonucleoprotein Helix in the Chromatoid Body of* Entamoeba invadens

Lake and Slayter (1970) applied the methods of three-dimensional reconstruction to the chromatoid body helix of *Entamoeba invadens*. Light micrographs of cysts of *Entamoeba invadens* show bodies called chromatoid bodies, which strongly absorb ultraviolet light. The chromatoid body contains helices which are composed of particles having the physical attributes of ribosomes but which have not yet been shown to be functional ribosomes. The three-dimensional structure of this molecule is of interest, however, as it seems to bear directly on the question of ribosome structure. (The ribosome is, of course, the factory molecule which reads a genetic message from molecules of messenger RNA and which manufactures proteins according to that genetic message.)

Figure 17(a) shows an electron micrograph of a longitudinal section through a chromatoid body and Fig. 17(b) shows a transverse section through the chromatoid body. The helices are 600 Å in diameter and are composed of units which appear as spherical beads about 300 Å across. An optical diffraction pattern of one of the helices masked from the array in Fig. 17(a) is shown in Fig. 18(a). The inner edges of the layer lines arise from the surface of the helix and index approximately on the helical net overlay in Fig. 18(b). The near side of this helix has been removed by sectioning and thus only the helical net corresponding to the far side of the helix is present. The layer lines have been labelled in Fig. 18(a). As we have stated previously, the diffraction

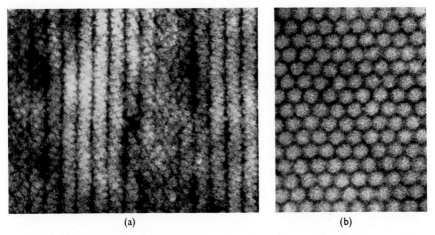

(a)   (b)

Fig. 17. (a), Longitudinal section through chromatoid body showing helices and asymmetric units comprising the helix; (b), transverse section through the chromatoid body of *Entamoeba invadens* showing packing of 600 Å diameter helices. Micrographs reproduced with the permission of Henry S. Slayter.

pattern from a helix consists not of discrete spots, but of diffraction spots which have been broadened into layer lines. However, if one plots along the abscissa the order of Bessel functions which are allowed to contribute to the

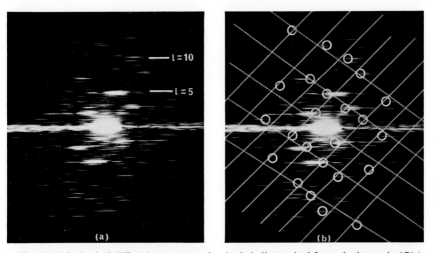

Fig. 18. (a), Optical diffraction pattern of a single helix masked from the image in 17(a). The diffraction pattern indicates that the far one-half of the helix has been removed by sectioning. This pattern is therefore easier to index than a two-sided pattern; (b), the diffraction pattern in (a) with the edge of each spot nearer the meridian circled. These spots lie approximately on the lattice drawn. Since the lattice has been drawn connecting the inner edges of diffraction lines it represents the structure at the outer edge of the helix.

diffraction pattern of a helix on a given layer line versus the layer line along the ordinate, one obtains a function which is described by a lattice. Such a plot is analogous to the diffraction pattern from a planar array corresponding to a flattened helix and is called a (n,l) plot (Klug et al., 1958). By measuring the positions of the broadened diffraction spots with respect to the meridian it is possible to determine a small range of orders of Bessel functions which may be contributing to the layer line. Thus measurement of the position of diffraction maxima in the optical diffraction pattern of a helix usually determines several (n,l) plots which are consistent with the measured positions of diffraction maxima. The positions of diffraction maxima which were measured from optical diffraction patterns of the helix of *Entamoeba invadens* are

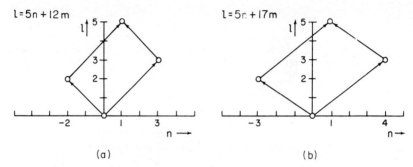

Fig. 19. The n, l plots for two possible indexings of the chromatoid body helix. l is the layer line number and n is the order of the Bessel functions contributing to the layer line. The circles indicate orders of Bessel functions allowed by the corresponding selection rules; (a), the n, l plot corresponding to the selection rule $l = 5n + 12m$, or twelve ribosomes per repeat five turns long; (b), the n, l plot corresponding to the selection rule $l = 5n + 17m$, or seventeen ribosomes per repeat five turns long.

consistent with either of the two (n,l) plots shown in Fig. 19. The helical lattices represented by these (n,l) plots are described by the selection rules $l = 5n + 17m$ and $l = 5n + 12m$, and contain 17 and 12 units per repeat respectively. In both helical lattices the units complete five turns per repeat. These two lattices cannot be distinguished by further measurements of the optical diffraction pattern. It is essential that the helical symmetry be determined since one must know the angles between equivalent views of the structure in order to perform the reconstruction. Fortunately, the phases, which can be calculated directly from an electron micrograph, contain information on the symmetry of the helix which is not contained in a photograph of the optical diffraction pattern. This information was used in the following manner to complete the identification of the helical parameters. It can be shown from the theory of helical diffraction that (see for example, Klug et al., 1958) in the region of the Fourier transform where Bessel functions of different order do not overlap (i.e., near the meridian), the magnitudes of the Fourier transform on a layer line are symmetrical about the meridian. The phases are symmetric if they arise from a Bessel function of even order; however, if they originate from an odd-order Bessel function then the phases on either side of the meridian will differ by 180°. Thus one can determine whether a Bessel function contributing to a layer line is of even or odd order by examining the symmetry of the phases across the meridian. In the study of the chromatoid body helix this was the information needed to index the helix completely. The

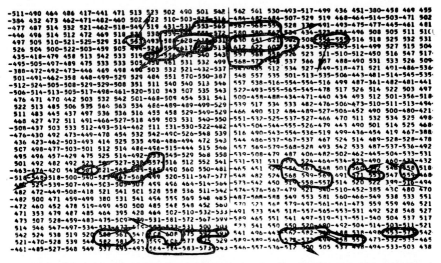

Fig. 20. Part of computer-generated output showing the Fourier transform calculated from a two-sided helix. The second, third, and fifth layer lines are shown. The length of the arrows indicate the magnitudes of the Fourier transform and their directions give the phase angles. For example, the interior spots on the third layer line has a phase of 170° on the right hand side of the meridian and 190° on the left hand side. The phase difference is 20° implying that an even-order Bessel function contributes to this layer line.

indexing procedure is better understood by examining Fig. 20; this shows a Fourier transform calculated on a digital computer from an electron microscopic image of a two-sided helix. This transform contains both magnitudes and phases. The length of a vector represents the magnitude of the transform at that point and the phase is represented by the direction of the vector. A phase angle of 0° corresponds to a vector pointing toward the right edge of the page. It is noted that phases on either side of the meridian on the fifth (top of Fig. 20) and second (bottom of Fig. 20) layer lines differ by approximately 180°, and the phases on the third layer lines are symmetrical about the meridian. This implies that odd-order Bessel functions contribute to the fifth and second lines and that an even-order Bessel function contributes to

7

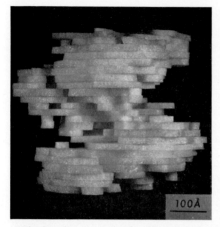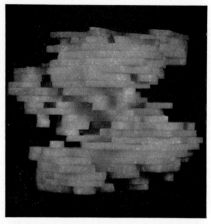

Fig. 21. A stereo pair of a "Styrofoam" model three-eighths of a repeat long representing the distribution of stain in the helix. Reproduced with permission from Lake and Slayter (1970).

the third layer line. By examining Fig. 19 we see that these conditions are satisfied only by the indexing $l = 5n + 17m$ (17 units in a repeat of five turns).

Knowing the mutual orientation of units within the helix, it was possible

Fig. 22. A diagrammatic representation of the three-dimensional structure of the chromatoid body helix. One entire helical repeat is shown. Reproduced with permission from Lake and Slayter (1970).

to reconstruct the three-dimensional structure. A stereo pair of a styrofoam model of the structure is shown in Fig. 21. The styrofoam indicates where the lead stain is specifically absorbed by the helix. From their studies, Lake and Slayter concluded that the dimensions of the unit (roughly 360 Å × 200 Å × 190 Å) corresponded reasonably well with the dimensions for 70s ribosomes given by small-angle X-ray measurements of ribosomes in solution (400 Å × 200 Å × 135 Å). If ribosomes are observed negatively contrasted their dimensions vary from 140 Å to 190 Å, which is not in agreement with the solution studies. This is understandable since negatively contrasted crystals frequently exhibit smaller dimensions than those given by X-ray diffraction measurements. Lake and Slayter were unable to find any structure reminiscent of the cap-like 30s ribosomal subunit which is frequently observed in electron microscopy. However a cleft-like region did seem to be present. Figure 22 is a diagrammatic tracing of photographs of the model shown in Fig. 21. A cleft which may represent a division between subunits was found and is visible in those units on the left side of the figure. Since *in vitro* crystallization of ribosomes has not yet been accomplished, optical and computer diffraction of electron micrographs will very likely continue to be the most productive method of analysing ribosome structure in an unambiguous three-dimensional manner.

C. Tomato Bushy Stunt Virus

All of the molecular arrangements reconstructed in three dimensions which have been discussed so far in this chapter have been helical arrays. There is nothing in the formal Fourier theory of three-dimensional reconstruction to require that only helical systems can be reconstructed; in fact, there is no requirement at all that a molecular structure contain rotational or screw symmetry elements. Any non-symmetric structure can be reconstructed provided that it can survive the damage incurred during the successive exposures to the electron beam required to photograph tilted views of the molecule. Crowther *et al.* (1970) used the method of Fourier three-dimensional synthesis of electron micrographs to demonstrate that viruses with non-helical symmetry could be reconstructed. They demonstrated the application of reconstruction methods by reconstructing both human-wart virus and tomato bushy stunt virus from electron micrographs. We shall discuss only their work with tomato bushy stunt virus.

Tomato bushy stunt virus is an icosahedral virus with a maximum diameter of about 330 Å. The major component of the coat protein is arranged on a $T = 3$ icosahedral surface lattice with the 180 major subunits clustered about two-fold axes so that electron micrographs give the appearance of 90 morphological units. Recent work has indicated the presence of a minor protein

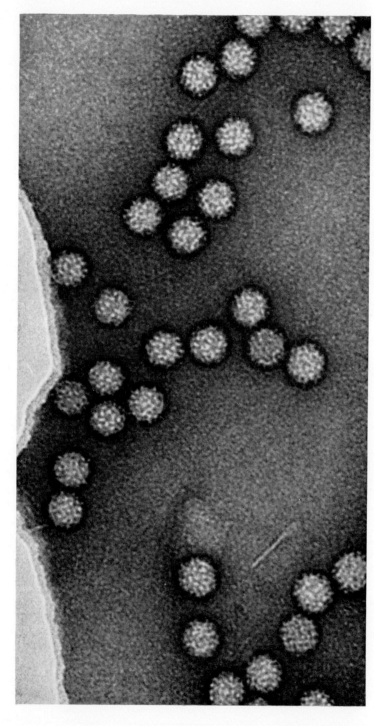

Fig. 23. A field of tomato bushy stunt virus particles embedded in negative stain. The particles used in the reconstruction shown in Fig. 25 are included in this image. Reproduced with permission from Crowther *et al.* (1970).

component and the three-dimensional reconstruction of the virus structure has shed some light on the placement of this minor protein component. Crowther *et al.* used some of the particles shown in the field reproduced in Fig. 23 in their reconstruction of the virus structure in three dimensions. Since a general view of a virus with icosahedral symmetry contains 60 independent, symmetry related views, except in the special cases of views down a symmetry axis in which cases there are fewer views, three independent views of virus particles were adequate to calculate the structure to a resolution of about 28 Å.

In determining the structure they faced several difficulties. The first problem was to identify the direction of view of the particles. Klug and Finch (1968) developed methods of comparing characteristic "views" of viruses with a gallery of computer-simulated views. By using this method Crowther *et al.* initially determined the orientation of viruses to within a few degrees. Using a "common lines" technique they refined the orientation of views. The common-lines technique is novel and utilizes the projection theorem in the following manner: the two-dimensional Fourier transform calculated from a micrograph of an icosahedral virus can be operated upon by the symmetry elements of the virus to generate 59 other identical planes throughout three-dimensional Fourier space. The original plane will, in general, be intersected by the other 59 planes. Crowther *et al.* searched for these common lines of intersection by computer calculations and thereby refined the orientation of the virus.

Unless one partially introduces the symmetry of the virus into the calculation by the use of Bessel functions, the problem is too large to be solved on present-day computers. To reduce the calculation to a manageable size, the virus is treated mathematically as a special type of helix. If one uses a five-fold axis as the z axis, an icosahedral virus can be formally represented as a degenerate helix with a rotation of $360°/5$ between repeating units and an infinite z translation between units. The transform of the virus is continuous in the Z direction and has no layer lines since there is an infinite repeat in the z direction. On planes of constant Z, only those Bessel terms can contribute which have orders that are multiples of five. The two-fold axes perpendicular to the five-fold axis can be put into the mathematics explicitly, but it is not possible to accommodate easily the three-fold axes. In order to introduce the symmetry of the three-fold axes into the structural determination, it is necessary to calculate the line of intersection of all 60 equivalent planes with each layer plane. The various allowed orders of overlapped Bessel terms contributing to each layer plane can then be determined from the values of the transform along the 60 lines of intersection. Determined in this way, the Bessel terms contain implicitly within them the entire symmetry of the 532 point group (including the symmetry of the three-fold axes). Once the contributions of the allowed orders of Bessel functions are calculated, the structural determination can be carried out as if the virus were a helix.

The three-dimensional structure of tomato bushy stunt virus calculated by Crowther *et al.* is shown in Fig. 24. The primary density is seen to lie on the local and strict two-fold axes. (Any molecule contained in the virus will coincide with an identical molecule if the molecule is rotated 180° about a strict two-fold axis; whereas, only those molecules which are "near" a local

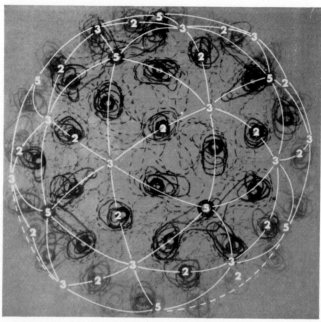

Fig. 24. Three-dimensional contour map of tomato bushy stunt virus with the $T = 3$ surface lattice superimposed. Contours indicate the absence of stain. Contours are clustered around the local (not indicated) and strict (indicated by 2's) two-fold axes. There are also contours to a lesser extent around the five-fold axes, indicated by 5's, but there seem to be no contours near the strict three-fold axes (indicated by 3's). Reproduced with permission from Crowther *et al.* (1970).

two-fold axis will be brought into coincidence with an identical molecule by a 180° rotation about the local two-fold axis.) The density on the local two-fold axes lies at a larger radius than the density on the strict two-fold axes (marked by 2's) resulting in the puckering of the rings of six seen in the stereo pair in Fig. 25. From the three-dimensional density map, it appears that the dimers on local two-fold axes are at a radius about 10 Å greater than the dimers on strict two-fold axes. The reconstructed model also shows an accumulation of density at the five-fold axes; this may be the minor protein component discussed earlier.

The evolution of structural studies from electron micrographs has been spectacular during the last 10 years. Although the techniques of optical diffraction have been particularly helpful in demonstrating the correspondences between electron microscopy and X-ray diffraction, major advances have come about by utilizing the phase information contained within electron micrographs. This has been possible by direct interpretation of electron micrographs, or by the more complicated techniques of optical filtering and three-dimensional reconstruction. These advances have been buttressed by the conceptual framework of X-ray diffraction. This framework has supplied us with theoretical and working concepts of the Fourier transform, symmetry

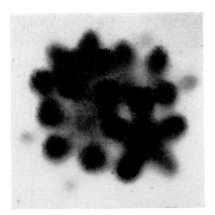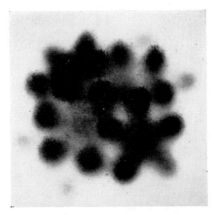

Fig. 25. A stereo pair of the density plot of the tomato bushy stunt virus reconstruction. High density indicates an absence of stain. Note that rings of five contain density within while rings of six seem to be empty. Reproduced with permission from Crowther *et al.* (1970).

operators, and helical diffraction theory. Optical diffraction by directing our attention to this natural framework has provided the conceptual impetus for many of the advances in analysing biological macromolecular structure by electron microscopy.

REFERENCES

Caspar, D. L. D., and Klug, A. (1962). *Cold Spr. Harb. Symp. Quant. Biol.* **27**, 1.
Caspar, D. L. D., Cohen, C., and Longley, W. (1969). *J. molec. Biol.* **41**, 87.
Crowther, R. A., Amos, L. A., Finch, J. T., DeRosier, D. J., and Klug, A. (1970). *Nature, Lond.* **226**, 421.
DeRosier, D. J., and Klug, A. (1968). *Nature, Lond.* **217**, 130.
Kiselev, N. A., and Klug, A. (1969). *J. molec. Biol.* **40**, 155.
Klug, A., and Berger, J. E. (1964). *J. molec. Biol.* **10**, 565.
Klug, A., and DeRosier, D. J. (1966). *Nature, Lond.* **212**, 29.
Klug, A., and Finch, J. T. (1968). *J. molec. Biol.* **31**, 1.

188   J. A. LAKE

Klug, A., Crick, F. H. C., and Wyckoff, H. W. (1958). *Acta Cryst.* **11**, 199.
Lake, J. A., and Slayter, H. S. (1970). *Nature, Lond.* **227**, 1032.
Longley, W. (1967). *J. molec. Biol.* **30**, 323.
Moody, M. F. (1967). *J. molec. Biol.* **25**, 167.

ADDITIONAL LITERATURE

Crowther, R. A., DeRosier, D. J., and Klug, A. (1970). *Proc. R. Soc. Lond.* A. **317**, 319.
DeRosier, D. J., and Moore, P. B. (1970). *J. molec. Biol.* **52**, 355.
Finch, J. T., and Gibbs, A. J. (1970). *J. Gen. Virol.* **6**, 141.
Finch, J. T., and Holmes, K. C. (1967). *In* "Methods in Virology", Vol. III (Ch. 9), Academic Press, New York.
Finch, J. T., and Klug, A. (1967). *J. molec. Biol.* **24**, 289.
Finch, J. T., Klug, A., and Stretton, A. O. W. (1964). *J. molec. Biol.* **10**, 570.
Glaeser, R. M., and Thomas, G. (1969). *Biophys. J.* **9**, 1073.
Hall, C. E. (1953). "Introduction to Electron Microscopy", McGraw-Hill, New York.
Hart, R. G., and Yoshiyama, J. M. (1970). *Proc. natn. Acad. Sci. USA* **65**, 402.
Heidenreich, R. D. (1964). "Fundamentals of Transmission Electron Microscopy," Interscience, London.
Huxley, H. E. (1968). *J. molec. Biol.* **37**, 507.
Kiselev, N. A., DeRosier, D. J., and Klug, A. (1968). *J. molec. Biol.* **35**, 561.
Klug, A., and Finch, J. T. (1968). *J. molec. Biol.* **31**, 1.
Krimm, S., and Anderson, T. F. (1967). *J. molec. Biol.* **27**, 197.
Labaw, L. W., and Rossman, M. G. (1969), *J. Ultrastruct. Res.* **27**, 105.
Moody, M. F. (1967). *J. molec. Biol.* **25**, 201.
Moody, M. F. (1968). *Nature, Lond.* **218**, 263.
Moore, P. B., Huxley, H. E., and DeRosier, D. J. (1970). *J. molec. Biol.* **50**, 279.
Reedy, M. K. (1968). *J. molec. Biol.* **31**, 155.
Thon, F. (1966). *Z. Naturforschg.* **21a**, 476.
Vainshtein, B. K., Barynin, V. V., and Gurskaya, G. V. (1968). *Doklady Akademii Nauk SSSR.* **182**, 569.
Yanagida, M., Boy de la Tour, E., Alff-Steinberger, C., and Kellenberger, E. (1970). *J. molec. Biol.* **50**, 35.

CHAPTER 6

Optical Fourier Synthesis

G. Harburn

*Physics Department, University College, Cathays Park,
Cardiff, South Wales*

I. Introduction

Image formation in terms of Fourier transformation has been discussed in Chapter 1. It can be shown in general (Goodman, 1968; Cowley and Moody, 1958) that image formation can be represented mathematically by two successive Fourier transformations. The physical meanings of these two transformations are, firstly, the diffraction pattern formed by scattering of radiation from the object and, secondly, the interference pattern resulting from subsequent recombination of the scattered wave (Section I.C, Chapter 1). It is well known that for optical wavelengths the recombination process can be effected by means of a lens; but for most other wavelengths, including X-rays, it is not possible to obtain the image from the diffraction pattern by any direct physical means.

When, for reasons such as improved resolution or more favourable scattering properties, a radiation that will not form direct images has to be used (Section I.A, Chapter 1) then the image must be obtained by calculation from data recorded from the scattered wave; i.e., the second Fourier transformation must be carried out purely by computation. Such calculations are usually performed on single-channel electronic computers but a physically satisfying alternative is to use optical diffraction as a multi-channel analogue computer. Since the intensity distribution in the Fraunhofer diffraction pattern of an object is proportional to the square of the modulus of its Fourier transform (Section I.C, Chapter 1) the required image can be obtained as the intensity distribution in the Fraunhofer pattern of a screen which represents the complex amplitude in the diffraction pattern of radiation scattered by the object.

Mathematically, the recombination process for the general case can be expressed by

$$\rho(\mathbf{r}) = \int_{-\infty}^{\infty} G(\mathbf{S}) \exp\{2\pi i \mathbf{r} . \mathbf{S}\} \, dV_s \qquad (6.1)$$

where $\rho(\mathbf{r})$ is the amplitude of the wave at the point in the image specified by vector $\mathbf{r}$ and $G(\mathbf{S})$ is the complex amplitude of the radiation scattered in the direction specified by the reciprocal-space vector $\mathbf{S}$ (compare Eqn. 1.8). For crystalline material the expression simplifies to

$$\rho(x, y, z) = \sum_{-\infty}^{\infty} \sum \sum F(hkl) \exp\{2\pi i(hx + ky + lz)\} \qquad (6.2)$$

where x, y and z are fractional coordinates in the image and h, k and l are the indices of the discrete orders in the diffraction pattern. In general $F(hkl)$ is complex and the phase angle may have any value in the range 0 to 2π (Section I.D, Chapter 1).

Thus if the various orders of diffraction are correctly represented in position, amplitude and phase by beams of light passing through apertures in a diffracting screen, the Fraunhofer diffraction pattern of the screen is an accurate representation of the object. In practice, since the screen is two-dimensional, only two of the indices h, k and l can be represented in it and the third order is zero. It can be seen from Eq. (6.2) that when one of the indices is zero the corresponding coordinate x, y or z is not represented in the image which appears as a projection of the unit cell contents on to a plane. Equation (6.2) then reduces to a form involving only two coordinates and two indices; for example if the diffracting screen is prepared from the $0kl$ X-ray data the image is governed by the equation

$$\rho(y, z) = \sum_{-\infty}^{\infty} F(0kl) \exp 2\pi i(ky + lz). \qquad (6.3)$$

It is a simple matter to position correctly the apertures in a diffracting screen. Methods of controlling the amplitude and phase of the light transmitted through each aperture are described in the two sections following.

II. AMPLITUDE CONTROL

It was shown in Section I that, in order for optical images to be produced from scattering data obtained at optical or other wavelengths, the holes in the mask representing the various diffracted orders must transmit a wave of the correct amplitude. Since, with ideal equipment, the mask is illuminated by a plane wave across which the amplitude distribution is uniform, some means must be available for adjusting the amplitude of the various transmitted beams to the desired value. Various methods have been described in the past but the three most important techniques are as follows.

A. Hole Size

When Bragg (1939) first put forward the idea of carrying out Fourier summations optically (Section I.F, Chapter 1) he proposed that each reflexion in the diffraction pattern of a crystal be represented by an aperture of area proportional to the X-ray amplitude. The Fraunhofer diffraction pattern of a hole is such that the amplitude at the origin is proportional to the area of the hole.

It is clear from Eq. (6.2) that the coefficients, $F(hkl)$, of the terms in the Fourier series should remain constant: they should not be functions of the position in the image at which the summation is being performed. This situation can be approximated to in the optical analogue procedure only if the holes in the mask are extremely small compared with their spacing; otherwise the coefficients will change appreciably because of the fall-off in amplitude in the diffraction pattern of a hole. However, it is not practicable to use extremely small holes in the mask because of the consequent low levels of illumination in the diffraction pattern. It would not matter if the coefficients changed from point to point in the image provided that they all changed in the same proportion—this would merely alter the overall level in different parts of the summation. The main criticism of Bragg's method of amplitude control by using apertures of different areas stems from the fact that the extent of the Airy disk in the diffraction pattern of a hole is inversely proportional to the radius of the hole. It follows that the fall-off in amplitude (and the manner in which the Fourier coefficients change) as the diffracting angle increases from zero varies considerably for apertures of different areas.

Figure 1 compares the amplitude-distribution curves for two diffracting apertures whose areas are in the ratio 10:1. Although the amplitudes at the origin of the diffraction pattern are in the ratio 10:1 this figure has decreased to 9:1 at a distance of 0·25 of the radius of the Airy disk of the large aperture. In practice it is unlikely that, for X-ray imaging by optical analogue, amplitude ratios of greater than 10:1 will be represented since, unless an extremely high-quality optical system is used with a very well characterized light beam, the quality of the images produced are not seriously affected by excluding contributions having amplitudes less than 0·1 of the maximum value. If errors of up to 10% in amplitudes are acceptable—and such amplitude variations will be commonly found in the wave incident on apertures in different parts of the mask—then optical summations may be carried out over this region

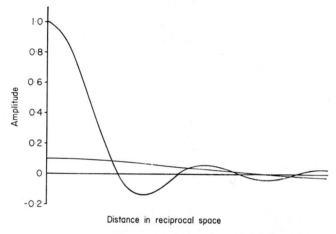

Fig. 1. Distribution of amplitude in the diffraction patterns of two circular apertures having areas in the ratio 10:1.

of diameter 0·25 of the diameter of the first dark ring in the diffraction pattern of the largest aperture in the mask.

Consider, as a simple example, the formation of an image from a reciprocal lattice (Section I.E, Chapter 1) in which the unit cell is a square of side a^*. The separation, a, of the corresponding lattice points in the image is proportional to $1/a^*$ or $a = k/a^*$. If the imaged contents of a complete cell are to fall within the region for which the amplitude representation satisfies the 10% condition then it is clear from Fig. 2 that half the diagonal of the unit cell should be less than 0·25 of the radius of the Airy disk due to the largest aperture in the mask. Thus, a^*/d, the ratio of the spacing of the orders in the mask to the diameter of the largest aperture, must be greater than 2·32. In spite of these limitations for anything approaching accurate work, this method of weighting remains an attractive one because of its simplicity.

A method of controlling amplitude by varying the size of a hole without altering the angular position at which the contribution first falls to zero has been described by de Vos (1948). He made use of annular apertures, the central opaque region serving to reduce the effective area of the opening. The

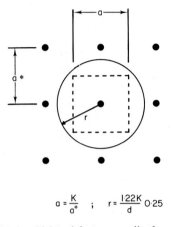

$$a = \frac{K}{a^*} \quad ; \quad r = \frac{1 \cdot 22K}{d}\, 0 \cdot 25$$

Fig. 2. Extent of Airy disk in which satisfactory amplitude representation is obtained.

introduction of this central obstruction shifts to a lower diffracting angle the position at which the amplitude first falls to zero, but this angle can be restored to its original value by reducing the diameter of the outer ring. Figure 3 shows the radii of the inner and outer rings which give required fractions of the amplitude contribution from an unobstructed circular hole of unit radius. In practice

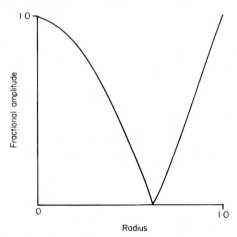

Fig. 3. Relation between the radii of annular apertures and the amplitudes at the origin of their diffraction patterns.

this method is not easy to use since the central portions of the annular apertures have to be supported. The simplest way of representing such an aperture is to print it on a photographic plate, which must then be placed in an "optical gate" to eliminate unwanted phase effects due to variations in thickness of the emulsion and its base (compare Section I.B, Chapter 11). Although the method has been used for preparing optical transforms of masks in which atoms of different atomic number are represented, the optical gate would be a serious embarrassment when the phases, as well as the amplitudes, of the various contributions have to be set to particular values.

B. Polarized Light

Hanson and Lipson (1952) described a method of amplitude control which depends on the effect produced by pieces of mica placed between crossed polarizer and analyser. The method can readily be understood by considering

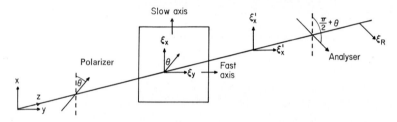

Fig. 4. Vector diagram for amplitude and phase control using plane-polarized light.

the behaviour of the light beam, represented by its electric vector, as it traverses the system. In Fig. 4 the polarizer is oriented so that the electric vector, of unit amplitude, incident on the mica sheet makes an angle θ with the x-axis. The components of this vector in the x and y directions can be represented by

$$\xi_x = \cos \theta \cos [\omega t - kz]$$
$$\xi_y = \sin \theta \cos [\omega t - kz].$$

If the piece of mica is positioned so that its slow and fast directions are parallel to the x and y axes respectively, the components in these directions after the wave has passed through the mica are given by:

$$\xi_x{}^1 = \cos \theta \cos [\omega t - k(z + d)]$$
$$\xi_y{}^1 = \sin \theta \cos [\omega t - kz]$$

where d is the difference in optical thickness of the mica for the two components. Each of these components has a contribution in the direction, specified by the analyser, making angle $(\pi/2 + \theta)$ with the x-axis. The resultant wave transmitted by the system is the sum of these two contributions and is given by:

$$\xi_R = \xi_x{}^1 \cos\left(\frac{\pi}{2} + \theta\right) + \xi_y{}^1 \sin\left(\frac{\pi}{2} + \theta\right)$$

$$= -\xi_x{}^1 \sin\theta + \xi_y{}^1 \cos\theta$$

$$= -\sin\theta\cos\theta\cos\left[\omega t - k(z+d)\right] + \sin\theta\cos\theta\cos\left[\omega t - kz\right]$$

$$= -\tfrac{1}{2}\sin 2\theta\{\cos\left[\omega t - k(z+d)\right] - \cos\left[\omega t - kz\right]\}$$

$$= -\sin 2\theta \sin\frac{kd}{2}\sin\left[\omega t - k(z + \tfrac{1}{2}d)\right].$$

Thus, at a given time and place along the z axis, the amplitude of the transmitted disturbance is proportional to $\sin 2\theta$ so that the amplitude can be adjusted to any desired value by altering the angle θ. The physical significance of the change in sign of $\sin 2\theta$ when θ exceeds $\pi/2$ is that the phase of the wave changes by π. The method therefore provides a means not only of varying the amplitude but also of setting the phase of the contribution to either 0 or π. The accuracy with which different amplitudes can be represented is reasonably good; for an accuracy of angular setting of $\pm 1°$ the error in the middle of the amplitude range is about $\pm 5\%$. However the accuracy depends on the relative amplitude level and is lowest at low levels, where, happily, the effect on a diffraction image is least.

To keep the intensity transmitted by the system as high as possible the mica sheet should be a half-wave plate so that $d = \lambda/2$ and the term $\sin kd/2 = 1$. However, since the sine function changes fairly slowly with angle in the region of its maximum value, the transmitted intensity remains high for small departures from this optimum thickness. Because of their length it is not convenient to use Nicol prisms for the polarizer and analyser and polaroid sheet is usually used. The high absorption of this material leads to a low level of illumination in the diffraction patterns, which is a disadvantage of this method of amplitude control. This factor would be of little significance if a high-power gas laser, emitting plane-polarized light, were used as the source of illumination.

An advantage of this method of amplitude control is that identical components are used for all amplitude levels, the required values being obtained by altering the orientation. The number of components that have to be constructed need not, therefore, exceed the largest number of diffracting orders that are likely to be represented in a mask.

C. Gauzes

The most satisfactory method of amplitude control is undoubtedly that in which all diffracting apertures have the same size and the amount of light transmitted by each is adjusted by covering the holes with gauzes of appropriate transmission factor. The transmission factor is defined by the clear area of gauze in unit total area of the sheet of material. Thus if the radius of the gauze hole is r and there are N holes within a limiting aperture of radius R, then the transmission is given by Nr^2/R^2.

The diffraction effects produced by a gauze of regularly sited holes are most easily appreciated if the diffracting object is broken down into the following three component functions (Section II.F, Chapter 11).

    (i) A two-dimensional array of points, infinite in extent, defining the centres of the gauze holes—a lattice function, the lattice cell usually being a square in practice.

    (ii) A single gauze hole—a transparency distribution function.

    (iii) The overall aperture which limits the number of effective gauze holes —a shape function.

The diffracting object is given by the product of the shape function and the result of convoluting the infinite net with the transparency distribution function. The complete optical transform is the result of the corresponding reciprocal-space combinations of the individual transforms of the three functions. Sections of the transforms of functions (i), (ii) and (iii) are shown in Figs. 5(a), (b) and (c) respectively; the amplitude a is plotted to an arbitrary scale for each diagram against s, the sine of the scattering angle, and the sections are taken along a line parallel to one side of the unit cell of the basic lattice and passing through the origin of reciprocal space. Figure 5(d) shows the result of multiplying the reciprocal lattice of Fig. 5(a) by the transform of a single gauze hole (Fig. 5b). The final transform (Fig. 5e) is obtained by convoluting the transform of the shape function (Fig. 5c) with the product function of Fig. 5(d). A photograph of a limited extent of the diffraction pattern of a gauze-covered hole is shown in Fig. 6. The diffraction pattern of the limiting aperture has been reproduced at points of the lattice reciprocally related to that of the gauze. The region of this diffraction pattern which is of interest for optical imaging is the Airy disk surrounding the origin because, when further gauze-covered apertures are included in the mask, the interference pattern is found within this area.

Since the spacing of the gauze holes is small compared with the spacings of the overall diffracting apertures in the mask, the phases of the contributions from gauze holes within a single aperture vary very little over the region of the Airy disk of the limiting aperture. The amplitude contribution within this Airy-disk region from any one aperture is, therefore, simply the sum of the contributions from the individual gauze holes. Since the contribution from each gauze hole is proportional to its area, the total contribution from the aperture, containing N holes each of radius r, is proportional to Nr^2, which is the transmission of the gauze. The amplitude contributions to a diffraction pattern from a number of apertures of the same size may, therefore, be adjusted by covering the apertures with gauzes having transmission factors proportional to the required amplitude contributions.

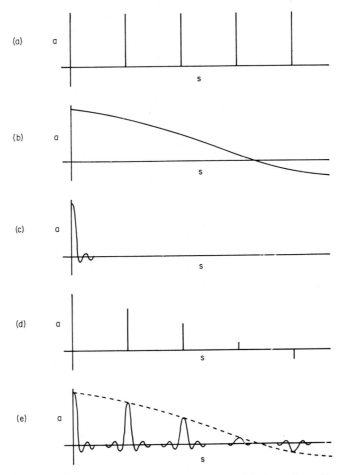

Fig. 5. The diffraction effects of a gauze: (a), reciprocal-lattice points; (b), diffraction pattern of a single gauze hole; (c), diffraction pattern of a limiting aperture; (d), the product of (a) and (b); (e), the convolution of (c) and (d).

It is an unfortunate disadvantage of the method that the full amount of light that could be passed by an aperture can rarely be used since there is a gap between the maximum transmission obtainable with a gauze (0·78 for a square mesh and 0·88 for a hexagonal mesh when the holes run together) and the transmission factor of 1·0 for an uncovered hole. Thus some lowering of the overall level of illumination in the diffraction images may have to be accepted. In most cases of X-ray imaging however, the magnitude of the $F(000)$ reflexion is so much larger than that for any other reflexion that $F(000)$ can be represented by a clear hole, all others being represented by holes of the same size covered with suitable gauzes.

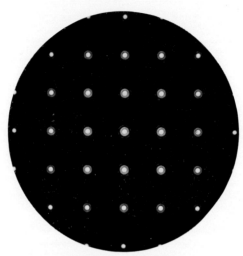

Fig. 6. Part of the diffraction pattern of a circular hole covered by a piece of gauze.

It is important, for the following reasons, to use a gauze in which the separation of adjacent holes is small compared with the diameter of the aperture that is to be covered.

  (i) The amount of light transmitted by an aperture should be independent of the position of the centre of the aperture relative to the centre of a gauze hole.

 (ii) The outline of the unobstructed gauze holes should be the same for each aperture and as nearly circular as possible. In practice this outline depends again on the position of the centre of the aperture relative to the centre of the gauze hole.

(iii) The "centre of gravity" of the transmission function for an aperture should coincide with the centre of the aperture.

Figure 7 shows examples of the violation of these three conditions in an extreme situation when the radius of the limiting aperture is only 1·5 times the spacing of the gauze holes. Clearly the smaller the separation of the gauze holes the better the conditions will be fulfilled. In practice satisfactory results are obtained if the limiting aperture contains about 30 gauze holes or more.

A disadvantage of the method is that each piece of gauze can be used to represent only one particular amplitude. A very large number of components

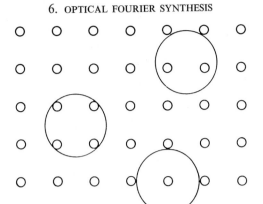

Fig. 7. Apertures in different positions on a gauze.

must therefore be available if new data are to be represented quickly without a lot of preliminary cutting and mounting of disks of gauze.

If the gauzes covering all the apertures in a mask are in parallel orientation and have the same spacing, additional images are found in the Airy disks surrounding the non-zero-order reciprocal-lattice points shown in Fig. 6.

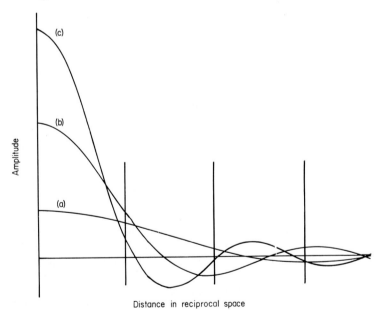

Distance in reciprocal space

Fig. 8. Distribution of amplitude in the diffraction patterns of gauzes of three different transmissions: (a), 0·15; (b), 0·45; (c), 0·75. The vertical lines show the positions of reciprocal-lattice points. The diagram is based on a gauze of 6 lines mm^{-1}.

However, it is easily seen from Fig. 8 that a set of gauzes having transmissions chosen to give correct amplitude representation in the neighbourhood of the origin does not give the same representation at any other reciprocal-lattice point. The additional images are therefore unlikely to be recognizable as such. It is shown in Section III.A.4 that gauzes can also be used in a mask to obtain contributions to a diffraction pattern that differ in phase by π. When this technique is used the image is formed not at the origin of the diffraction pattern but in the neighbourhood of one of the other reciprocal-lattice points.

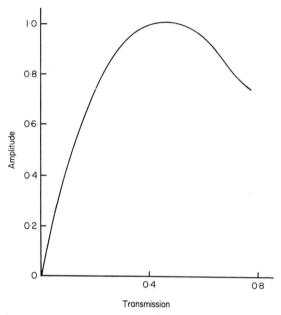

Fig. 9. Relation between transmission factor for a square-mesh gauze and amplitude contribution at one of the four reciprocal-lattice points adjacent to the origin.

It is then necessary to adjust the transmissions of the gauzes covering the various diffracting apertures so that the desired amplitude contributions are obtained at the appropriate reciprocal-lattice point. Figure 9 shows the relationship between gauze transmission and amplitude contribution at one of the four reciprocal-lattice points adjacent to the origin. It is assumed that the gauze is based on a square mesh.

III. Phase Control

It was stated in Section I that the phases of the beams in the Fraunhofer diffraction pattern of an arbitrary object can have any value between 0 and 2π. Consequently, for an image of such an object to be produced by optical analogue, some means must be available for setting the phases of the light beams representing the Fraunhofer diffraction pattern of the object to any required value in the range $0-2\pi$. There is a simplification if the object being imaged is centrosymmetric, since the diffracted beams then have phases of either 0 or π, and only these values need be represented in the diffracting mask. In addition there is a further simplification in a few particular cases when all the diffracted beams have the same phase. This occurs when a centrosymmetric structure has an atom at the origin which scatters X-rays so strongly that its contribution to the diffraction pattern dominates the contributions of all the other atoms in the unit cell.

Several methods are available for obtaining phases of 0 and π but only one satisfactory technique has been developed for the full 0 to 2π range.

A. Phases 0 and π

1. Mica

The simplest method of setting the discrete phases 0 and π was described by Bragg (1942). Every hole in the diffracting mask is covered with a piece of mica cut from the same optically uniform sheet, which must be a half-wave plate for light of the wavelength being used. All the holes for which the relative phase of the transmitted light is 0 are covered by pieces of mica having the same arbitrary orientation in the plane of the screen. The pieces covering holes required to transmit beams of phase π are oriented at right angles to the arbitrary direction for phase 0.

Figure 10 clarifies the physics of the method. The unpolarized beam of light incident on the screen can be regarded as made up from beams having all possible planes of polarization. For any one plane of polarization the wave can be resolved into components along the fast and slow directions for the mica. By virtue of the mica's being a half-wave plate the difference in optical path traversed by these two components is $\lambda/2$, corresponding to a phase difference of π. It can be seen from the diagram that for a wave polarized in a particular plane traversing two half-wave plates oriented at right angles, the two components in both the x and the y directions have a phase difference of π. Since both pairs of components specifying the beams have this phase difference, the difference also exists between the two beams themselves. The same argument applies to the beams polarized in all other directions which together make up the main beam. The method works well in practice.

2. Tilted Mica

Buerger (1950) has described a method of adjusting the phase at each aperture in the mask by altering the optical path lengths traversed by the various beams. Apertures requiring phase 0 remained unaltered while those of phase π were covered with a thin mica slip. The mica was tilted to an angle such that the optical path difference between beams of light transmitted by the covered and the uncovered holes was an odd integral multiple of half the wave length of the incident light.

It is better to modify Buerger's technique slightly and cover all the apertures with slips of mica because typical mica samples, tilted at a small angle, may transmit as little as 80% of the light incident on them. The difference in transmission for a piece perpendicular to the direction of propagation of the light beam and one tilted at an angle of about 30° is usually 5% or less so that, by covering all the holes, the transmitted amplitude is made nearly independent of the phase. If necessary, an allowance can be made for any small

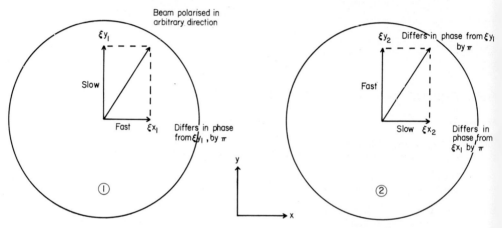

Fig. 10. Vector diagram for setting phases of 0 and π using half-wave plates.

loss of amplitude when setting the overall contribution of the hole. A more important advantage of covering all the holes with mica is that the optical path difference between the two types of beam can be $\lambda/2$ instead of $(n + \frac{1}{2})\lambda$ where n is some unknown integer, probably of the order of 100. In this way, any possible ill effects due to dispersion of the small range of wavelengths used in the instrument may be avoided. From Buerger's results it would seem that this point is of no practical significance but mica has been used in just this way (Taylor and Lipson, 1964) to obtain temporal incoherence between contributions from different parts of a diffracting mask. This problem will not arise when a laser, with its highly monochromatic output, is used as the source of illumination.

The angle of tilt which increases the optical path length by $\lambda/2$ can be found in two ways. First, if two of the pieces of mica are placed over the apertures of a two-hole grating the diffraction pattern is the well-known Young's fringe system with one fringe passing through the centre of the pattern (Fig. 11a). If one of the pieces of mica is tilted, so that the phase of the light transmitted by the hole which it covers is altered, the pattern shifts along a line perpendicular to the length of the fringes. When the phase difference between the contributions from the two holes is π, the system shifts a distance equal to half of the fringe spacing (Fig. 11b). The half-way position can be found by comparing the shifted fringes with an unmodified set of the same spacing produced by two other holes in a remote part of the mask, the illumination between the pairs of holes being spatially incoherent. Alternatively the required amount of shift can be defined by an adjustable line in a traversing

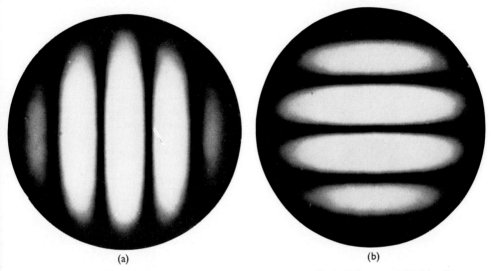

(a)   (b)

Fig. 11. Fringes formed by a grating of two holes. For (b) the light transmitted by the holes differs in phase by π.

eye-piece attached to the viewing microscope. By setting the fringes to the half-way position, first with the normal to the mica to one side of the perpendicular to the plane of the aperture and then to the symmetrical position on the other side of that perpendicular, a reasonably accurate value of the required angle of tilt can be found.

Secondly, a four-hole grating can be used and this gives a rather more accurate method of determining the required tilt. The diffraction pattern of such a mask consists of two subsidiary fringes between adjacent main fringes, one of which passes through the centre of the pattern. As the phase of the contributions from two adjacent holes at one end of the row is changed, by tilting the mica, the intensity distribution in the pattern alters. When the phase difference between contributions from the two pairs of holes is π it can easily be demonstrated that the pattern should again be symmetrical and as shown in Fig. 12(a). If the phase difference is not quite π, the intensity of one of the main fringes adjacent to the origin increases while that of the other decreases (Fig. 12b). The condition of equal intensity of these fringes can be judged very critically by eye and so the required angle of tilt can be determined accurately.

In practice the thickness of the mica should be such that the angle of tilt is large enough for setting errors to be unimportant, but not so large that the proportion of light that is reflected, rather than transmitted, becomes unacceptable. Angles between 15° and 20° have proved satisfactory: in this range

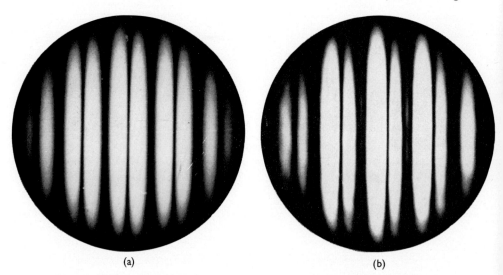

(a)   (b)

Fig. 12. Fringes formed by a grating of four holes: (a), the light transmitted by pairs of holes differs by π; (b), the light transmitted by pairs of holes differs by nearly π.

an error of 0·5° in tilt corresponds to an error in the phase angle of about 10° and the tilt can be determined empirically to rather better than 0·5°. On account of the biaxial nature of mica all the pieces used in a mask must be cut from the same sheet, of constant optical thickness, and arranged in parallel orientation.

3. *Plane-polarized light*

It was shown in Section II.B that if the orientation of a piece of mica, situated between crossed polaroids, is altered then the amplitude of the transmitted light changes and there is a phase reversal every 90° throughout the rotation. Thus a simple method is available for setting phases to the particular values 0 and π with complete precision.

4. *Gauzes*

The additive property of Fourier transforms of functions referred to a common origin (see Fig. 7, Chapter 10) can be used to represent the discrete phases 0 and π in a diffracting mask. Consider a mask which consists of a gauze of holes repeated regularly on a square mesh. The diffraction pattern of this mask was discussed in Section II.C and consists basically of a set of reciprocal-lattice points. The amplitude at the reciprocal-lattice point h, k is given by the Fourier summation expression

$$F(h,k) = \sum_{n=1}^{N} f_n(h,k) \exp\{2\pi i (hx + ky)\}. \tag{6.4}$$

If the gauze holes are very small, the scattering function $f_n(h,k)$ for each hole is scarcely dependent on h and k and can conveniently be written as unity. For larger holes the magnitude of the amplitude contribution at the various reciprocal-lattice points drops off in the manner shown in Fig. 5(d). For a particular reciprocal-lattice point the summation must be carried out over the unit-cell contents which, for a regular gauze, is just one hole. When the arbitrarily chosen origin of the diffracting mask coincides with the centre of a gauze hole, the fractional coordinates x, y of the unit-cell contents are both zero and the contribution to all reciprocal-lattice points is +1. If the gauze is shifted, without altering its orientation, so that its holes lie half-way along the unit-cell diagonals for the first position, the fractional coordinates of the unit cell contents are now $\frac{1}{2}$, $\frac{1}{2}$. It can easily be seen that at reciprocal-lattice points for which $h + k = 2m + 1$, for integral values m, the contribution is -1 while at all other reciprocal-lattice points it is +1. Figure 13 demonstrates the method. Figure 13(a) shows a small portion of a square lattice and Fig. 13(b) is its diffraction pattern, the reciprocal lattice. In Fig. 13(c) the appropriate translation of the lattice has been made and, as expected, its diffraction pattern has the same appearance as that shown in Fig. 13(b) since, although the transform is different in phase, the square of its modulus, which is the recorded quantity, is unchanged. In Fig. 13(e) the two lattices have been added together giving a new lattice of smaller pitch oriented at 45° to the original two: the new reciprocal lattice (Fig. 13f) consists of those points in the original reciprocal lattices at which the contributions from the two real-space lattices have the same sign. At the "missing" reciprocal lattice points the two contributions have the same magnitude but opposite signs and destructive interference has taken place.

To represent antiphase regions in a diffracting screen it is therefore necessary merely to cover one type of aperture with a suitable gauze and to cover apertures whose contributions are to have the opposite sign with the same gauze translated by the appropriate amount. A diffraction pattern incorporating the

G. HARBURN

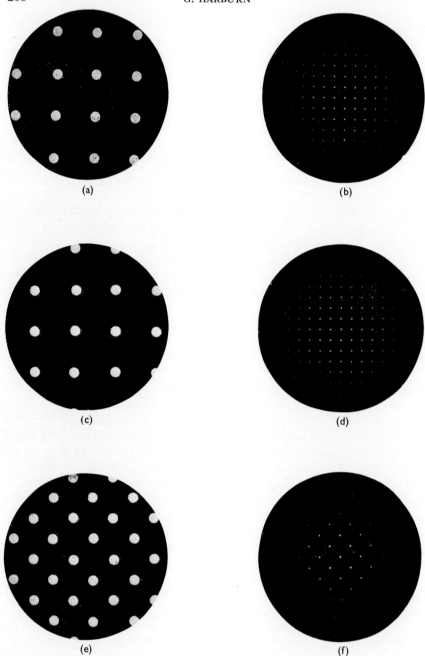

(a)

(b)

(c)

(d)

(e)

(f)

Fig. 13. Representation of phases 0 and π by shifting a sampling lattice.

required antiphase effects can be observed in the region surrounding any of the "missing" reciprocal-lattice spots. However, it is best to choose one of the four adjacent to the origin where the overall level of illumination is greatest and where any ill effects which might arise [for example, due to the light not being satisfactorily monochromatic (Harburn *et al.*, 1965)] from working at relatively high diffracting angles are minimized. The diffraction pattern of the array of apertures, without phase representation, can be observed around the origin and in the neighbourhood of the other non-disappearing reciprocal-lattice points. Figure 14 is a simple example of this technique. The apertures of the two-hole grating have been covered with separate pieces of gauze shifted

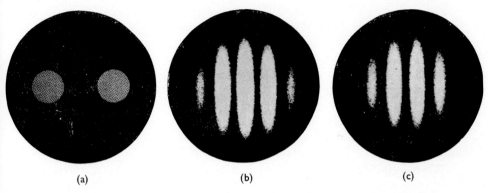

(a)   (b)   (c)

Fig. 14. (a), a two-hole grating. The holes are covered by identical gauzes shifted relative to one another by half the unit cell diagonal; (b), the diffraction pattern of (a) in the neighbourhood of the origin; (c), the diffraction pattern of (a) around one of the reciprocal-lattice points adjacent to the origin.

by the appropriate amount relative to one another. The apertures are shown in Fig. 14(a); Fig. 14(b) is an enlargement of the region near the origin and Fig. 14(c) shows the neighbourhood of the 01 reciprocal-lattice point for the gauze to the same scale.

In the above discussion the gauze was shifted by half the unit-cell dimensions in two directions at right angles. Phase control can also be obtained by other simple shifts. For example, if the gauze is shifted half the unit cell dimension in the direction of one axis, say the x-axis, Eq. (6.3) shows that antiphase effects will be found at all reciprocal-lattice points for which h is odd.

When this method of phase control is employed the gauzes can also be used to provide amplitude control. In fact the method is the only convenient one available for producing masks representing diffraction patterns which contain extended regions over which the amplitude varies, but the phase is constant at either 0 or π.

B. Phases 0 to 2π

1. Tilted Mica

In principle, Buerger's method of phase control, discussed in Section III.A.2 can be extended to represent any desired phase in the range 0 to 2π. The control is achieved by tilting a piece of mica to such an angle that the increased optical path length corresponds to the required phase delay. The method suffers from several disadvantages. First, the accuracy with which the phase can be set varies throughout the range since the relationship between phase angle and tilt is not a linear one. It is doubtful if phase errors could be kept below 20° at the higher angles of tilt and it is known that such errors could have a serious effect on the diffraction pattern. Secondly, for reasonably accurate work, the different amounts of light transmitted by mica slips tilted to different angles would have to be taken into account. Although this effect could be allowed for in setting the amplitudes it would involve extra work and, at the least, be a source of annoyance. Thirdly, because of the large number that would be required, it would scarcely be practicable to make components that were permanently set to produce a single phase value. The alternative, a relatively small number of components in which the tilt could be set to any required value to a high degree of accuracy, leads to serious design problems.

There is in fact no published account of work in which this method has been used and, in view of the satisfactory nature of the technique described in the next section, there seems to be no reason to adopt it. The only advantage of the tilted mica technique would be the higher levels of illumination found in the diffraction patterns.

2. Circularly-polarized Light

Harburn and Taylor (1961) have described a method of setting the relative phases of the beams of light transmitted by the apertures of a diffracting screen to any desired value. The technique depends on the relative orientation of half-wave plates in a beam of circularly-polarized light. Consider a plane-polarized wave, specified by $\xi = a \sin \omega t$, incident on a quarter-wave plate for which the fast and slow directions coincide with the x and y axes respectively in the coordinate system of Fig. 15. If the angle between the plane of polarization of the incident wave and the x-axis is 45° the beam emerging from the quarter-wave plate is circularly-polarized and may be specified by the components

$$\xi_x = a' \cos \omega t$$
$$\xi_y = a' \sin \omega t$$

where $a' = a/\sqrt{2}$.

Suppose that this circularly-polarized beam is incident on a half-wave plate whose fast direction is at angle θ to the x direction, then the components of ξ_x and ξ_y in the fast and slow directions are

$$\xi_F = \xi_x{}^F + \xi_y{}^F = a' \cos \omega t \cos \theta + a' \sin \omega t \sin \theta$$
$$= a' \cos (\omega t - \theta)$$
$$\xi_s = \xi_x{}^s + \xi_y{}^s = -a' \cos \omega t \sin \theta + a' \sin wt \cos \theta$$
$$= a' \sin (\omega t - \theta).$$

If, after traversing the half-wave plate, the beam in the fast direction is considered to be unaltered ($\xi_F' = \xi_F$) then the slow beam has undergone a phase

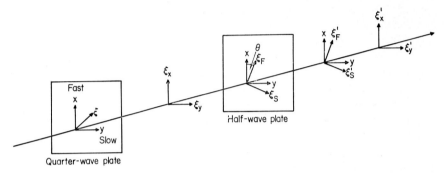

Fig. 15. Vector diagram for full phase control using circularly-polarized light.

change of π so that $\xi_s' = -\xi_s = -a' \sin (\omega t - \theta)$. Combination of the components of ξ_F' and ξ_s' in the x direction gives

$$\xi_x' = a' \cos (\omega t - \theta) \cos \theta + a' \sin (\omega t - \theta) \sin \theta$$
$$= a' \cos (\omega t - 2\theta)$$

and the phase of the original vibration in the x direction has been changed by 2θ. Similarly $\xi_y' = a' \sin (\omega t - 2\theta)$ and, since both components specifying the beam have suffered the same delay, the phase of the whole beam is altered by 2θ. If the fast directions of two half-wave plates make angles θ and θ' with the x direction the beams emerging from the half-wave plates will differ in phase by $(\omega t - 2\theta) - (\omega t - 2\theta') = 2(\theta' - \theta)$, which depends only on the relative orientations of the plates. The amplitude of the beam is not altered and the phase change is linear and continuous, the range 0 to 2π being covered by rotation of the mica from 0 to π. Since the relative orientations can be set to about $1°$ the error in the phases is very small. The illumination level in a diffraction pattern obtained from such components is lower than would be obtained by Buerger's method on account of the necessary polaroid filter.

This objection would be removed if the plane-polarized output of a laser were used.

IV. APPARATUS

A. Mica

Several of the techniques described in the two previous sections make use of mica. It is possible that other materials could be used instead but, in addition to its useful birefringent properties, mica has the great advantage that it cleaves easily and is readily available in good quality, optically-clear sheets. With a little practice it is possible to cleave mica into sheets which have uniform optical thickness over an area of up to, say, 50 cm². The uniformity of the sheets can be checked by viewing them between crossed polaroids with white-light illumination. Regions of constant optical thickness then appear as areas having uniform tint: sharp changes in colour correspond to cleavage steps on the surfaces. If the mica is very thin it has to be viewed at large angles of incidence and it should be kept reasonably flat to avoid confusion from gradual changes of colour across optically uniform regions.

When mica of a particular optical thickness is required it can be made up as a composite sheet from two pieces of arbitrary thickness suitably oriented relative to one another. Details of this technique are given by Childs (1956). In practice, however, such a composite sheet would be difficult to handle when cut into the small pieces needed to cover individual diffracting apertures. With a certain amount of trial and error single mica sheets of thickness sufficiently close to any desired value can be produced by cleaving thicker pieces. The method of continuous phase control described in Section III.B.2 uses circularly-polarized light. This is most easily produced by passing a plane-polarized beam through a quarter-wave plate, the plane of polarization bisecting the fast and slow directions in the birefringent material. When much work of this type is envisaged it is well worth while obtaining a commercially produced quarter-wave plate. If, however, a mica sheet is required the approximate thickness can be calculated from the expression

$$t = \frac{\lambda}{2\pi} \frac{\alpha}{(n_1 - n_2)}$$

where n_1 and n_2 are the principal refractive indices for the material and α is the phase delay required. Assuming typical values of n_1 and n_2 for mica the thickness of a quarter-wave plate is about $3 \cdot 5 . 10^{-2}$ mm.

Preliminary sorting of possible sheets can be made with a micrometer screw gauge prior to the following optical check. The piece of mica under test is placed between crossed polaroids, illuminated with parallel light of the relevant wavelength and oriented to give maximum transmitted intensity as measured by a simple photo-electric photometer. If the analyser is subsequently rotated and the sheet is a quarter-wave plate there is no change in the intensity of the light transmitted by the system. In practice it has been found that a sheet will give satisfactory results if the intensity does not vary by more than 5%.

To make a unit for producing circularly-polarized light the quarter-wave plate and a piece of polaroid sheet must be mounted in a device carrying a scale which enables their relative orientation to be set to an accuracy of, say, 0·25°. To ensure that they are correctly set a graph of the maximum and minimum intensity transmitted by an analysing polaroid as it is rotated through 180° should be plotted for a small range of relative orientations of the two components in the neighbourhood of the approximately correct value. It is then a simple matter to deduce the correct setting from the graph. For greatest precision a similar procedure should be used in the original selection of the

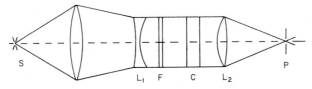

Fig. 16. Modified condenser system. S, source; L_1 and L_2, auxiliary lenses; F, filter; C, circularly-polarizing unit; P, pinhole.

quarter-wave plate. To produce circularly-polarized light the quarter-wave plate must be illuminated with a parallel beam. The condenser system of the optical diffractometer can be modified in the manner shown in Fig. 16 to incorporate a section where the light is parallel. To satisfy coherence requirements (Taylor and Thompson, 1957) the lenses L_1 and L_2 have focal lengths which give a cone of light at the pinhole of the same semi-angle as in the unmodified system.

To select a half-wave plate of mica one arbitrary plane in the circularly-polarized beam of light emerging from the unit just described is picked out with an analysing polaroid. If a half-wave plate is rotated in the circularly-polarized beam there is no change in the intensity of the light transmitted by the system. It has been found that a mica sheet gives satisfactory results in the phase-changing technique of Section III.B.2 if the transmitted intensity does not vary by more than 10%.

Thin sheets of mica can be cut with scissors or a guillotine. If a firm but unhurried action is used, shattering at the edges is confined to a very narrow strip. When a sheet of mica is cut into smaller pieces it is advisable to mark a predetermined corner of each piece before cutting so that, later, there is no confusion about its orientation relative to other pieces.

B. Gauzes

The most convenient gauzes for amplitude control are undoubtedly those produced by etching holes in copper foil. The etching process requires a photographic plate carrying black dots of high contrast corresponding to the holes required in the copper sheet. This plate is obtained from a cross-line screen, such as is used in commercial printing processes, by Fresnel diffraction. The screen consists of two diamond-ruled plates, the grooves produced being filled with opaque material. Rulings of 150 and 60 lines per in. have been used to prepare gauzes. The ruled surfaces are in contact with their lines at right angles to form a square mesh. The screen is uniformly illuminated with white light and supported above a photographic plate at such a distance that, in the shadowed image, the corners of the mesh are rounded off by diffraction effects. Provided that a suitable emulsion is used (for example, Kodalith) the size of the circular dots obtained can be altered by changing the exposure.

The procedure for obtaining etched gauzes from this photographic plate is a straightforward application of well-established photomechanical techniques such as are used to produce printed circuits for the electronics industry. The process will be outlined briefly; further details are given by Eisler (1959). In the past, gauzes were made on standard printed-circuit boards consisting of thin sheets of copper bonded to a laminated plastic backing. The copper was removed from the backing at the conclusion of the etching process. Recently, most manufacturers have improved the bond between the copper and the backing to such an extent that it is now extremely difficult to separate the two by chemical action. It is more convenient, therefore, to fasten a suitably sized sheet of copper on to a glass plate with an adhesive that can later be readily dissolved. Kodaflat solution has proved to be a satisfactory adhesive; it can be broken down with ethyl acetate. The copper sheet should be about 0·001 in. thick; plain-quality electro-deposited copper foil of $\frac{3}{4}$ oz per sq. ft (200 gm m^{-2}) is very suitable. The electro-deposition process leaves one side of the foil in a slightly rough condition which provides an excellent key for the adhesive.

The copper is rolled down on to the glass sheet (which may be lightly ground to assist adhesion) and the metal surface cleaned with fine domestic scouring powder, well washed and allowed to dry. A small amount of photo-sensitive resist (for example, Kodak P.C. Resist) is poured in the centre of the copper sheet which is then spun on a horizontal turntable at about 100 r.p.m. to spread out the resist into a very thin film. When the film is dry and hard the sheet is removed from the turntable and a contact print with ultra-violet light is made from the pattern of black dots on the photographic plate. It is important that, during the exposure, extremely good contact is maintained between the photographic emulsion and the film of resist; a vacuum printing frame should be used. The regions of resist protected from the ultra-violet light are dissolved away with a suitable developer (for example, Kodak P.C. Developer) leaving bare copper exposed. The sheet is placed, with the copper surface downwards, in a bath of ferric chloride which should have a specific gravity of at least 1·35. At any stage of the etching process the plate can be rinsed with water and examined under a microscope. The etching is terminated when holes of the desired size are obtained and the copper is then removed from the glass sheet.

A full range of gauze transmissions can be obtained by using different etching times in conjunction with, say, three photographic plates giving meshes of dots of three different sizes. In addition the hole size varies slightly over every piece of gauze so that it is easy to pick out a region of precisely the required transmission. Such a region is most easily selected with a photo-electric photometer in which the beam of light has a cross-section corresponding to the shape of the piece of gauze that is required. Gauzes based on a square mesh have a maximum transmission of about 0·78. This figure is increased to 0·88 for a close-packed hexagonal mesh. In practice square-mesh gauzes are too fragile to handle if their transmission exceeds about 0·7.

C. Supporting Plugs

When pieces of gauze or mica are incorporated in a diffracting screen some means must be available for securing them in the correct position and orientation. It is possible to fasten the basic material in place with adhesive tape, but, particularly if much work is envisaged, it is much easier to mount the gauze or mica on small brass plugs which can be placed directly into correctly sited holes in a board. Figure 17 shows sections of some plugs that have been used.

Hanson and Lipson (1952) secured their disks of mica with circles of spring wire against ledges machined in brass cylinders. The orientation of the mica was altered by means of a screwdriver blade which engaged with a slot cut across the top of the cylinder. Harburn has found it simpler to cut square pieces of mica and has mounted these on square-section brass rod. Each piece of rod is bored along its length with a hole—the diffracting aperture—and a short length of the square section is turned down to a cylinder of convenient diameter to fit in the holes in the mask. The square section of this type of plug is useful for adjusting the orientation of the piece of mica glued on to it (mica can be fastened to brass with an epoxy resin such as Araldite). A knob, carrying on its circumference a pointer which moves over a graduated circular scale, is

8

used to turn a rod which has a slotted end. The slot fits snugly over the square section of the plug. The knob is carried at the end of a parallel-motion linkage so that the relative orientations of a number of plugs in a mask can be quickly and accurately adjusted by using the slotted rod on them in turn.

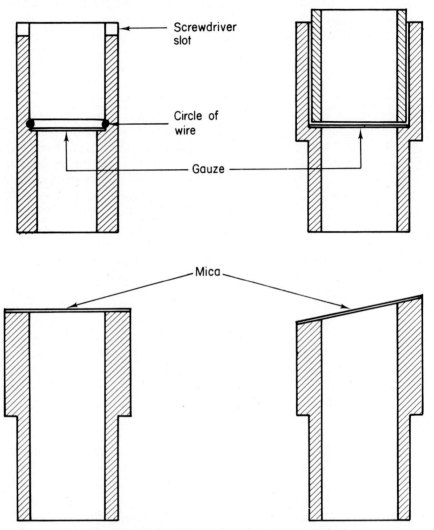

Fig. 17. Sections of brass supporting plugs for gauze and mica.

Disks of gauze can easily be removed from the copper sheet with a punch and die. The disks can be secured with a circle of wire or a thin-walled brass cylinder in plugs that are counterbored to provide a locating ledge.

V. OPTICAL FOURIER SYNTHESIS

A. *No Phase Representation*

Fourier summations for which no phase representation is needed are of little practical importance since they are very rare. It is unusual for an object or a unit cell to contain such a large concentration of material (heavy atom) that

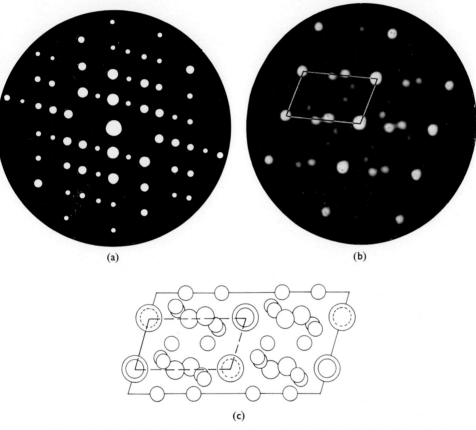

(a)   (b)

(c)

Fig. 18. Diopside: (a), the *h0l* section of the reciprocal lattice; (b), optical Fourier synthesis prepared from (a), showing the unit cell for the projection; (c), atomic sites and unit cells.

the wave scattered from this region dominates the contributions to the diffraction pattern from the rest of the object so that all the diffracted beams have the same phase. However, a few cases, such as the simple uranium salts studied by Zachariasen (1949), do occur and Bragg (1939) demonstrated the technique for diopside, $CaMg(SiO_3)_2$.

In the (010) projection of diopside the calcium and magnesium atoms are superimposed giving the necessary high concentration of electron density. Figure 18(a) shows a mask representing the $h0l$ section of the reciprocal solid

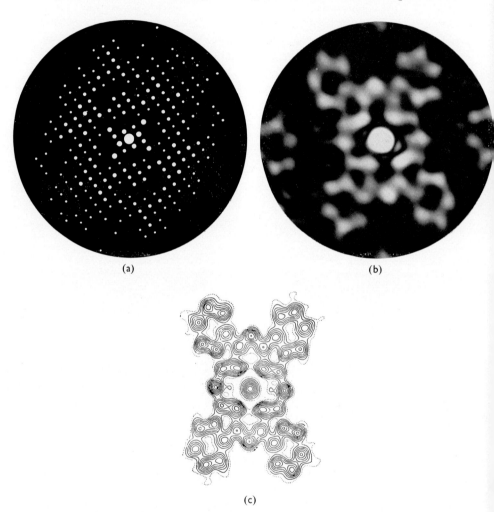

(a)   (b)

(c)

Fig. 19. Nickel phthalocyanine: (a), reciprocal-lattice section; (b), optical Fourier synthesis; (c), contour map of electron density distribution.

for diopside, the amplitudes of the various beams diffracted by the mask being controlled by the areas of the apertures as described in Section II.A. The optical summation given by this mask is shown in Fig. 18(b) and may be compared with the map of the atomic sites for the complete unit cell reproduced in Fig. 18(c). In this projection the complete unit cell reduces to a smaller cell of sides $a/2$, $c/2$ containing one superimposed pair of the heavy atoms. This smaller cell is also indicated in Fig. 18(c) and corresponds to the cell shown in Fig. 18(b). It should be noted that the optical summation shows the high concentration of scattering material at the origin of the pattern. This disposition is a consequence of assuming that all the phases in the diffraction pattern are the same. If the origin shown for the complete unit cell of Fig. 18(c) were used to calculate the diffraction pattern, pronounced fringing effects, with discrete phases of 0 and π, would be found. It is well known from Fourier-transform theory that a change of origin in a diffracting object gives rise to a change in the distribution of phase in the diffraction pattern of the object.

The technique may sometimes usefully be extended to centrosymmetric structures that do not contain a heavy atom. The amplitude of all the diffracted beams must be increased by a positive amount equal in magnitude to the amplitude of the largest negative term in the diffraction pattern. All the diffracted beams then have the same sign and the synthesis corresponds to a new structure which is identical to the original one except that a heavy atom has been added at the origin. Dunkerley and Lipson (1955) demonstrated this approach by producing an optical image of modified nickel phthalocyanine in which the nickel atom is at the origin. The consequence of making all the phases the same is to overemphasize the nickel atom and, when the image is recorded, the high intensity near the origin tends to obscure details of other, nearby atoms. Figure 19(a) shows the reciprocal-lattice section after the amplitudes, again represented by hole size, have been adjusted to make all the phases the same and Fig. 19(b) shows the optical Fourier synthesis obtained from the mask. The calculated electron-density map is shown in Fig. 19(c).

The method is relevant only for a very few genuine cases and the extended method would be tedious to use for an unknown structure. For every set of signs that was proposed a new set of amplitudes would have to be calculated and apertures of corresponding size made at the reciprocal-lattice points in a screen.

B. Centrosymmetric Structures

Buerger (1950) published the first example of an optical Fourier synthesis in which phases of both 0 and π were represented in the mask. He placed tilted mica plates over those holes which had to transmit a wave of phase π and controlled the amplitudes by choice of diameter for the diffracting apertures. He produced an excellent image of a projection of the six atoms in the unit cell of marcasite, FeS_2. Harburn improved the method by covering all the apertures with mica (see Section III.A.2) and controlling the amplitudes of the diffracted beams by means of gauzes. Figure 20(a) shows Harburn's synthesis for the eight atoms in the (100) projection of the unit cell of urea $CO(NH_2)_2$; Fig. 20(b) shows one molecule further enlarged. The contributions to the original diffraction pattern from the hydrogen atoms are so small that they are not reproduced in the image.

Two interesting qualitative points, which can be demonstrated for the

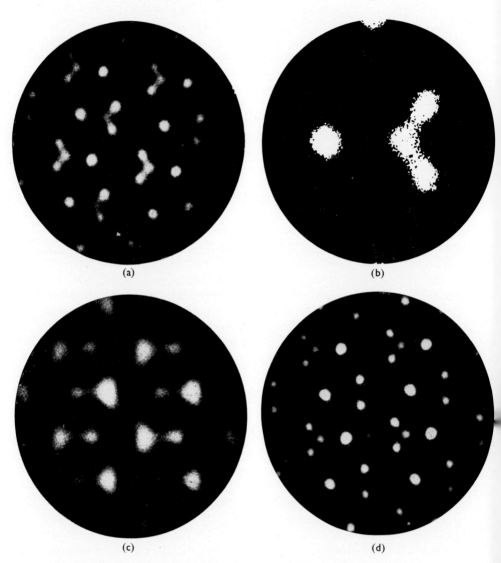

(a)   (b)

(c)   (d)

Fig. 20. Urea: (a), thirteen independent terms; (b), one molecule from (a); (c), six independent terms; (d), thirteen independent terms, all of the same amplitude.

structure of urea, emerge from the work that has been done on optical Fourier synthesis. First, a recognizable image of an atomic structure can be obtained from a surprisingly small number of reflexions. Figures 20(a) and (c) show

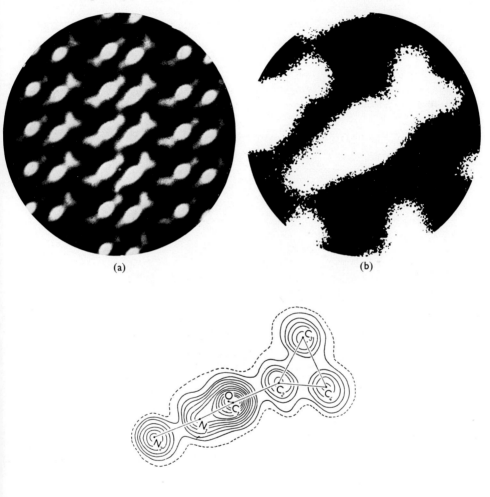

(a)

(b)

(c)

Fig. 21. (a), Optical Fourier synthesis for cyclopropane-carbo-hydrazide; (b), one molecule from (a) enlarged; (c), electron density distribution for one molecule.

the synthesis obtained for urea using, respectively, the 13 and the six strongest reflexions from the experimental X-ray data. The former is an excellent image and, knowing the chemistry of the molecule, the latter could probably be used to find the signs of other reflexions. Only 16 sets of signs would have to be examined to obtain Fig. 20(c). Secondly, the phases of the reflexions are more important than their amplitudes in determining the positions of the atoms in an image of a unit cell. For the synthesis of Fig. 20(d) the 13 reflexions used for (a) were given the same amplitude as $F(00)$. The figure is an extreme example of a sharpened Fourier synthesis in which the resolution has been appreciably improved at the expense of a very small amount of spurious detail. It is clear that, for Fourier summations in the early stages of a crystal structure determination, the amplitudes need not be set with any great precision.

The resolution in the best image of urea is excellent in spite of the small number of terms used because the unit cell contain few atoms and these do

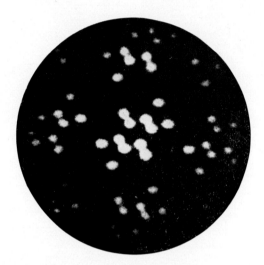

Fig. 22. Optical Fourier synthesis for hexamethylbenzene.

not overlap. However, for a unit cell containing more atoms, some of which overlap in projection, the resolution deteriorates. Figure 21(a) shows the optical Fourier synthesis for the (010) projection of cyclopropane-carbo-hydrazide prepared from the data of Chestnut and Marsh (1958). Figure 21(b) is a further enlargement of one molecule which may be compared with the calculated electron-density contour map (Fig. 21c). The resolution is not good but, in conjunction with chemical knowledge, it should be possible to deduce a structure that would readily refine.

Hanson and Lipson (1952) and Hanson *et al.* (1951) have published examples of optical syntheses for hexamethylbenzene (Fig. 22) and durene respectively. They controlled both amplitude and phase by the relative orientation of half-wave plates of mica in plane-polarized light as described in Section. II.B. Hanson and Lipson give a full account of the use of this method and state that, once the reciprocal-lattice plate had been set up, they were able to examine all the possible combinations (128) of signs for a determination of the structure of hexamethylbenzene in about two hours. Each set of signs consisted of 10

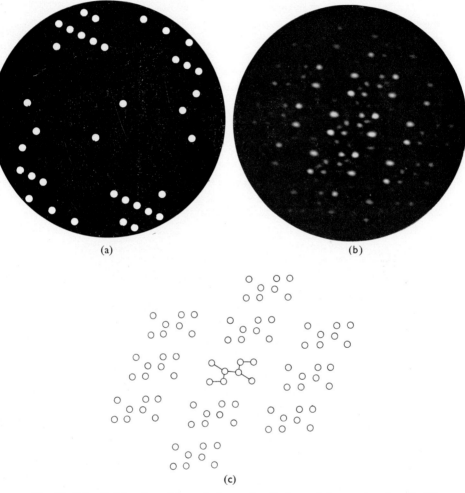

Fig. 23. Dimethylglyoxime: (a), mask for weak reflexions of phase π (number 6); (b), optical synthesis; (c), atomic sites.

independent reflexions and this number proved sufficient to produce a plausible image which refined to the accepted structure.

A synthesis for which both amplitude and phase were controlled by gauzes is shown in Fig. 23. The data used were those of Merritt and Lanterman (1952) for the (001) projection of dimethylglyoxime $(CH_3CONH_2)_2$. For this demonstration the final calculated structure factors were used in the manner shown in Table I. The reflexions were divided into six groups according to their amplitudes being strong, medium or weak and their phases being positive or negative. Masks were then prepared for each group of reflexions: Fig. 23(a) shows mask number 6. Since the phase control was to be achieved by a diagonal shift of the sampling lattice, as described in Section III.A.4, amplitude contributions to the 01 reciprocal-lattice point in the diffraction pattern of the gauze had to be considered. Figure 9 shows that, for the 150 line-per-inch square-mesh gauze that was to be used, only the 0–0·45 transmission range could be utilized to represent the full range of structure factors. Since it is not easy to make an etched gauze of this pitch with a transmission less than about 0·1 the range was further reduced and, for the present synthesis, transmissions of 0·45, 0·3 and 0·1 were used to represent the strong, medium and weak groups of reflexions respectively.

The complete diffracting mask was built up by making six separate exposures through the line screen, on which was placed each of the six masks in turn, on to a photographic plate (see Section IV.B). The exposures required to give the three different transmissions had been determined previously and the line screen was shifted by the appropriate amount after printing through the three masks representing the positive orders. The photographic plate, when processed, was suitable for use in the etching process described in Section IV.B. A contact print would have to be made if the diffracting screen was required in the form of a photographic plate. The optical synthesis obtained is shown in Fig. 23(b) and this may be compared with the atomic positions given by Merritt and Lanterman which are shown in Fig. 23(c). The amount of spurious detail is not surprising considering the way in which the weak reflexions were overemphasized in the mask.

The technique has several disadvantages. First, it is not easy to represent a wide range of amplitudes and each particular group of reflexions having the same amplitude requires a separate mask and exposure. Secondly, to examine the effect of changing the phase of any order at least two new masks have to be made and the photographic process repeated. Thirdly, because the image is formed round the 01 reciprocal-lattice point, which may lie at a relatively high diffracting angle, highly monochromatic light must be used—preferably from a laser. This precaution is necessary to avoid serious degradation of the image due to elongation of the spots representing the atoms in the direction away from the 00 reciprocal-lattice point. An example of this effect is shown below.

The main advantage of the method is that it requires neither optically uniform sheets of mica nor polarized light. In addition it is the only technique available for forming an image from a diffraction pattern containing a

TABLE I

$h\,k$	F_{calc}	Group	Mask No.	$h\,k$	F_{calc}	Group	Mask No.
0 0	620	$+s$	1	0 5	29	$+m$	2
1 0	172	$+s$	1	1 5	−22	$−m$	5
2 0	−58	$−m$	5	2 5	−15	$−w$	6
3 0	27	$+m$	2	3 5	18	$+w$	3
4 0	42	$+m$	2	4 5	−20	$−w$	6
5 0	−23	$−m$	5	5 5	−47	$−m$	5
6 0	−28	$−m$	5	6 5	−18	$−w$	6
7 0	−20	$−m$	5				
				0 6	−6	$−w$	6
0 1	90	$+s$	1	1 6	7	$+w$	3
1 1	−4	$−w$	6	2 6	−11	$−w$	6
2 1	−119	$−s$	4	3 6	−19	$−w$	6
3 1	7	$+w$	3	4 6	−17	$−w$	6
4 1	9	$+w$	3	5 6	−19	$−w$	6
5 1	−76	$−s$	4				
6 1	−65	$−m$	5	1 $\bar{1}$	−125	$−s$	4
7 1	−3	$−w$	6	2 $\bar{1}$	141	$+s$	1
				3 $\bar{1}$	28	$+m$	2
0 2	−169	$−s$	4	4 $\bar{1}$	−31	$−m$	5
1 2	30	$+m$	2	5 $\bar{1}$	15	$+w$	3
2 2	59	$+m$	2	6 $\bar{1}$	31	$+m$	2
3 2	−59	$−m$	5				
4 2	−67	$−m$	5	1 $\bar{2}$	−51	$−m$	5
5 2	−25	$−m$	5	2 $\bar{2}$	77	$+s$	1
6 2	−11	$−w$	6	3 $\bar{2}$	13	$+w$	3
7 2	6	$+w$	3	4 $\bar{2}$	−52	$−m$	5
				5 $\bar{2}$	−15	$−w$	6
0 3	12	$+w$	3				
1 3	76	$+s$	1	1 $\bar{3}$	−19	$−w$	6
2 3	62	$+m$	2	2 $\bar{3}$	−19	$−w$	6
3 3	−44	$−m$	5	3 $\bar{3}$	−18	$−w$	6
4 3	−22	$−m$	5	4 $\bar{3}$	−2	$−w$	6
5 3	40	$+m$	2	5 $\bar{3}$	−18	$−w$	6
6 3	18	$+w$	3				
7 3	−15	$−w$	6	1 $\bar{4}$	84	$+s$	1
				2 $\bar{4}$	−5	$−w$	6
0 4	93	$+s$	1	3 $\bar{4}$	−16	$−w$	6
1 4	14	$+w$	3	4 $\bar{4}$	18	$+w$	3
2 4	6	$+w$	3				
3 4	23	$+m$	2	1 $\bar{5}$	43	$+m$	2
4 4	4	$+w$	3	2 $\bar{5}$	24	$+m$	2
5 4	−5	$−w$	6	3 $\bar{5}$	19	$+w$	3
6 4	−11	$−w$	6				
7 4	−19	$−w$	6	1 $\bar{6}$	−17	$−w$	6

G. HARBURN

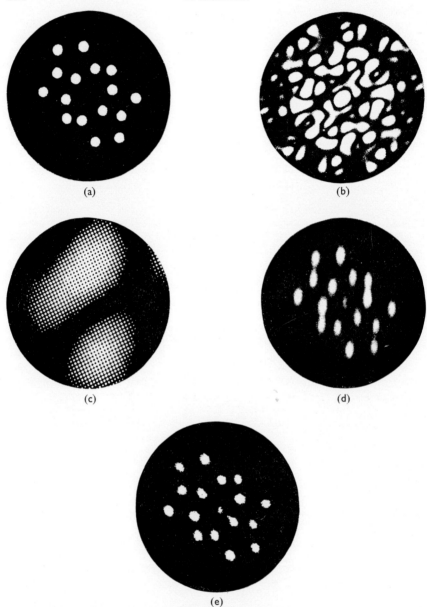

Fig. 24. (a) Mask with holes representing the atoms of a hypothetical, centrosymmetrical molecule; (b), optical diffraction pattern of (a); (c), part of the half-tone mask made from (b). The displacement between the sampling lattices representing regions of opposite phase can be seen; (d), image showing elongation due to illumination containing a small range of wavelengths; (e), image obtained using a laser as the source of light.

continuous distribution of amplitude rather than discrete beams. This situation is interesting but currently of no practical importance in X-ray crystallography. Figure 24 illustrates the possibilities; (a) shows a mask containing a centro-symmetrical array of holes representing a single hypothetical molecule; (b) is the diffraction pattern of the mask; the pattern contains discrete extended regions each of which has phase either 0 or π. The phase of each region was determined by comparing this diffraction pattern with that for the same mask with an extra hole added at the centre of symmetry. The regions which were enhanced were assigned phase 0 and those in which the intensity diminished, phase π. A contact print from an enlargement of the diffraction pattern was used, in conjunction with two masks to isolate the positive and negative regions, to print through the line screen on to a photographic plate. The screen was translated the requisite amount between the two exposures. Part of the final mask which was a photographic plate is shown in Fig. 24(c). Further details of the procedure have been given by Harburn et al. (1965).

The final image formed by a diffractometer using light from a filtered mercury-vapour lamp is reproduced in Fig. 24(d). The smearing out of the spots due to the range of wavelengths passed by the filter is clear. The improvement obtained when a laser is used is shown in Fig. 24(e). It is evident from Fig. 24(c) that some of the gauze holes have run together so that transmissions in excess of 0·78 have been used to represent the strongest parts of the transform. It follows, from Fig. 9, that the amplitude representation in the neighbourhood of the 01 reciprocal-lattice point is not good. It is not surprising, therefore, that incorrect detail has appeared at the centre of symmetry in the final image.

C. Non-centrosymmetric Structures

Figure 25 shows an optical Fourier summation prepared by Harburn and Taylor (1962) for the non-centrosymmetric (100) projection of sodium nitrite using the data of Carpenter (1952). Amplitude control was achieved with gauzes and phase control by adjusting the relative orientation of half-wave mica plates in a beam of circularly-polarized light (Section III.B.2). The image, which was built up from nineteen independent terms, is extremely good but that is to be expected for a unit cell containing such a small number of atoms which do not overlap in the projection chosen. Nevertheless, and in spite of the fact that no further demonstrations exist for known structures, it can be said that the technique is simple, effective and rapid. Once the gauzes representing the amplitudes of the diffracted X-ray beams have been mounted on a reciprocal-lattice plate, the image given by any trial set of, say, 50 different phases could be evaluated in about 15 minutes.

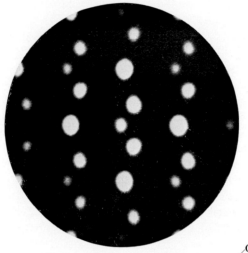

Fig. 25. Optical synthesis for sodium nitrite Na=N⟨ $_O^O$.

VI. Conclusion

The techniques for controlling the phases of light beams that have been described in this chapter can be used in other crystallographic applications apart from the production of images from X-ray data. An obvious extension to Fourier summation is the production of error syntheses and Hanson and Lipson (1952) have demonstrated the procedure for durene. Taylor and Underwood (1959) have shown that the labour of preparing structure-factor graphs (Bragg and Lipson, 1936) can be considerably reduced by using optical methods. Finally, Harburn and Taylor (1961) have used their method of phase control for examining non-central sections of the optical reciprocal solid for the contents of a unit cell.

With one exception all the examples of optical Fourier syntheses for crystal structures that have been shown in this chapter were originally produced at least ten years ago. This fact reflects the advances that have been made in computer design and production over the last 15 years. When Buerger (1950) and Hanson and Lipson (1952) published their work on optical means of carrying out Fourier summations their techniques promised a very real saving in time over the traditional hand methods of computation. In addition, at the time the optical techniques were introduced, many crystal structure determinations were carried out by physicists who were at least as interested in the method by which the structure was solved as in the final result itself. Many such people found the optical analogue procedure attractive from a physical point of view.

Now, however, the vast majority of scientists who work out crystal structures are interested only in the result and not at all in the method by which it is achieved. They prefer to interpret the numerical output of large fast computers, to which they have easy access, rather than the intensity distribution in an optical diffraction pattern. Thus Fourier summation by optical analogue has no serious part to play in contemporary X-ray crystallography. Nevertheless optical Fourier synthesis of experimental X-ray data remains an excellent example of the idea that image formation is a two-stage diffraction process (Section I.C). Experience with the technique could make a valuable contribution to the training of a crystallographer. It is quite certain that the methods described in this chapter, and particularly that of Harburn and Taylor (1961) for continuous phase control, could be developed from their present positions of interesting demonstrations to the point where they could rapidly provide high-quality "images" of the contents of unit cells in crystals. It seems equally certain however that there is no demand for such a system, unless for economic reasons the essentially free access to the computer has to be abandoned; then the large cost will become apparent and the methods described in this chapter may be given a new lease of life.

REFERENCES

Bragg, W. L. (1939). *Nature, Lond.* **143**, 678.
Bragg, W. L. (1942). *Nature, Lond.* **149**, 470.
Bragg, W. L. and Lipson, H. (1936). *Z. Krist.* **95**, 323.
Buerger, M. J. (1950). *J. appl. Phys.* **21**, 909.
Carpenter, G. B. (1952). *Acta Cryst.* **5**, 132.
Chestnut, D. B. and Marsh, R. E. (1958). *Acta Cryst.* **11**, 413.
Childs, W. H. J. (1956). *J. Sci. Inst.* **33**, 298.
Cowley, J. M. and Moody, A. F. (1958). *Proc. phys. Soc.* **71**, 533.
Dunkerley, B. D. and Lipson, H. (1955). *Nature, Lond.* **176**, 81.
Eisler, P. (1959). "The technology of printed circuits," Heywood, London.
Goodman, J. W. (1968). "Introduction to Fourier optics," McGraw-Hill.
Hanson, A. W. and Lipson, H. (1952). *Acta Cryst.* **5**, 362.
Hanson, A. W., Taylor, C. A. and Lipson, H. (1951). *Nature, Lond.* **168**, 160.
Harburn, G. and Taylor, C. A. (1961). *Proc. R. Soc. A*, **264**, 339.
Harburn, G. and Taylor, C. A. (1962). *Nature, Lond.* **194**, 764.
Harburn, G., Taylor, C. A. and Yeadon, E. C. (1965). *Brit. J. appl. Phys.* **16**, 1367.
Harburn, G., Walkley, K. and Taylor, C. A. (1965). *Nature, Lond.* **205**, 1095.
Merritt, L. L. and Lanterman, E. (1952). *Acta Cryst.* **5**, 811.
Taylor, C. A. and Lipson, H. (1964). "Optical transforms," Bell, London.
Taylor, C. A. and Thompson, B. J. (1957). *J. Sci. Inst.* **34**, 439.
Taylor, C. A. and Underwood, F. A. (1959). *Acta Cryst.* **12**, 336.
Vos, P. J. de (1948). *Acta Cryst.* **1**, 118.
Zachariasen, W. H. (1949). *Acta Cryst.* **2**, 94.

CHAPTER 7

Low Energy Electron Diffraction (LEED)†

W. P. ELLIS

University of California, Los Alamos Scientific Laboratory,
Los Alamos, New Mexico, U.S.A.

I. INTRODUCTION

A. Principles of LEED

Davisson and Germer in 1927 demonstrated that low energy electrons were diffracted by crystalline nickel surfaces. Their experiment and the observation by Thomson (1928) of diffraction of high energy electrons confirmed the deBroglie relation, $\lambda = h/p$, between wavelength λ, momentum p, and Planck's constant h. For nonrelativistic electrons accelerated from rest through a voltage difference, V, the relation is $\lambda = \sqrt{150/V}$ Å. In the energy range, 10–500 eV, electrons are scattered elastically by only two to four atomic layers and thus are a sensitive probe for studies of surface structure on faces of single crystals.

Although low energy electron diffraction, LEED, was discovered at the same time as high energy electron diffraction, HEED, it did not immediately

† Work performed under the auspices of the U.S. Atomic Energy Commission.
229

become a universal method for studies of surface structure because of the difficulties with contamination. At a gas pressure of 10^{-6} torr, a monolayer adsorbs on a reactive surface in about one second. Until clean ultrahigh-vacuum systems that operate in the 10^{-10} torr region became readily available, only a few researchers, such as Professor H. Farnsworth at Brown University, studied surfaces with low energy electrons prior to the preceding decade. Within the past few years, because of the greatly improved variety of instrumental techniques, including LEED, efforts toward an understanding of clean surfaces and the reactions that occur there have expanded remarkably (Germer, 1965; MacRae, 1963; Somorjai, 1969).

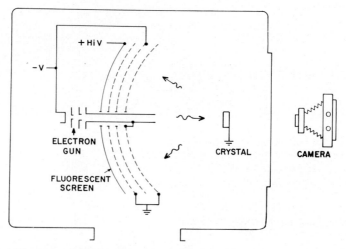

Fig. 1. Schematic of the principal features of the fluorescent-screen LEED apparatus, not to scale.

In the diffraction mode of operation, the LEED instrument displays a diffraction array produced by the outermost atomic layers. With minor modifications to the external electronics, changes in work functions and secondary emission ratios can also be measured, plasmon and other characteristic losses can be observed (Scheibner and Tharp, 1967), and surface chemical constituents can be detected by the presence of Auger transitions (Harris, 1968) and ionization losses (Ellis and Campbell, 1970) in the secondary electron energy distribution. In this chapter, only the diffraction pattern and its relation to surface structure are considered.

Optical transforms (Taylor and Lipson, 1964) are especially pertinent and useful in LEED studies because the simulated diffraction pattern from a two-dimensional grating corresponds to the kinematical pattern of the outermost layer of surface atoms, and is presented in the same format as with the fluorescent-screen apparatus. The method is particularly useful in the study of complex structural features which may be expected to occur on surfaces: ordered or statistically disordered arrays of antiphase domains, the nucleation and growth of surface films, misaligned crystallographic faces, and adsorption

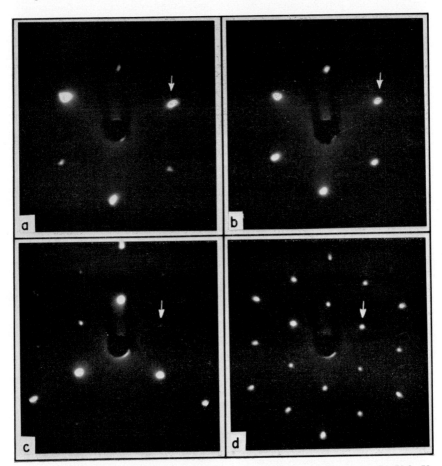

Fig. 2. LEED patterns from $UO_2(111)$ at primary energies of (a) 65 eV; (b) 70·3 eV; (c) 113·8 eV; and (d) 139·2 eV.

which is statistical on the atomic scale (Campbell and Ellis, 1968; Ellis and Campbell, 1968; Fedak *et al.*, 1968).

   Figure 1 is a schematic of the principal features of a 3-grid, ultrahigh vacuum, fluorescent-screen LEED apparatus. The crystal face under examination is positioned at the centre of curvature of a series of spherically concentric grids. Electrons are emitted by the cathode at a variable negative voltage, $-V$, focused on the crystal by the gun, and are reflected back to the fluorescent screen. The inelastically scattered electrons, which constitute 90–95 % of the secondary electron current, are suppressed by applying a negative bias at cathode potential to the second grid. Two or four grids can be used instead of three, and clearly the voltage bias can be arranged in many different ways depending upon the signal processing and sort of information desired. In the diffraction mode as shown, the crystal is usually at ground potential, the negative cathode potential is varied between 10–500 V, and only the elastically scattered electrons which carry diffraction information are displayed on the screen.

Figure 2 shows photographs of the type of diffraction pattern observed with the fluorescent screen, also called the post-acceleration, LEED instrument. They are from a carefully prepared $UO_2(111)$ planar surface free of detectable contaminants, at primary beam energies, E_p, of (a) 65 eV; (b) 70·3 eV; (c) 113·8 eV; and (d) 139·2 eV. The patterns demonstrate that the surface is single crystalline, perpendicular to the [111], and that the surface crystallography is simply related to that in the bulk. As the voltage is increased, the diffraction beams blink on and off smoothly as they converge toward the specular reflection, as shown by the arrow pointing to a given beam in Fig. 2.

Intensities are best measured with a Faraday collector, but as will be discussed briefly in later sections, the interpretation of these intensities in terms of surface structure has been a major difficulty in LEED.

B. Nomenclature of Surface Crystallography and Diffraction

It has become conventional to discuss LEED in terms of the two-dimensional mesh of outermost surface atoms (Germer, 1965). As shown in Fig. 3(a), **a** and **b** are the lattice vectors of the two-dimensional unit mesh from which the reciprocal net, i.e., the diffraction array, $\mathbf{a}^*$ and $\mathbf{b}^*$, is derived in the usual manner (Brillouin, 1946), by putting $\mathbf{a} \cdot \mathbf{a}^* = \mathbf{b} \cdot \mathbf{b}^* = 1$, $\mathbf{a} \cdot \mathbf{b}^* = \mathbf{b} \cdot \mathbf{a}^* = 0$. To obtain the proper relative directions between $\mathbf{a}^*$ and $\mathbf{b}^*$, it is convenient to use a unit vector $\hat{\mathbf{c}}$ perpendicular to the ab plane, and follow the conventional rules of vector multiplication for obtaining the three-dimensional reciprocal lattice, i.e., $\mathbf{a}^* = \mathbf{b} \times \hat{\mathbf{c}}/\mathbf{a} \cdot \mathbf{b} \times \hat{\mathbf{c}}$, $\mathbf{b}^* = \hat{\mathbf{c}} \times \mathbf{a}/\mathbf{a} \cdot \mathbf{b} \times \hat{\mathbf{c}}$.

Figure 3(b) shows an example of the indexing of LEED beams from the (100) face of a cubic lattice; the same convention will be used in the following optical simulations. The integral-order beams, e.g., 01, 11, are indexed according to an extension of the bulk crystallography to the unit mesh of the outermost atomic layer of the cleaned crystalline substrate, e.g., Cu(100). The fractional-order beams, e.g., 1/2 1/2, represent a surface structure that bears an integral relation to the substrate, e.g., as seen upon exposure of Cu(100) to O_2 (Simmons et al., 1967). According to the nomenclature of Wood (1964), this latter structure is designated $Cu(100)\sqrt{2}(1 \times 1) - R45° - O_2$ because as shown in

Fig. 3(c), the unit vectors of the oxide layer, **a′**, **b′**, are $\sqrt{2}$ times as great as the substrate and are rotated 45°. With structures more complicated than the ones shown here, the matrix formulation of Brillouin (1946) and Park (1969) is more convenient in relating the reciprocal and direct lattices. More commonly though, the particular structure in Fig. 3(c) is referred to simply as

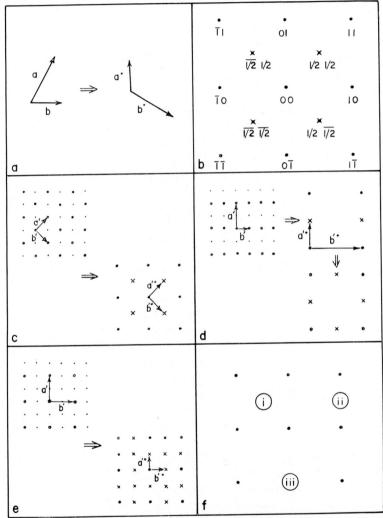

Fig. 3. Conventions in the nomenclature of two-dimensional diffraction. (a) **a**, **b** real unit mesh vectors, **a***, **b*** reciprocal net unit vectors; (b) indexing of LEED beams; (c) the $c(2 \times 2)$ grating and pattern; (d) the $P(2 \times 1)$ grating and pattern; (e) the $P(2 \times 2)$ grating and pattern; (f) positions on the surface, "i"—centred bond, "ii"—bridge-bond, "iii"—place exchange, or directly above substrate atom.

$c(2 \times 2)$, meaning that the surface structure is two substrate unit vectors long with an additional central atom.

Figure 3(d) shows the frequently observed primitive (2×1), i.e., $P(2 \times 1)$, LEED pattern, so called because $\mathbf{a}'^* = \frac{1}{2}\mathbf{a}^*$ and $\mathbf{b}'^* = \mathbf{b}^*$. With equal coverage in both directions, that is, also with $\mathbf{a}'^* = \mathbf{a}^*$ and $\mathbf{b}'^* = \frac{1}{2}\mathbf{b}^*$ coverage on half of the surface, the diffraction pattern appears as shown in the lower right-hand corner. This pattern is similar to the $P(2 \times 2)$ shown in Fig. 3(e) except that in Fig. 3(d) the 1/2 1/2 beam is missing.

It should be emphasized that from beam positions alone atomic positions within the unit mesh cannot be determined. The extra atom may be geo-metrically centred on the substrate unit mesh in the centred-bond positions as shown by "i" in Fig. 3 (f). It may also occupy a bridge-bond position "ii" in Fig. 3(f), or it may be positioned directly above a substrate atom or undergo place exchange and occupy a regular lattice site as shown by "iii" in Fig. 3(f). There may be lattice distortion both parallel and normal to the surface plane, and other positions are possible. Although the surface structures in Figs. 3(c), (d), and (e) are shown for "iii", without good intensity measurements and an equally good theoretical analysis, it may be impossible on the basis of LEED data alone to designate atomic positions within the unit mesh.

C. Reciprocal Lattice for Weakly Penetrating Radiation

Figure 4(a) shows a section through the reciprocal lattice of an infinitely extended perfect crystal; the Fourier transform leads to infinitely small points in reciprocal space (Lipson and Taylor, 1958; Vainshtein, 1964). The incident wave-vector is denoted by κ_0. In kinematical treatment applied to penetrating radiation, for example X-rays and neutrons, diffraction beams appear where the Ewald sphere of radius $1/\lambda$ intersects the reciprocal-lattice points. The 60° reflection cone of observation for LEED is also shown. However, Fig. 4(a) does not apply to LEED because the penetration is through 2–4 atomic layers and dynamical effects are important.

As shown in Fig. 4(b), if a crystal three atomic layers thick is probed by penetrating radiation, the reciprocal-lattice points elongate into rods and secondary points of scattering appear between the major Bragg reflections (see for example Françon, 1966). This effect, variously referred to as the crystal form factor, the shape transform, Laue interference function, or reciprocal lattice for weakly penetrating radiation, broadens the diffraction beams and gives rise to secondary maxima in the intensity measurements as $1/\lambda$ is varied, i.e., with LEED in the intensity-versus-voltage curves. As will be discussed later, the preceding concept is only partially valid for LEED because of dynamical effects but some general features may be explained in this fashion (Duke and Tucker, 1969). For example, as shown in Fig. 5(a), the reciprocal lattice for weak penetration into three atomic layers of a misaligned surface gives rise to subsidiary beams in the diffraction pattern. For deeply penetrating back-reflected radiation, e.g., X-rays, pattern 5(b) would be observed. Instead, the LEED pattern for $UO_2(553)$ at $E_p = 75$ eV, Fig. 5(c), shows the type of beam splitting expected from a reciprocal lattice such as Fig. 5(a). The kinematical pattern calculated for scattering from only the outermost layer

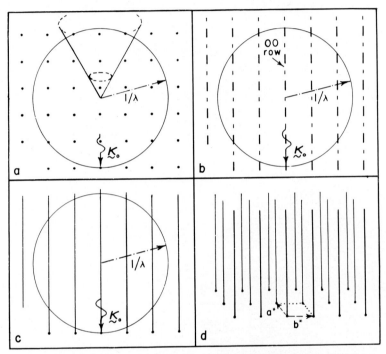

Fig. 4. Reciprocal lattices and the Ewald sphere, with incident plane-wave vector, κ_0, for (a) penetrating radiation in perfect, infinite crystal; (b) penetrating radiation in crystal 3 atomic layers thick; (c) the planar 2-dim. reciprocal lattice; (d) isometric view of the 2-dim. rods of #c.

of an array of equally spaced steps of minimum height is shown in Fig. 5(d) (Ellis and Schwoebel, 1968). Agreement between calculation and observation is shown in Fig. 5(d) by encircled beams.

Of central concern to us in this chapter is the comparison between LEED and the optical analogue. In the Ewald construction for diffraction by a two-dimensional array, the reciprocal lattice consists of continuous uniform rods as shown in Figs. 4(c), (d) rather than the simple planar dots of Fig. 3. Thus, except for the angular dependence of scattering amplitude, beams in the analogue are not modulated in intensity as they are in LEED, but they do display quite nicely in reciprocal space the equivalent diffraction beams.

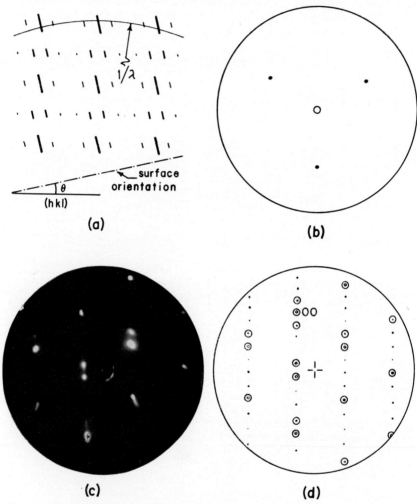

Fig. 5. Misaligned surfaces. (a) reciprocal lattice for penetration into 3 layers of a misoriented $fcc(\sim111)$ surface; (b) back-reflected pattern for X-rays; (c) observed LEED pattern from $UO_2(553)$ at $E_p = 75$ eV; (d) calculated kinematical LEED pattern.

D. Comparison with High Energy Electron Diffraction (HEED)

The transmission of a high-energy electron beam through a foil of 100 or so atomic layers in thickness contains a vanishingly small amount of information about the outermost atomic layers. The radius of the Ewald sphere of course is much greater than in LEED because the energies may be a thousand times as great. In the forward scattering direction, the sphere may intersect several reciprocal-lattice points for the bulk crystal as shown in Fig. 6(a). The back-scattering amplitudes are too low to be of much value.

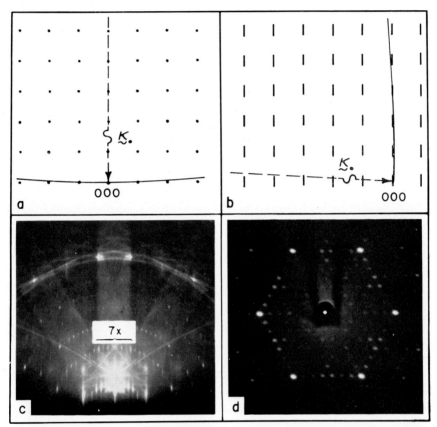

Fig. 6. Comparison of LEED with HEED: (a) reciprocal lattice for transmission HEED, beam normally incident on thin foil; (b) grazing incidence HEED; (c) grazing incidence 50 keV HEED pattern from Si(111), photo by N. Taylor, Varian Assoc.; and (d) LEED pattern of the Si(111)P(7 × 7) structure, $E_p = 54$ eV.

Grazing-incidence diffraction however does contain information about the outermost atomic layers (Simmons *et al.*, 1967; Siegel and Menadue, 1967). At a grazing angle of 1°, a 50 keV beam may penetrate only a few atomic layers, and the pattern appears on a diffraction photograph as streaks at the angles where the Ewald sphere intersects the reciprocal-lattice rods of the surface layers, as shown in Fig. 6(b). A direct comparison between LEED and HEED is seen in Figs. 6(c), (d). Figure 6(c) shows a 50 keV grazing incidence pattern from the Si(111) surface. Between the Laue beams are six superlattice reflections, outlined by the brackets, which indicate a surface lattice periodicity seven times as great as in the bulk. The same diffraction array is seen in the normal incidence LEED pattern taken at $E_p = 50$ eV shown in Fig. 6(d); between the brighter Bragg reflections, the less intense beams of the Si(111) $-$ P(7 $\times$ 7) pattern are displayed. In the high-energy apparatus, inelastically scattered electrons are not suppressed with the result that Kikuchi bands appear prominently in the photograph.

Among the possible advantages of HEED are that the atomic scattering amplitudes are known much better for high energies and the techniques for intensity analysis are more highly developed (Cowley, 1967). A disadvantage, or perhaps another advantage depending upon the intent of the experiment, is that at grazing incidence protrusions may be the only parts of the surface probed. Also, an experimental bother is the requirement that in order to obtain the complete reciprocal lattice for the crystal face, the specimen must be rotated about an axis normal to the surface. But in conjunction with LEED, HEED appears to be yet another powerful tool to investigate surface structure.

E. Difficulties in the Interpretation of LEED Observations

1. Diffraction Intensities

Ironically, the principal difficulty in the interpretation of LEED arises from the same causes that make low energy electrons such a sensitive surface probe. Both the elastic and inelastic scattering amplitudes are so high that the coherently scattered electrons penetrate only a few atomic layers, and the intensities are not related in a simple manner to structure. The kinematical effect of diffraction by three layers can be demonstrated readily by the optical transform of a row of holes. The three-dimensional reciprocal lattice of an infinitely extended linear row of mathematical points is a stack of equally spaced, infinitely thin sheets as shown in Fig. 7(a). The intersections of the sphere of reflection with the reciprocal lattice sheets, Fig. 7(b), appear as lines. For a row of 38 holes, the Fraunhofer pattern indeed is a series of sharp lines as shown in Fig. 7(c). If the incident wave-vector, κ_0, were from left to right in Fig. 7(c) and $1/\lambda = \sqrt{V}/150$ Å^{-1} were changed as in LEED, the hypothetical intensity-versus-voltage curve of the specular reflection would be a series of sharp spikes at the Bragg reflections as shown in Fig. 8(a) (cf. Fig. 4a). In the Fraunhofer pattern of three holes, Fig. 7(d), because of the shape transform the principal maxima are broadened and secondary peaks appear. The corresponding hypothetical I—V curve for the three-layer case would be as shown in Fig. 8(b) (cf. Fig. 4b). For comparison, a typical LEED experimental I—V

curve observed for example with Ni or Cu is shown in Fig. 8(d). Although Fig. 8(b) bears a resemblance to 8(d) in that there are principal and secondary Bragg maxima, the correspondence is not very good. Herein lies a major difficulty in the interpretation of LEED results—the analysis of intensity data

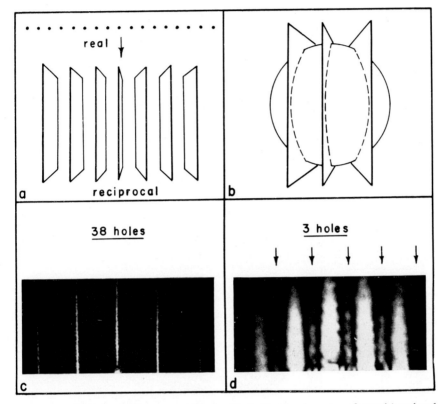

Fig. 7. Diffraction by linear rows of holes illustrating the shape transform: (a) real and reciprocal lattice for an infinitely extended row of holes; (b) intersection of the sphere of reflection with the reciprocal lattice sheets. Fraunhofer patterns from (c) 38 holes and (d) 3 holes.

to obtain defined structural information. Considerable theoretical efforts from several points of view are being applied by researchers to this aspect of the low energy dynamical diffraction problem (Seah, 1969; Beeby, 1968; McRae, 1968; Boudreaux and Heine, 1967).

2. The Diffraction Pattern

As a result of the difficulty of intensity analysis, the highly developed convolution techniques used in X-ray analysis (Hosemann and Bagchi, 1962), have not been applied to low energy electron diffraction. For a given diffraction array there may be several structures all equally possible on an *a priori* basis. In our effort at Los Alamos to interpret LEED data we have found that optical transforms are extremely valuable as a preliminary to, and as a secondary check on, computations. In addition, transforms have indicated types of as yet unobserved diffraction patterns which may be expected for complex surfaces. In discussing the application of optical transforms to LEED, it should be remembered that, although the corresponding beams are observed,

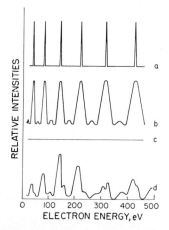

Fig. 8. Intensity-versus-incident beam energy curves for (a) penetrating radiation, kinematical case, infinite perfect crystal; (b) 3 atomic layer case; (c) the 2-dim. analogue, and (d) typical LEED *I–V* curve.

the intensities do not vary as in Fig. 8(d). Instead, the hypothetical two-dimensional optical transform *I—V* curve is a straight line as seen in Fig. 8(c) (compare with Figs. 4c, d).

II. APPLICATION OF TRANSFORM METHODS

A. Transforms of Simple Surface Structures

Except for the subsequent addition of lenses to sharpen the patterns, the experimental procedures used in taking the optical transforms for this chapter were the same as reported by Ellis and Campbell (1968). The diffraction grating, consisting of about 1200 scattering centres, was prepared by first piercing the desired array in an 8 1/2″ × 11″ sheet of opaque paper with a fine needle. The paper was then mounted in front of an illuminated, trans-lucent screen and photographed with transparency film to produce a grating *ca.* 4 mm × 6 mm. Coherent light from a 1 mw, 632·8 nm He–Ne cw laser

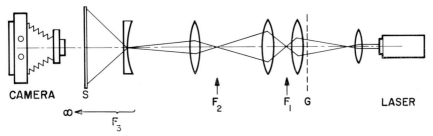

Fig. 9. Optical system for projecting the diffraction pattern: *G*—grating; *S*—display screen; $F_{1,2,3}$—Fraunhofer zones.

was then passed through the grating and the Fraunhofer, i.e., far-field, pattern was projected onto a frosted-glass screen in the fashion shown schematically in Fig. 9.

Examples of simple transforms taken in this manner are shown in Fig. 10.

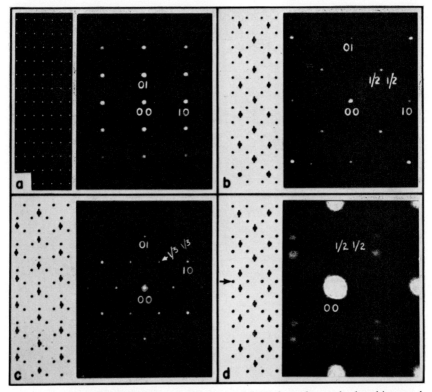

Fig. 10. Simple optical transforms: (a) simulated (110) face of an orthorhombic crystal; (b) the square $c(2 \times 2)$ pattern; (c) cubic $(111)\sqrt{3}(1 \times 1)R30°$ surface structure and pattern; (d) antiphase splitting in the $c(2 \times 2)$ pattern.

In this and most of the subsequent illustrations the grating is shown to the left of the diffraction pattern in the same relative orientation as used to take the transform. In Fig. 10(a) we see the grating and simulated LEED pattern of the (110) face of an orthorhombic crystal. As known and expected, the pattern is reciprocal to the grating. Figure 10(b) shows the square $c(2 \times 2)$ grating and pattern discussed earlier. For greater clarity, in Figs. 10(b), (c), and (d) drawings rather than photographs are shown. The substrate atoms are represented by small dots; the centred-bond adsorbed atoms are indicated by larger dots with vertical bars. In Fig. 10(c) we see a simulation of a $(111)\sqrt{3}(1 \times 1) R30°$ pattern from a crystal in the cubic system.

Figure 10(d) shows a simulation of the sort of structure and diffraction effect that may be expected upon exposure of a crystal surface to gas: splitting of the fractional-order beams by the growth of random nuclei into antiphase domains. The antiphase boundary in the $c(2 \times 2)$ structure is indicated by the arrow. As will be discussed later, such splitting depends upon domain size and the way the nuclei grow together.

B. Ordered Steps and Kink Sites

From high-index UO_2 surfaces, such as the $UO_2(553)$, the Bragg reflections split into doublets as shown in Fig. 5(c) (Ellis and Schwoebel, 1968). Such splittings are also seen on Nb (Haas, 1966), W (Taylor, 1968), cleaved Ge (Henzler, 1970), and Re (Feinstein and Macrakis, 1969). In its simplest representation, the $UO_2(553)$ surface consists of (111) terraces separated by equally spaced ⟨110⟩ steps one atomic layer in height. A similar configuration of the same general type is illustrated in Fig. 12(a) for the (140) simple cubic surface. However, there is no *a priori* reason to believe that an ordered surface of this sort is thermodynamically stable. Instead, one might expect such a surface to decay into randomly spaced microfacets of multiple height with the low-thermodynamic-energy planes exposed. To test the conclusion based upon kinematical calculations that the observed doublet pattern represents a surface with equally spaced steps of minimum height, optical simulations were used.

Figure 11 shows six simulated diffraction patterns. Figure 11(a) is the pattern from a two-dimensional hexagonal array representing beam positions either for the perfectly oriented $UO_2(111)$ LEED pattern, or the back-reflected pattern for penetrating radiation where κ_0 is parallel to ⟨111⟩ (*cf.* Figs. 2, 5). Figure 11(b) shows the simulated pattern from a single step on the (111) plane. Each dot above the pattern represents a row of atoms parallel to the ⟨110⟩ in the surface and the grating, partially shown only for Fig. 11(c), is the flat projection of this configuration on the (111) plane. There is characteristic out-of-phase splitting of the integral-order beams. The diffraction pattern in

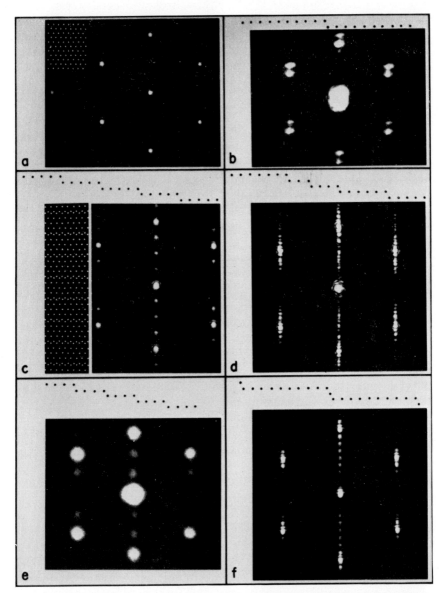

Fig. 11. Surface steps: (a) well oriented $fcc(111)$, compare with Fig. 2; (b) antiphase splitting of the integral-order beams by a single step; (c) $fcc(553)$ surface with equally spaced steps of min. ht.; (d) $fcc(553)$, statistically disordered steps of min. ht.; (e) $fcc(442)$ surface with equally spaced steps of min. ht.; (f) $fcc(553)$, steps of double height, equally spaced.

Fig. 11(c) is from a simulated (553) surface with parallel, equally spaced steps of minimum height, and 11(d) is from a (553) surface with statistically spaced steps of minimum height. For the grating of Fig. 11(c) the sequence is 5 rows of atoms forming a (111) terrace, then a step, 5 rows then a step, and so on. For Fig. 11(d) the sequence was 5 rows then a step, 7 rows then a step, 3 rows, 4, 6, 8, 3, 7, 4, and 1 row between steps for a total of 10 terraces with varied step spacings. Figure 11(d) has a good deal more fine features than the others, which the LEED instrument undoubtedly would not resolve. Instead, in a LEED pattern of this sort we would expect to observe unresolved streaks and a few discrete beams seemingly out of sequence. Figure 11(e) is from a simulated (442) surface, and 11(f) is from a (553) surface with equally spaced steps of double height. Among these simulations, Fig. 11(c) agrees most closely with the observed and calculated doublet pattern from $UO_2(553)$ heated in O_2, and agreement for surfaces heated in vacuum is best for 12(d), (f). Thus we take these results as additional support for the contention that the doublet LEED pattern of Fig. 5(c) displays the simplest case.

These results suggest that even more complicated high-index planes can be simulated and that the optical transforms may serve as a guide in the interpretation of experimental LEED patterns from those surfaces. Figure 12(a) is a sketch of the type of surface discussed above: equally spaced steps of minimum height parallel to [001], with (010) terraces on the simple cubic (140) face. Instead, if the surface is cut parallel to, for example, the (3 12 1) plane, the steps are no longer straight lines but have kinks, also called half-crystal sites, as shown by the shaded cubes in Fig. 12(b). In Fig. 12(b), the kink sites are ordered in one particular crystallographic array. There is no reason to assume however that for a given set of experimental conditions the kinks will form one array in preference to another. It is instructive to simulate the diffraction effects for different distributions of half-crystal sites on such vicinal surfaces. The term "vicinal" has been applied by Rhead and Perdereau (1969) because the surface normal is in the vicinity of a crystallographic pole. Figure 12(c) shows the pattern and a portion of the grating for one particular ordering of a high-index $fcc(111)$ vicinal surface. The dark lines to the left are traces of the ledges, i.e., the out-of-phase boundaries in the grating. In Fig. 12(d) the kink sites along a given ledge are equally separated but between different ledges they are statistically disordered. The patterns in Figs. 12(c) and 12(d) show similar features: the subsidiary beams are aligned in rows perpendicular to the antiphase boundaries. Similar LEED patterns have been observed from Cu(100) vicinal surfaces by Rhead and Perdereau (1969) and from $UO_2(\sim111)$ by us here at Los Alamos (Ellis, unpublished). The two patterns in Figs. 12(c), (d) can be distinguished easily by the positions of the extra beams, and with longer exposures Fig. 12(d) shows considerably more streaking. These two examples, chosen from several configurations simulated

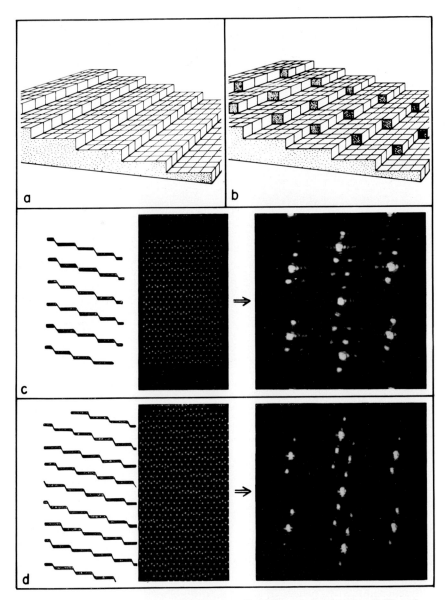

Fig. 12. Vicinal surfaces: (a) ordered steps of min. ht. on simple cubic (140) surface; (b) ordered kink sites on simple cubic (3 12 1); (c) ordered vicinal *fcc*(~111) surface; (d) disordered kink sites on *fcc*(~111).

9

for this surface, clearly demonstrate that indeed the ordering of half-crystal positions can be determined by optical transforms, which implies that in principle the same information is obtainable in LEED studies of real vicinal surfaces.

To date we have not analysed our LEED results for vicinal surfaces, but the approach merits the attention which will soon be given. Since the surface defects shown in Fig. 12(b) are important in theories of surface diffusion (Burton *et al.*, 1951; Schwoebel, 1967), catalysis, vaporization and condensation processes, and reactivities of interfaces, it seems pertinent to investigate

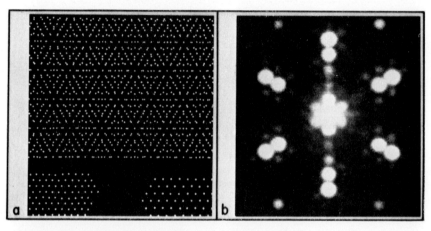

Fig. 13. Hexagonal coincidence lattice: (a) grating at top from superposition of two hex meshes at bottom, spacings ratio 4: :5; (b) diffraction pattern from #a, top showing intense beams from 2 sublattices and extra coincidence beams.

these vicinal surfaces more thoroughly and to design LEED experiments with the preceding concepts in mind.

C. Coincidence Lattices

The somewhat complicated coincidence-lattice diffraction pattern is shown in Fig. 13(b). The grating, shown at the top of Fig. 13(a), was produced by the superposition of two hexagonal arrays at the bottom with lattice parameters in the ratio 4: :5. The pattern consists of the more intense integral-order beams from the two hexagonal lattices, and satellite beams from the difference in the two reciprocal lattices. Although the pattern is not common in LEED, it has been seen for oxide layers on Ni(111), and has been attributed to multiple scattering between a hexagonal monolayer of NiO and the Ni(111) substrate (Bauer, 1965). However, the transform grating is two-dimensional and multiple scattering cannot occur. Thus, although multiple scattering probably does occur, it does not have to be invoked to explain the observed pattern and the simple appearance of fractional-order beams in a coincidence-lattice display cannot be used to support multiple-scattering theories.

D. The $UO_2(111) + O_2$ Ring Pattern

Another example of the applicability of optical transforms to LEED relates to the appearance of third-order rings around the integral-order beams from

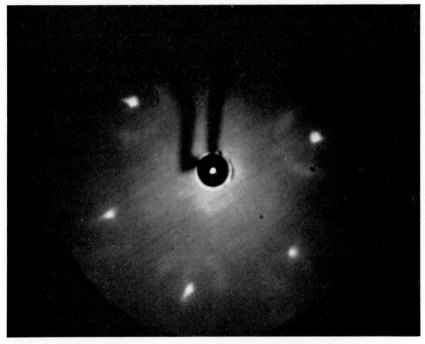

Fig. 14. The $UO_{2+x}(111)$ LEED ring pattern at $E_p = 49.2$ eV.

the $UO_{2+x}(111)$ surface. The LEED pattern at $E_p = 49$ eV is shown in Fig. 14. Similar patterns are seen in high-energy transmission electron diffraction and are attributed to Debye-Scherrer scattering of the bulk diffracted beams by an amorphous layer on the surface. But LEED differs from the high energy case because of limited penetration of the elastically scattered electrons. Furthermore, the rings do not follow the same curves of intensity-versus-voltage as the encircled beams. The qualitative analogy seemed correct however in that there appears to be some random surface structure.

It was not until we began optical analogues that possible solutions emerged. Prior to being successful we simulated several plausible structures including diffusely adsorbed O_2, mosaic microdomains, double-beam diffraction by first passing the light through the regular hexagonal array and then through a diffuse layer, reversal of the preceding, dislocations, other antiphase boundaries, and so forth, none of which reproduced the experimental LEED pattern. One possible configuration which did reproduce the ringed pattern consisted of triangular clusters occupying regular lattice sites but randomly

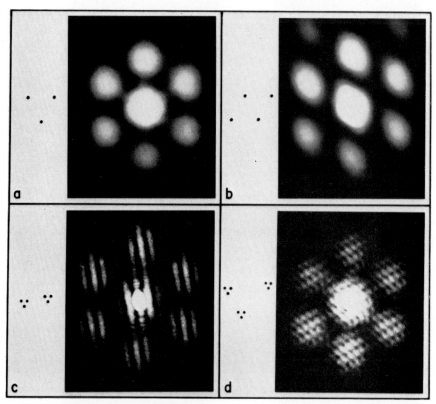

Fig. 15. Ring formation: (a) transform of triangular cluster; (b) transform of 4 scattering centres; (c) transform of 2 separated triads; (d) transform of 3 separated triads.

separated from neighbouring clusters by approximately three lattice spacings (Ellis and Campbell, 1968). It is instructive to demonstrate here just why the rings do appear in the transform.

Figure 15a shows the diffraction pattern obtained by passing laser light through only three scattering centres arranged in a triangular cluster. The beams are very broad and form a hexagonal array reciprocal to the cluster. Addition of a fourth centre to the cluster distorts the beams as shown in Fig. 15(b). Two triangular clusters separated by about four lattice parameters, but

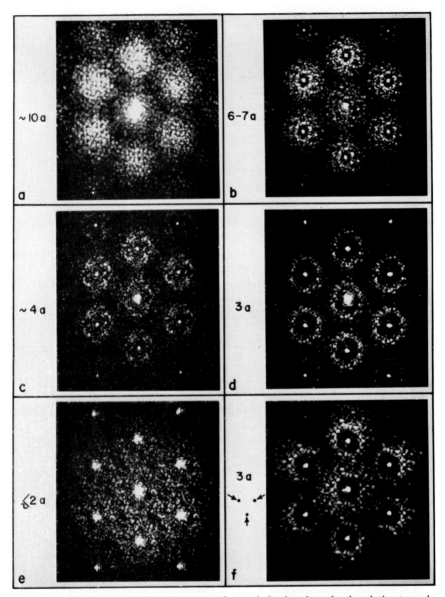

Fig. 16. Development of the ring pattern from triads placed randomly relative to each other, but approximately equally separated. Separation: (a) ~10 lattice spacings, i.e., ~10a; (b) 6–7a; (c) ~4a; (d) 3a; (e) <2a; (f) 3a with central relaxation.

bearing integral multiple translations of the substrate lattice vectors, lead as expected from Fourier analysis to a splitting of the diffuse beam into dark and light bands perpendicular to the separation vector. Three triangular clusters, again separated by integral translations of the hexagonal lattice vectors, lead to splittings in directions perpendicular to the three separation vectors. With a suitable statistical distribution of such triads, it should be possible then to simulate the ring pattern of Fig. 14.

Figure 16 illustrates the development of discrete rings as the separation of randomly distributed clusters is decreased. In Fig. 16(a), the clusters are separated in random directions by approximately ten lattice parameters. The beams are broad, and although they show granularity, in an equivalent LEED pattern they would be uniformly diffuse. At a separation of 6–7a, the central integral-order beam has a dark ring around it, Fig. 16(b); at ~4a the dark ring has increased in diameter, Fig. 16(c). When the triads are statistically separated by 3a, the ring pattern is well developed and corresponds quite closely to the observed LEED pattern shown in Fig. 14. At a separation of $\leqslant 2a$, Fig. 16(e), the rings overlap and the dark inner ring disappears. If there is central relaxation of the triads in the plane of the surface, as shown in Fig. 16(f), the integral-order beams retain the same scattering angle but are no longer concentric with the rings. This latter pattern is not observed experimentally with UO_{2+x}. From these simulations, we conclude that the rings may possibly display a disordered non-stoichiometric uranium oxide surface in which there is perturbation of the lattice by triangular clusters randomly separated from neighbouring clusters by approximately three atomic spacings. This interpretation may not be unique of course, and multiple scattering may still be an important consideration. A more detailed description could not be deduced without greater knowledge of atomic scattering factors and a refined analysis of diffraction intensities.

As a beginning in the computational analysis of the ring pattern, Feber and Grimmer (1969) modified the kinematical diffraction model for beam attenuation and out-of-plane atomic displacements, and on a large modern computer calculated the LEED diffraction pattern for $UO_{2+x}(111)$ shown in Fig. 17. The basis for computation was the plane-wave amplitude term

$$\mathbf{A} = A_0 \sum_l f_l \exp\left[i(\kappa_0 - \kappa)\cdot\mathbf{r}_l\right]$$

summed over a grating 20×20 atoms arranged statistically as described previously for Fig. 16(d). Beam attenuation was accounted for in f_l, the atomic scattering amplitude, and atomic displacements through $\mathbf{r}_l$, the separation vector. The wave-vectors, κ_0 and κ, have the usual meanings (Vainshtein, 1964). In the illustration, the heights of the contour lines are proportional to intensity. Although spotty because of limited statistics, the rings are definitely repro-

duced. On the basis of these initial calculations, they have tentatively concluded that the interplanar spacing of the outermost oxygen on stoichiometric
$UO_2(111)$ may be greater than in the bulk and that in $UO_{2+x}(111)$ random
triads of uranium ions may be displaced toward the surface. The computer
results suggest that the rationale for varying the scattering factor in the gratings
of Figs. 15 and 16, i.e., making larger and smaller holes, may in fact be a
phase factor resulting from atomic displacements normal to the (111) plane.
In addition to producing the rings, these displacements also perturb the
curves of intensity-versus-voltage, which suggests that better intensity
measurements be compared with a more refined theoretical treatment.

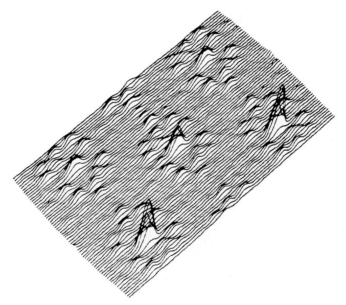

Fig. 17. Modified kinematical diffraction pattern calculated for $UO_{2+x}(111)$.

Before proceeding to the next topic, another possibility for the ring structure
should be considered. As suggested previously by Fig. 11, antiphase domains
in a $P(1 \times 1)$ structure can produce splitting of the integral-order beams.
Furthermore, if the domains are randomly oriented, approximately equal in
size, and small, the splitting may produce observable rings. Oxygen may
conceivably adsorb on an $fcc(111)$ plane in two out-of-phase positions as shown
by the triangles in Fig. 18(a). The extended substrate (111) planar lattice
positions are represented by solid dots and adsorbed atoms by open circles.
The antiphase boundary is represented by a broken line. A statistical antiphase
$P(1 \times 1)$ array, with a domain size ~3–4 lattice spacings, is shown in Fig. (18b),
which is drawn to the same scale as the actual grating of Fig. 18(c), left. The

diffraction pattern seen in the right of Fig. 18(c) indeed does show rings around a hexagonal array of beams. But not all beams are encircled: the 00 and alternate {11} beams with arrows pointing to them have no rings. In the actual LEED pattern as well as the simulations of Fig. 16, all beams are encircled.

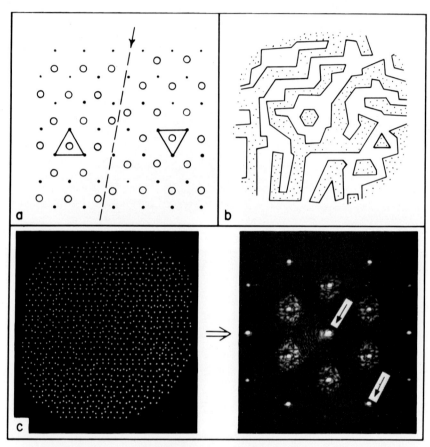

Fig. 18. $fcc(111)P(1 \times 1)$ diffraction rings: (a) antiphase boundary in $P(1 \times 1)$ array; (b) antiphase domain array outlined for (c) the grating; and (d) Fraunhofer pattern from (c).

Accordingly, although the possibility of a diffusely adsorbed monolayer of oxygen is an appealing concept, we conclude that a statistical arrangement of this sort is not responsible for the observed diffraction rings from $UO_{2+x}(111)$. This conclusion is also supported by computations in which variations of this sort of the atomic scattering factor, f_i, did not reproduce the rings.

Thus, in attempting to solve the $UO_{2+x}(111)$ ring diffraction problem, we proceeded as follows: 1. observed the LEED rings; 2. searched with optical analogues for possible solutions, culminating in the array of random triads; 3. performed an initial computer analysis, which in turn suggested 4. better intensity measurements; and 5. dynamical multiple-scattering treatment which contains corrections for 2, 3, or 4 atomic layers, phase shifts, beam attenuation, resonance scattering and dispersion effects, and so on. In the above sequence, it is pertinent to note that without step 2 the subsequent analyses would have been seriously delayed if not impossible.

E. Antiphase Boundaries in the Square $c(2 \times 2)$ Surface Structure

1. Growth of Random Nuclei

Another pertinent application of optical transforms to LEED studies concerns the diffraction patterns produced by random nucleation centres in the initial stages of a gas-solid reaction. Such a situation is partially simulated in Fig. 19(a). The grating comprises a continuous square lattice with small $c(2 \times 2)$ nuclei placed randomly. In all of Fig. 19, the "i" domains are in phase with each other, the "o" regions are also in phase with each other, but the "o"'s are out of phase with the "i"'s. Centred bonds were used as per Fig. 10(b). The pattern of Fig. 19(a) shows diffuse half-order beams that are granular because of poor statistics. With a larger grating and random separation, the fractional-order beams would be broadened, but smooth in intensity without discrete splitting.

As the nuclei grow together, they may do so in several different ways of which one is simulated in Fig. 19(b). In this grating, the boundaries between adjacent "i" and "o" regions are the perpendicular bisectors of the lines joining nucleation sites, i.e., they are immobile boundaries formed by initially circularly symmetric domains. The sites of nucleation, indicated by the i's and o's, are approximately equally separated at random orientations. The pattern was taken with an optical system improved over the earlier unit (Ellis and Campbell, 1968), by the addition of projection lenses to sharpen the Fraunhofer display. In a fashion similar to the preceding sections, the 1/2 1/2 beams, rather than simply being broadened, formed rings. With another grating, not shown but similar to Fig. 19(b) with smaller domain size, the ring diameter was larger. Thus, although such a $c(2 \times 2)$ pattern has not been reported in the literature, with a suitable statistical arrangement of immobile antiphase domains, the fractional-order beams can form rings. If the boundaries were to rearrange along a low-index direction, such as the [11], the result would be as shown in Fig. 19(c). The nuclei in 19(b) and 19(c) are the same, but in 19(c) rather than forming rings, the 1/2 1/2 beams display indistinct antiphase splitting. The observations taken with such gratings lead to the conclusion that discrete antiphase splitting appears only when there is ordering of the domains, either

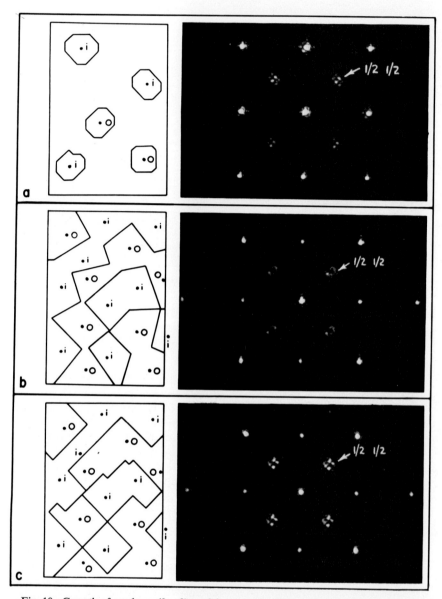

Fig. 19. Growth of random $c(2 \times 2)$ nuclei on square (100) surface: (a) initial stage with random isolated nuclei; (b) fully covered surface with immobile antiphase boundaries; and (c) fully covered surface with rearranged boundaries parallel to [11] and [1$\bar{1}$], same nuclei as (b).

into approximately equally spaced parallel arrays, or into rectangular regions of approximately equal size with a predominance of 90° intersections. However, there are instrumental limitations, which may vary between units, that may obscure rings and splittings if the domain size is larger than 20 or so atomic diameters. Such general considerations as the preceding demonstrate that antiphase domains may be present on surfaces to a greater extent than is generally realized.

2. Ordered Parallel Arrays

A good example of antiphase splitting occurs with equally spaced "i" and "o" domains whose interfaces are parallel to a crystallographic direction, e.g., [10] as shown in Fig. 20(a). The grating is composed of bridge-bond $c(2 \times 2)$ domains three substrate atomic spacings wide. The 1/2 1/2 beam splits perpendicularly to the antiphase boundaries into rows of fractional-order beams, having the indices 1/2 0, 1/2 1/3, 1/2 2/3, 1/2 1, et cetera, with the more intense beams at 1/2 1/3 and 1/2 2/3. If the domains are five substrate spacings wide, as in Fig. 20(b), the 1/2 1/2 beam splits into 1/2 0, 1/2 1/5, 1/2 2/5, and so on with the more intense beams at 1/2 2/5 and 1/2 3/5, i.e., those closest to the 1/2 1/2 position. An interpretation of this sort has been advanced by Estrup and Anderson (1966) to explain the LEED patterns produced by hydrogen on W(100). An alternative bridge-bond array five spacings wide is shown in Fig. 20(c) with zig-zag boundaries. The 1/2 1/2 beams in the pattern of Fig. 20(c) split similarly to those in Fig. 20(b), but dim beams also appear at 1 1/5, 1 2/5, 1 3/5, and so on. Centred-bond arrays also lead to such splittings as shown in Fig. 20(d), which is identical in beam positions to Fig. 20(b).

The beams shown in Figs. 20(a), (b), (c), and (d) are sharp because the gratings, which are only partially shown, are large and the surfaces are fully covered. On a real surface in the early stages of a reaction there may be nucleation arrays consisting of only two antiphase domains such as illustrated in Fig. 20(e) for centred-bond $c(2 \times 2)$. Each domain is three substrate lattice spacings wide. The pattern again shows splitting of the 1/2 1/2 beams, but this time more diffusely. In addition, the 10 and 01 beams are also split. If such a structure were to exist on a real surface, the LEED pattern would show an indistinct, broadened splitting of the 1/2 1/2 beams and a diffuse linear broadening of the 10 and 01 beams, depending of course upon the extent of coverage, relative arrangements of the domains, what is on the rest of the surface, and so on.

On an actual crystal face, it is expected that splittings such as shown in Figs. 20(a–e) would occur with equal probability in the other direction rotated by 90° and that the 1 mm² LEED beam would cover many such domains. In such a case the observed LEED pattern would then appear as shown in Fig. 20(f).

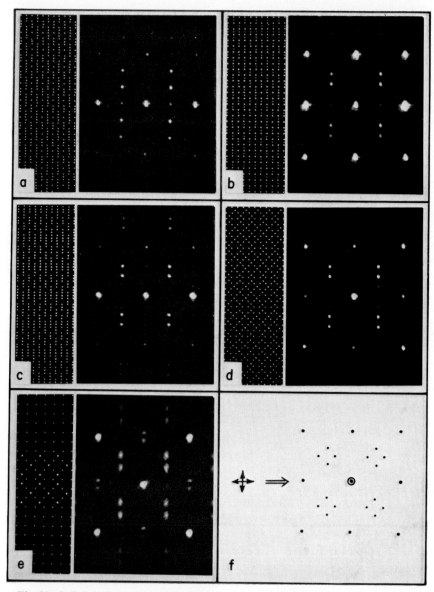

Fig. 20. Splitting of the $c(2 \times 2)$ 1/2 1/2 beam by parallel antiphase domains: (a) bridge bonds, domains 3 substrate lattice spacings wide; (b) same as (a), 5 spacings wide; (c) bridge bonds, zig-zag boundaries, domains 5 spacings wide; (d) centred bonds, 5 spacings wide; (e) centred bonds, only 2 antiphase domains, each 2 substrate spacings wide; and (f) pattern with equal coverage in both directions.

3. *The Cu(100) + O₂ "Four-spot" Structure*

The splitting of the 1/2 1/2 beam into four spots which Simmons *et al.* (1967) and Lee and Farnsworth (1965) observed upon exposure of the Cu(100) face to oxygen affords another good illustration of the applicability of optical transforms to LEED studies. The reported pattern is shown in Fig. 21(a); it is identical to Fig. 20(f) but is reproduced again to avoid the suggestion that the parallel antiphase arrays of Fig. 20 are the sole possibilities. In fact there are several alternatives as will be seen. One structure proposed by Fedak *et al.* (1968) is illustrated in Fig. 21(b). The grating consists of $c(2 \times 2)$ domains 3×3 substrate spacings square alternating in an antiphase relation with smaller $c(2 \times 2)$ domains 2×2 substrate spacings square. The continuous

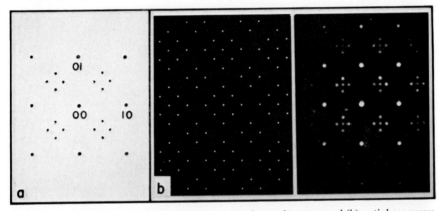

Fig. 21. Cu(100) + O₂ "four-spot" pattern: (a) observed pattern and (b) antiphase array of nuclei and pattern (Fedak *et al.*, 1968).

substrate scattering centres have been omitted. In the pattern, shown on the right in Fig. 21(b), the 1/2 1/2 beam shows splitting into the four-spot pattern. The presence of the central 1/2 1/2 beam at about one-half the intensity of the four observed beams does not rule out this grating. For reasons discussed earlier, LEED intensities vary with primary-beam energy in a dynamical, somewhat unpredictable fashion, and it may be that this beam is obscured by the relatively high background described by Simmons *et al.* (1967).

Two other possibilities to consider for the "4-spot" structure are shown in Fig. 22. The grating shown in Fig. 22(a) consists of a $c(2 \times 2)$ antiphase nucleus; each of the four domains within the nucleus is 3×3 substrate spacings square. Again, the substrate atoms have been omitted. The transform, Fig. 22(b), of a grating with domains 2×2 spacings square displays broadened integral-order beams and a skewed "4-spot" pattern. For a nucleus with 3×3 domains, i.e., Fig. 22(a), the pattern, Fig. 22(c), shows sharper integral-order

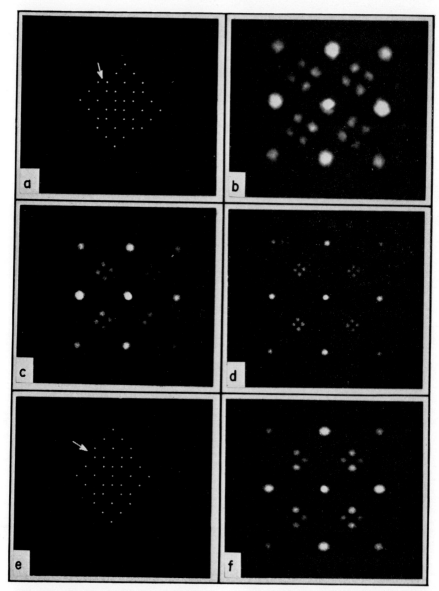

Fig. 22. Simulated "four-spot" patterns from isolated antiphase nuclei, antiphase kink sites indicated by arrows: (a) spiral grating, 3×3 spacings square; (b) pattern from spiral grating, 2×2 spacings square; (c) pattern from spiral grating, 3×3 spacings square; (d) pattern from spiral grating, 5×5 spacings square; (e) alternative diamond-shaped antiphase nucleus; and (f) its transform.

beams and again skewed but less separated "4-spots". For a nucleus with
5×5 domains the beams are even sharper and the splitting is smaller as seen
in Fig. 22(d). Another possible antiphase nucleus of this same general type
is shown in Fig. 22(e) with its "4-spot" transform in Fig. 22(f). An innovation
of these structures is the presence of antiphase kink sites shown by arrows
in Figs. 22(a), (e), where further growth of the nucleus is facilitated if the
interaction at the arrow is attractive. If the formation of such nuclei were
random, the mirror image of Fig. 22(a) would occur with equal probability
so in that the observed pattern the "4-spots" would be smeared as indicated
in an exaggerated fashion in Fig. 23(a). With equal distributions of the nucleus

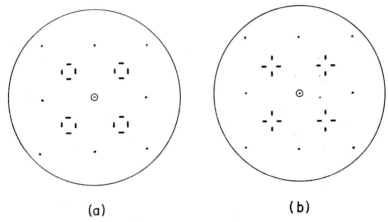

(a)   (b)

Fig. 23. Expected patterns from surface covered equally by the nuclei of Fig. 22 and
their images: (a) spiral nuclei of Fig. 21(a); and (b) diamond nuclei of Fig. 21(e).

of Fig. 22(e) and its $90°$ rotational image, the pattern would be smeared as
shown in Fig. 23(b).

Thus there are several configurations that relate to and possibly may account for the $Cu(100) + O_2$ "4-spot" pattern: 1. extended parallel and equally
spaced antiphase arrays and their $90°$ rotational image such as Fig. 20(a); 2,
parallel antiphase nuclei such as Fig. 20(e); 3, the grating of Fig. 22(b); 4,
the antiphase nucleus of Fig. 22(a); or 5, the nucleus of Fig. 22(e). The decision
as to which of these if any is most probable will undoubtedly be made when
intensity data are analysed by dynamical means. Even then several uncom-
fortable assumptions will have to be made about the extent of surface coverage,
the presence of undissociated, diffusely scattering oxygen, and so forth.

4. Beam Broadening

Figures 22(b), (c), and (d) clearly show the effect of domain size on beam
broadening. As discussed earlier in connection with the shape transform and

Fig. 8, as the domain size increases, the beams become sharper. With a given distribution Γ of domains on the surface, by kinematical theory the diffraction profile is $F\{\Gamma\phi\} = F'$, where F is the Fourier transform operator, ϕ is the scattering potential, and F' is the broadening effect.

Again, however, a dynamical treatment of the total LEED information appears necessary for a thorough description. The advantage of using optical analogues in this aspect of LEED has been the ease with which simple broadening can be distinguished from ring formation and discrete beam splitting.

F. The Low Energy Electron Microscope

Although high energy electron microscopy constitutes an extensive and highly developed area of research, the low energy equivalent has received only scant attention (Bauer, 1965; Turner and Bauer, 1966). The optical difficulties

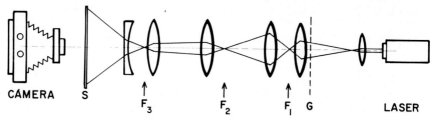

Fig. 24. Optical analogue of the electron microscope: G-grating; $F_{1,2,3}$-Fraunhofer planes; S—display screen.

with a beam of low energy electrons of course are considerably more severe than at high energies. For example: 1. the scattering amplitudes and angles are much greater; 2. this sharply increases the demands upon lens design regarding spherical aberration; 3. mutual electrostatic repulsion defocuses a beam of slow electrons; 4. stray magnetic fields have to be minimized to the extreme; and 5. the low energy method most certainly has to be a reflection technique. Some of these obstacles can be removed by high energy methods, i.e., by biasing the crystal at 100 V or so relative to a high-voltage cathode, and processing the beam with a conventional high energy optical train. But to date the full potential value of low energy electron microscopy has not been explored. Optical simulations can be of great assistance in this area because they readily demonstrate the type of information which may be obtained and thus can act as a guide in instrumental design as well as serve as an analytical method in the interpretation of results.

Figure 24 is a schematic of an optical analogue to the electron microscope (Park and Houston, 1968). Coherent light from a small cw He–Ne laser is decollimated by a condenser, passed through the grating, G, through a series of objective and projector lenses and finally onto the screen, S, where the image of the grating is displayed. Fraunhofer planes are indicated by $F_{1,2,3}$; F_2 is the position where a dark-field, DF, aperture is inserted to select a single diffraction beam for the image display (see also Section VII.A, Chapter 10).

Figure 25 shows the simulated microscopy of the same centred-bond $c(2 \times 2)$ grating used in Fig. 19(a); Fig. 25(a) is a bright-field, BF, photograph that is taken with no aperture in F_2, and Fig. 25(b) is the 00 dark-field image (Section VI.C.2, Chapter 10) taken by excluding all but the 00 beam. More striking is Fig. 25(c) which shows the dark-field image produced with only the 1/2 1/2 beam. The latter exemplifies the most promising area of low energy electron microscopy: dark-field microscopy with fractional-order beams to display the general features of nuclei on surfaces. At the present stage of development of low energy techniques, it is inconceivable that structural detail can be displayed on an atomic scale such as simulated in Fig. 25(a). All one could possibly hope for in a BF LEED photograph such as Fig. 25(a), even with good, as yet undeveloped lenses, is an indistinct difference in intensity indicating areas of $c(2 \times 2)$ coverage, somewhat similar to the 00 DF photo of Fig. 25(b). Other DF modes also display the nuclei as shown in Figs. 25(d), (e). The 1/2 1/2 DF image has the greatest contrast and may possibly be within the attainable capabilities of present-day instrumentation. The problem of spherical aberration is not as restrictive in the dark-field mode. Resolutions of 50 Å are easily achieved on commercial high-voltage electron microscopes, 20 Å are common, and 5 Å are reported so that diffuse areas 20–30 Å across as in Fig. 25(c) should be resolvable with suitable alterations to existing high-voltage systems (VIII.B, Chapter 10). Instrumental development with the general high-vacuum and electronic capabilities has been under way at Argonne National Laboratory and the University of Chicago for the past several years by Crewe and his associates (Crewe, 1966; Crewe et al., 1970).

Somewhat surprisingly the optical analogue shows quite clearly that it may be possible to distinguish between bridge bonds and centred bonds in $c(2 \times 2)$ nuclei. Figure 26 has the same arrangement of nuclei as Figs. 25 and 19(a), but as seen in the BF image in Fig. 26(a), the bonding is of the bridge type (cf. Fig. 3f). The 00 and 1/2 1/2 DF photographs in Figs. 26(b), (c) are the same as in Figs. 25(b), (c), but in the 11 DF images, the nuclei subtract intensity with bridge bonds, Fig. 26(d), whereas they add with centred bonds, Fig. 25(d) Furthermore, for the 10 DF image in 26(e), the nuclei with bonds parallel to this text appear dark and the areas with bonds perpendicular appear light. For the 01 DF image of Fig. 26(f) the reverse is true.

Thus, optical analogues have shown that a dark-field low energy electron microscope of even modest resolution, when perfected, may be invaluable in the study of the nucleation and growth of surface structures and may possibly be able to distinguish between bonding types.

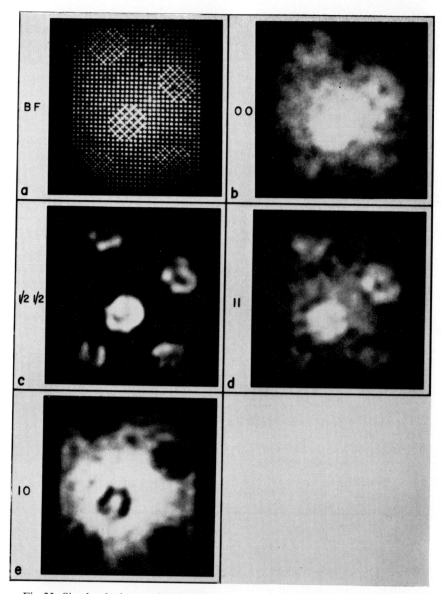

Fig. 25. Simulated micrographs of random $c(2 \times 2)$ nuclei, centred bond: (a) bright-field image. Remainder are dark-field photos taken by imaging only the beams indicated, e.g., (b) 00 beam; (c) 1/2 1/2 beam, and so on.

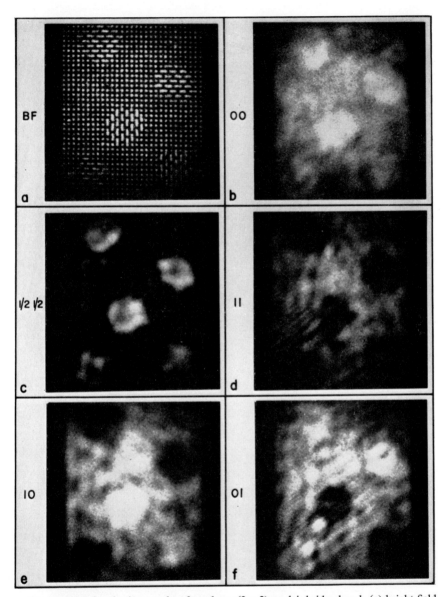

Fig. 26. Simulated micrographs of random $c(2 \times 2)$ nuclei, bridge bond: (a) bright-field image. Remainder are dark-field photos taken by imaging only the beams indicated, e.g., (b) 00 beam; (c) 1/2 1/2 beam, and so on.

REFERENCES

Bauer, E. (1965). *In* "Adsorption et Croissance Cristalline" (M. R. Kern, ed.), p. 19–48. Proc. of the Int'l. Conf., Editions du Centre National de la Recherche Scientifique, Paris.

Beeby, J. L. (1968). *J. Phys. C (Proc. phys. Soc.) Ser. 2*, **1**, 82.

Boudreaux, D. S. and Heine, V. (1967). *Surface Sci.*, **8**, 426.

Brillouin, L. (1946). "Wave Propagation in Periodic Structures". Ch. 6, McGraw-Hill Book Co., Inc., New York, U.S.A.

Burton, W. K., Cabrera, N. and Frank, F. C. (1951). *Phil. Trans. R. Soc. Lond.*, **243A**, 299.

Campbell, B. D., and Ellis, W. P. (1968). *Surface Sci.*, **10**, 124.

Cowley, J. M. (1967). *Prog. Mat'ls. Sci.*, **13**, 267.

Crewe, A. V. (1966). *In* "Sixth International Congress for Electron Microscopy, Kyoto" (R. Uyeda, ed.), Vol. I, p. 625–632, Maruzen Co. Ltd., Tokyo.

Crewe, A. V., Wall, J. and Langmore, L. (1970). *Science*, **168**, 1338.

Davisson, C. J. and Germer, L. H. (1927). *Phys. Rev.*, **30**, 705.

Duke, C. B. and Tucker, C. W., Jr. (1969). *Surface Sci.*, **15**, 231.

Ellis, W. P. and Campbell, B. D. (1968). *In* "Proceedings of the Symposium on Low Energy Diffraction", Am. Cryst. Ass'n., Tucson, Feb. 4–7. Polycrystal Book Service, Pittsburgh.

Ellis, W. P. and Campbell, B. D. (1970). *J. appl. Phys.* **41**, 1858.

Ellis, W. P. and Schwoebel, R. L. (1968). *Surface Sci.*, **11**, 82.

Estrup, P. J. and Anderson, J. (1966). *J. chem. Phys.*, **45**, 2254.

Feber, R. C. and Grimmer, D. P. (1969). "Computer Simulation of LEED Patterns from the (111) Planar Surface of UO_2. I. Initial Studies", Los Alamos Scientific Laboratory Report No. LA-4152, U.S. Dept. of Commerce, Springfield, Virginia.

Fedak, D. G., Fischer, T. E. and Robertson, W. D. (1968). *J. appl. Phys.*, **39**, 5658.

Feinstein, L. G. and Macrakis, M. S. (1969). *Surface Sci.*, **18**, 277.

Françon, M. (1966). "Diffraction Coherence in Optics", Pergamon, New York.

Germer, L. H. (1965). *Scient. Amer.*, **212**, 32.

Harris, L. A. (1968). *J. appl. Phys.*, **39**, 1419.

Haas, T. W. (1966). *Surface Sci.*, **5**, 345.

Henzler, M. (1970). *Surface Sci.*, **19**, 159.

Hosemann, R. and Bagchi, S. N. (1962). "Direct Analysis of Diffraction by Matter," North Holland Pub. Co., Amsterdam.

Lee, R. N. and Farnsworth, H. E. (1965). *Surface Sci.*, **3**, 461.

Lipson, H. and Taylor, C. A. (1958). "Fourier Transforms and X-ray Diffraction," G. Bell and Sons, Ltd., London.

MacRae, A. U. (1963). *Science*, **139**, 379.

McRae, E. G. (1968). *Surface Sci.*, **11**, 479.

Park, R. L. (1969). In "The Structure and Chemistry of Solid Surfaces" (G. A. Somorjai, ed.) p. 28-1, John Wiley and Sons, Inc., New York.

Park, R. L. and Houston, J. E. (1968). Presented at "1968 LEED Theory Seminar," Cornell Univ., July 1, unpublished.

Rhead, G. E. and Perdereau, J. (1969). "LEED Studies of Vicinal Surfaces," to be published as Proceedings of the Int'l. Conf., Editions du Centre Nat'l. de la Recherche Scientifique, Paris.

Scheibner, E. J. and Tharp, L. N. (1967). *Surface Sci.*, **8**, 427.

Schwoebel, R. L. (1967). *J. appl. Phys.*, **38**, 672.

Seah, M. P. (1969). *Surface Sci.*, **17**, 181.

Siegel, B. M. and Menadue, J. F. (1967). *Surface Sci.*, **8**, 206.

Simmons, G. W., Mitchell, D. F. and Lawless, K. R. (1967). *Surface Sci.*, **8**, 130.

Somorjai, G. A. (ed.) (1969). "The Structure and Chemistry of Solid Surfaces," Proceedings of the Fourth International Materials Symposium, Berkeley, John Wiley and Sons, Inc., New York.

Taylor, N. (1968), unpublished research.

Taylor, C. A. and Lipson, H. (1964). "Optical Transforms." G. Bell and Sons, Ltd., London.

Thomson, G. P. (1928). *Proc. R. Soc. Lond.*, **117**, 600.

Turner, G. and Bauer, E. (1966). In "Sixth International Congress for Electron Microscopy, Kyoto" (R. Uyeda, ed.), Vol. I, p. 163. Maruzen Co. Ltd., Tokyo.

Vainshtein, B. K. (1964). "Structure Analysis by Electron Diffraction." The MacMillan Co., New York.

Wood, E. A. (1964). *J. appl. Phys.*, **35**, 1306.

CHAPTER 8

Optical Data Processing

B. J. Thompson

The Institute of Optics, University of Rochester,
College of Engineering and Applied Science, Rochester, New York, U.S.A.

I. Introduction

Optical data processing, signal processing, information processing, pattern recognition, spatial frequency filtering are all fields that can mean different things to different people. Thus, the contents of this chapter have to be defined; the subject matter will be limited to be consistent with the title of the book and hence will be restricted to those areas that use optical transforms. Specifically, only coherent optical processing will be described—a process in which the Fourier transform of a given input is produced optically and operated upon to change the optical image of that input in a predetermined way.

Coherent optical data processing became a serious subject for study in the 1950's following the impact of the work of Duffieux (1946) on the Fourier integral and its application to optics and the subsequent use of communication theory in optical research (Elias *et al.*, 1952; Elias, 1953; Rhodes, 1953). The work was initiated in France (Maréchal and Croce, 1953) and also researched by Cheatham and Kohlenberg (1954) and O'Neill (1956). Since then a considerable literature has been developed on specific processes and applications. General review material can be found in two texts (O'Neill, 1963; Goodman, 1968) and a recent book has been devoted to the "general technology of the subject" (Shulman, 1970).

Starting in the early 60's a number of conferences have been held on optical processing and related topics and the proceedings provide an excellent record of progress. The first of these was in 1962 (O'Neill) and contains papers by Inglestam, Clark-Jones, Maréchal, Lohmann, Steel, Pancharatnam, Reza, Pole, Lamberts and Gamo. Other meetings followed (Pollack *et al.*, 1963; Tippett *et al.*, 1965) with two unedited proceedings in 1969 (NASA) and 1970 (AGARD).

New impetus has recently been given to the field of optical processing once it was realized that the required filters could be made holographically. The first statement of this idea was in a paper by Vander Lugt (1964) on signal detection by complex spatial filtering.

Today a variety of problems can be attempted by the technique of optical processing; these include removal of raster and half-tone, contrast enhancement, edge sharpening, enhancement of periodic (and isolated) signals in the presence of additive noise, aberration balancing, spectrum analysis, cross-correlation of recorded data, and matched and inverse filtering. After a brief historical introduction to the subject these applications will be described in some detail, organized in terms of the type of filter required.

A. Historical Background

The idea of using coherent optical systems in such a way as to allow for the manipulation of optical images is not new. The basic ideas are included in Abbe's theory of vision in a microscope first published in 1873 (Section I.B, Chapter 1). More specifically the illustrative experiments of Porter (1906) are certainly prophetic of much of present-day optical processing. To interpret Abbe's ideas somewhat loosely it is true to say that he realized that image formation in a microscope was more correctly described as a coherent process rather than the more familiar incoherent process. Abbe pointed out that the light illuminating the object of interest in a microscope would be diffracted (a coherent-light phenomenon). To form an image this diffracted light must be collected by the objective lens and the nature of the image and the resolution would be affected by how much of that diffracted light was collected. (By comparison we may recall that in incoherent image formation the nature of the image is not changed; only the resolution is affected by the change in numerical aperture of the imaging system). As an example we can consider an object consisting of a periodic variation in amplitude transmittance; the light diffracted by this object will exist in a series of discrete directions (or orders of diffraction). This series of orders contains a zero order propagating along the optical axis and a symmetric set of orders on either side of this zero order. Abbe correctly speculated upon what would happen as the microscope objective accepted different combinations of these orders. What information is carried by the zero order alone? What is the nature of the image as other orders are included? For instance if the zero order and one first order are collected then the information obtained from the image will be that the object consisted of a periodic distribution (compare Section VII.A, Chapter 10); a measurement will then give the correct periodicity. However, the spatial location of the periodic structure is not correctly ascertained. If now the other first-order diffracted light is included, then the correct spatial location of the periodic structure is also included. Additionally, however, a small spurious intensity peak will become visible between the main intensity peaks defining the periodic structure. The nature of the image is dependent upon the diffracted light collected. As more orders are included the image more closely resembles the object. We note that for a minimum acceptable image we must include the zero order and the two first-order diffracted beams and even these will show only the periodicity (Section VII.A, Chapter 10). We must comment again that by comparison the incoherent image under similar circumstances will not show the type of spurious effects discussed above. Finally we may examine the nature of the image if the first orders are collected but the zero order is omitted. The image appears to have twice the frequency of the original object! Abbe's theory of vision in a microscope was a description of image formation in terms of the addition of interfering beams, these beams being the diffracted light from the object.

Porter showed that these ideas were correct in an excellent series of experiments to determine the effects on the image of a periodic object when various combinations of orders of diffraction are removed. The Porter experiments are excellent examples of simple blocking filters and are often repeated for illustrative purposes.

These two pieces of work were really the forerunners of the process that is now called coherent optical data processing or spatial-frequency filtering. The name spatial-frequency filtering (or just spatial filtering) arises from the fundamental step of forming the Fourier transform; the transform is the two-dimensional spatial frequency spectrum of the object (Section III.A, Chapter 1). Hence low-frequency information is displayed near the optical axis and the higher frequencies further away from the axis. Thus the object itself can be thought of as a superposition of sine-waves of various frequencies and amplitudes.

B. Optical Schlieren Systems

Schlieren systems have been known and used in many forms for a variety of purposes such as the display of refractive index variations caused by gas flows, etc. This type of object is, of course, purely a phase object since its only effect is to produce a difference in optical path between various portions of the incident wave-front. Schlieren systems are related to the Foucault knife-edge test that was designed for the study of phase defects in lenses and mirrors. These systems clearly had an influence on the invention of the phase-contrast microscope by Zernike (Section VI.C, Chapter 10), which is also an example of spatial filtering. The phase-contrast microscope is applicable to the display of small phase differences in the object as an intensity distribution in image space. We shall discuss the knife-edge test and the phase-contrast microscope as special examples of spatial filtering.

C. Two-lens Coherent Image Formation

The basic system required for coherent optical processing (or optical spatial filtering) is that shown as a two-lens imaging system in Fig. 1(a). A collimated beam of quasi-monochromatic coherent light is used to transilluminate the object. Originally the illumination had to be derived from an arc source filtered for one of the prominent lines and then imaged onto a small aperture and the light allowed to propagate. Today of course the use of the laser is an attractive alternative (Section IV.C, Chapter 1). The first lens produces an exact Fourier transform of the object amplitude transmittance in its rear focal plane. We recall that this Fourier transform is just the complex amplitude distribution in the Fraunhofer diffraction pattern associated with the object. The second lens then transforms this distribution into an image. Again the image amplitude distribution is equivalent to the object amplitude with no phase errors and the image lies on a plane. Purely from geometrical-optical considerations this system is easily described as a simple imaging device that produces a magnification in the ratio of the focal lengths. If we depart from the exact location of the object then, of course, an image of the same magnification is still formed in a different plane and the image amplitude has associated with it a quadratic phase factor. The original systems of this type for image processing were not in this exact configuration, however the two-lens systems were used. Maréchal called the process a "double diffraction" experiment; today it is simply called an optical spatial-filtering system. The word spatial filtering arose from the operation of the optical device. The Fourier transform of the object is its spatial frequency spectrum. This spectrum may thus be operated upon by a filter and the nature of the resulting image changed in a controlled way.

Figures 1(b) and (c) show two other alternative systems that can be useful when phase considerations are no longer important. In (b) the transform plane is in the focal plane of the lens but the transform is multiplied by a quadratic

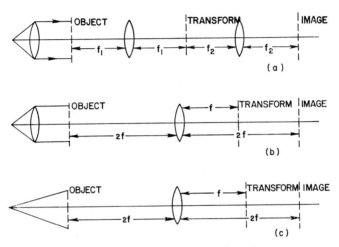

Fig. 1. Various optical data processing schemes.

phase factor; the same lens is also used to form an image of the object. In the third system one lens does the complete operation; the object is illuminated with a diverging beam and its transform, with additional phase factor, is formed in the image plane of the source. Hence these last two systems are useful when operations on the square modulus of the transform are required.

II. COHERENT OPTICAL PROCESSING

A. Basic Operations

To examine the process in a little more detail we shall use the system outlined in Fig. 1(a) and define the amplitude transmittance of the object as $A(x,y)$, the distribution in the transform plane as $U(\xi, \eta)$ and the final image amplitude as $\Psi(x',y')$. The coordinate system is indicated in Fig. 2. The coordinates $x'y'$ are in a reflected geometry to avoid the negative signs that occur because of the double-transformation process; the image is, of course, inverted.

The first lens produces an exact Fourier transform and hence

$$U(\xi, \eta) = C_1 \iint A(x,y) \exp\left[-ik(x\xi + y\eta)/f_1\right] dx\, dy, \qquad (8.1)$$

where $k = 2\pi/\lambda$ and λ is the mean wavelength of the illumination. We will use the following shorthand notation

$$U(\xi, \eta) = C_1 \tilde{A}(\xi/\lambda f_1, \eta/\lambda f_1); \qquad (8.2)$$

the $\sim$ indicating the Fourier transformation.

A general operation is then performed on $\tilde{A}$ to achieve some desired result. This filtering operation will be denoted by $F(\xi, \eta)$, where this filter may be

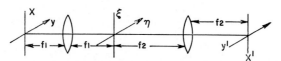

Fig. 2. Schematic diagram of two-lens coherent image forming systems.

a simple opaque blocking filter or may operate on either the amplitude and/or phase of the complex optical field in that plane. The complex amplitude distribution immediately after the filter $U'(\xi, \eta)$ is given by

$$U'(\xi, \eta) = U(\xi, \eta)\, F(\xi, \eta) = C_1 \tilde{A}(\xi/\lambda f_1, \eta/\lambda f_1)\, F(\xi, \eta). \qquad (8.3)$$

The second lens now forms an image by transformation of U'. Hence,

$$
\begin{aligned}
\Psi(x', y') &= C_1 C_2 \iint \tilde{A}(\xi/\lambda f_1, \eta/\lambda f_1)\, F(\xi, \eta) \exp\left[\frac{-ik}{f_2}(x'\xi + y'\eta)\right] d\xi\, d\eta \\
&= C_2 C_2\, A\!\left(\frac{x' f_2}{f_1}, \frac{y' f_2}{f_1}\right) \circledast F\!\left(\frac{x'}{\lambda f_2}, \frac{y'}{\lambda f_2}\right), \qquad (8.4)
\end{aligned}
$$

where the $\circledast$ means a convolution. Note that the magnification factor f_2/f_1 arises naturally. The resultant intensity distribution in the image is

$$\Psi(x', y')\, \Psi^*(x', y'),$$

the star here meaning a complex conjugate.

B. Filter Realization

The filters can be grouped into a number of types.

1. Blocking Filters

Here certain spatial frequencies are removed completely. These filters are simple to make and use but problems can occur because of their sharp cut-off which can produce "ringing" in the same way that a sharp cut-off filter in an electrical circuit produces ringing.

2. Amplitude Filters

These filters operate in a continuous way on the amplitude (but not the phase) of the transform. The required variation in amplitude transmittance is often produced by recording the appropriate distribution on film or by controlled evaporation of metal onto a glass substrate.

3. Phase Filters

Here only the phase is operated upon as in some phase-contrast microscopes. Phase distributions are difficult to manufacture in a controlled way especially if continuously varying phase is required.

4. Complex Filters

In this situation the amplitude and phase of the optical field is operated upon. Clearly this is the most general type of filter and is the most difficult to construct. Originally the amplitude and phase portions were made separately with the amplitude being made as indicated above. The phase filter can be manufactured by using an evaporated layer of material such as magnesium fluoride. Today the most efficient technique for fabrication of the filter is by an interferometric method in which the required complex amplitude is recorded as a hologram. Very often the required filter can be generated by the system without specific knowledge of the filter specifications.

C. Simple Blocking Filters

The experiments conducted by Porter are excellent examples of the use of blocking filters. Portions of the transform are completely eliminated and the effect on the image determined. This is particularly attractive when periodic objects are considered which have a transform consisting of a discrete set of diffraction orders; the effects of removing diffraction orders is readily demonstrated. As an example we will consider an object consisting of a set of six long slits approximately 1/2 mm wide and 20 mm long separated by a centre-to-centre space of 2 mm. The transform consists of an infinite series of diffraction orders under an envelope determined by the diffraction pattern of the single slit. Figure 3 shows the resultant images as various orders of diffraction are allowed to pass through the filter and proceed to the image plane. Figure 3(a) is the image formed when all the orders are allowed to pass. Figure 3(b) is the image produced when only the two first orders are passed and all other orders blocked; the image contains eleven slits with half the spacing of the original slits since the fundamental periodicity of the transform is blocked.

An image with three times the periodicity can be produced (Fig. 3c) by passing the first order on one side of the optical axis and the second order on the other side of the optical axis. Finally the centre portion of the transform consisting of the zero order and the two first-order peaks can be removed and the remaining orders allowed to continue to form the image (Fig. 3d); the fundamental spacing is retained from adjacent higher orders.

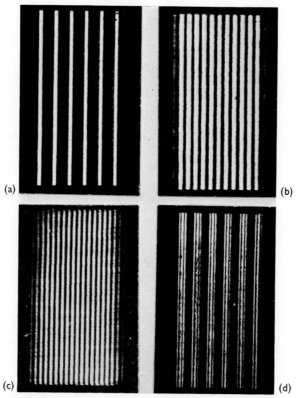

(a) (b) (c) (d)

Fig. 3. Illustration of the effect of a blocking filter on the image of a periodic object (a) all orders passed; (b) first orders only; (c) two orders passed—first order on one side of optical axis and the second order on the other side of the optical axis; (d) all orders except the zero and the first orders.

A similar set of slits was cut in a transparent plastic sheet to produce a phase object (Section VI.C.1, Chapter 10). The resulting diffraction pattern seen in the transform plane is shown in Fig. 4(a). Two separate patterns are seen. The sheet had a slight wedge angle to it and so the light passing through the transparent material is deviated and the transform centred at one side of the normal optical axis. However, the light passing through the slits in the material is not deviated and hence forms a light distribution centred on the normal optical axis. Because of this effect the two portions of the transform can be separated out quite easily. If the light associated with the transmission through the plastic is removed then a high-contrast image of the slits is seen,

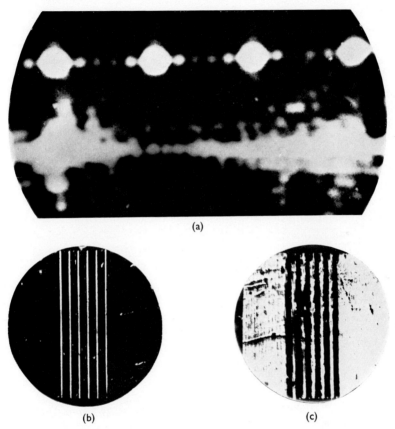

Fig. 4. Transform of six slits cut in a transparent plastic sheet that had a slight wedge shape; (b) image of slits only; (c) image formed with light passing through the plastic and on-axis contribution removed.

(Fig. 4b). Conversely the image formed by the light transmitted through the plastic is shown in (Fig. 5c)—a good image of a phase object with the slits appearing as dark bars.

1. *Raster Removal*

A popular and sometimes useful application of a blocking filter is to remove a raster from a television type picture. We can consider the input as a periodic function whose envelope is the continuous scene; or equivalently the periodic function samples the scene. Hence the transform consists of a periodic distribution in one direction with a periodicity reciprocally related to the raster periodicity. Centred at each of these locations is the transform of the envelope (i.e. the transform of the scene). If the scene is correctly sampled then the individual transforms of the scene do not overlap each other. Hence, if the filter selects one single order and the area surrounding it, the information about the periodicity will be lost and the scene information retained, thus yielding a continuous-tone image without a raster. Figure 5 shows a typical result. The top left hand corner of each photograph shows, in enlargement, a portion of the picture to make the effect more visible. The raster is quite clearly removed in the processed photograph. Other examples are given by Shulman (1970).

2. Half-tone Removal

The two-dimensional equivalent of the example discussed above is the removal of the half-tone to produce a continuous image again. This time the two-dimensional transform consists of a two-dimensional array of orders with the scene transform centred at each order. Selection of one order alone will produce an image without the half-tone—but with considerable loss of energy. A method to overcome this is to add sufficient additional path difference to a selection of orders so that the coherence length of the radiation is exceeded between the light reaching the image plane from these orders. Now the various orders will each produce the same image but no interference will take place and the images simply add in intensity. To achieve this result the illumination of the object has to be of short coherence length (say less than a millimetre) —this is readily produced from a high-pressure arc lamp. It is then only necessary to use small pieces of microscope cover-slip glass to introduce the necessary path differences between the orders.

There are, of course, other methods that remove the raster or half-tone. The photograph can be "squinted" at. This is equivalent to stopping down the lens until the impulse response is large enough that the half-tone is not resolved. Another method is to defocus the imaging optics slightly to "blur" out the half-tone. In both of these examples scene information is lost as well as the raster or half-tone. Excellent examples of half-tone removal can be found in Cagnet et al. (1962).

The question naturally occurs, "why remove the half-tone at all since no extra information is added?" Cosmetically, of course, the image is more pleasing without the half-tone. However, there is a more important reason. When looking at the image under high magnification so that the field of view contains only a few half-tone dots (say 10×10) it becomes very difficult to

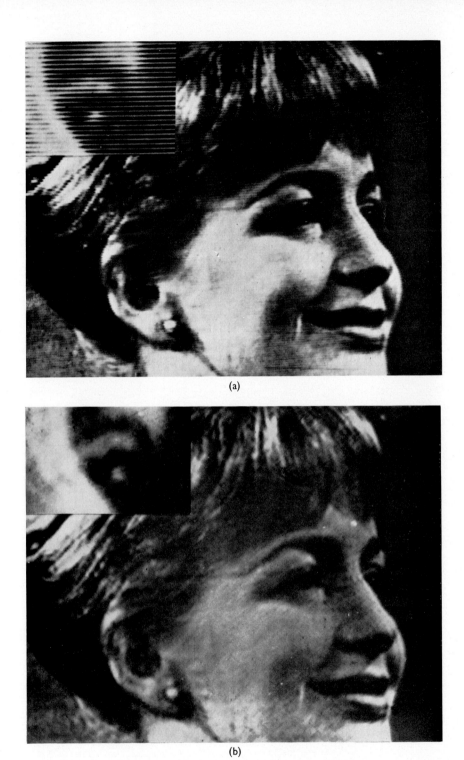

(a)

(b)

Fig. 5. Optical processing for the removal of a raster.

"see the image" and hence interpret what is there. We can't see the image for the half-tone!

3. *Series-termination Effects—An Illustration*

An optical spatial-filtering system provides an excellent means for demonstrating series-termination effects in Fourier synthesis. The [010] projection of the molecule of bishydroxydurylmethane, as determined by Chaudhuri and Hargreaves (1956) was chosen as a suitable example (Thompson, 1959; Lipson and Thompson, 1959; Taylor and Lipson, 1964). An object representing 28 molecules of bishydroxydurylmethane was made by punching holes in an opaque sheet. The intensity distribution associated with the transform (top right-hand corner) is shown in Fig. 6 together with the object (top left-hand corner). The transform shown is the portion that is contained inside the envelope of the central maximum of the diffraction pattern of the individual hole that represents the atom. The calculated electron-density map is shown in the lower half of the figure together with the appropriate portion of the weighted reciprocal lattice. One distinct difference exists that must be stressed: optically the atoms are represented as circular holes having sharp edges, whereas the information in the weighted reciprocal lattices is derived from real atoms. In Fig. 6(b–f) the optical diffraction pattern is limited by various circular apertures centred on the optical axis so that an increasing portion of the transform is removed. For each optical example the equivalent calculated electron-density map is also given together with the portion of the weighted reciprocal lattice used in the calculation. A spurious effect is noted in Fig. 6(b) at the centre of the molecule. The remaining illustrations are self-explanatory and need no further comment.

4. *Band-pass Filters*

The example given above can also be thought of as an example of a low-pass filtering, i.e., the higher frequencies are removed completely and the lower frequencies continue untouched. High-pass filtering can also be achieved by the reverse of the above process. The lower frequencies are removed by a stop and the higher frequencies remain unchanged. In general any band of frequencies can be allowed to continue to the final image plane by the use of the appropriate annular aperture as a filter.

5. *Removal of Additive Noise*

If we have a record of an object together with an additive noise contribution then the transform consists of an addition of the transform of the object and the transform of the noise. It is sometimes possible partially to separate out the object from the noise because the object transform may exist in known

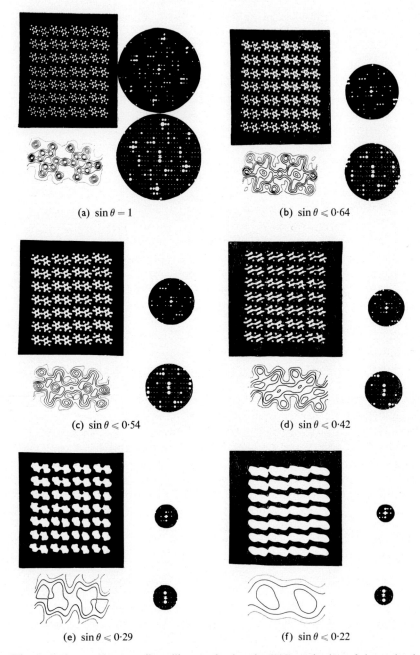

(a) $\sin \theta = 1$

(b) $\sin \theta \leqslant 0.64$

(c) $\sin \theta \leqslant 0.54$

(d) $\sin \theta \leqslant 0.42$

(e) $\sin \theta \leqslant 0.29$

(f) $\sin \theta \leqslant 0.22$

Fig. 6. Series termination effects illustrated using the (010) projection of the molecule of bishydroxydurylmethane.

locations in the transform plane (O'Neill, 1956). For example, if the object consisted of a periodic distribution such as a grating then its transform is located in a discrete set of orders. On the other hand, the noise spectrum is some broad distribution determined by the nature of the noise (i.e. if the noise is a Gaussian function then its transform is also Gaussian). If now a filter is used that lets through the discrete orders from the periodic object and blocks the rest of the transform, then the periodic object will be enhanced with respect to the noise. It is important to state this operation quite precisely: we enhance a *known* signal in the presence of an *additive* noise (compare Section I.E, Chapter 11 and Section III, Chapter 5).

Similar results can be obtained with an isolated object in the presence of additive noise (O'Neill, 1956). It is necessary to know what the signal is, but not its location in the object space. The spatial-frequency spectrum of the object is centred on the optical axis independent of its xy location in object space. However, the orientation of the object is important, but the filter could be rotated to select all possible orientations of the object.

Kozma and Kelly (1965) have shown how signals may be selected from additive noise by using a complex filter. The filter was made as the record of the real part of the resultant of putting the required function on a spatial carrier; the amplitude is recorded as density variations and the phase as spacing variations in the periodic structure of the carrier. Several workers have attempted to use these techniques for the removal of so-called "grain noise" in photographic images (see, e.g. Thiry, 1964). This is quite a different problem. Firstly the noise is not noise in the usual sense since the grains are the image; it is also clearly signal-dependent. A number of possible models exist; for example the grain can be thought of as a multiplicative function in object space. Hence the resulting transform is the transform of the object convoluted with the transform of the grain (Section II.F.3, Chapter 10). A convolution is much harder to undo than is mere addition. The presence of the multiplicative grain or signal dependent grain is the cause of great difficulties in optical processing of photographically recorded images and considerably more effort is required in this area before any more complicated optical processing of photographically recorded images can be considered.

6. *Signal Separation by Orientation*

Very often in optical processing the object that is required to be enhanced has to be known. Sometimes, however, it is equally valuable to know something about the noise or unwanted portion of the object. An example of this, which also illustrates the importance of having a two-dimensional spectrum, is that of selecting a signal from a set of seismic type traces. Figure 7(a) shows a set of 25-pen recorder traces, one of which contains a signal of importance. It will be noted that the quiet traces are essentially information in the horizontal

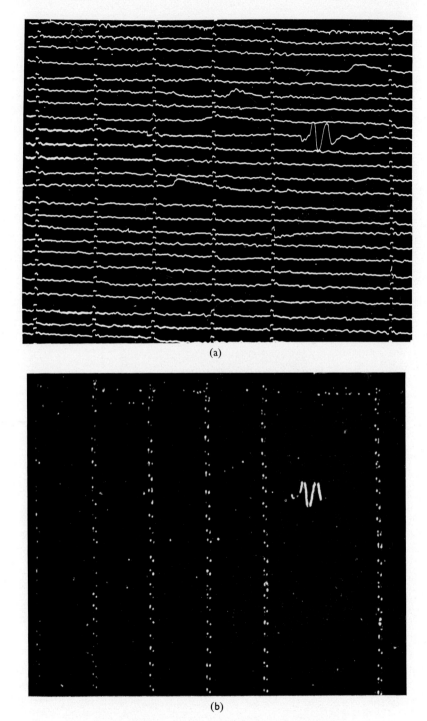

(a)

(b)

Fig. 7. Illustration of signal separation by orientation.

direction whereas when a signal is present the pen traces contains a strong vertical component. Hence a simple filtering operation that removes all the transform in one direction (that corresponding to the quiet trace) and retains the remainder of the transform is required. The filter is then simply an opaque strip. The processed image is now shown in Fig. 7(b). Clearly the signal remains

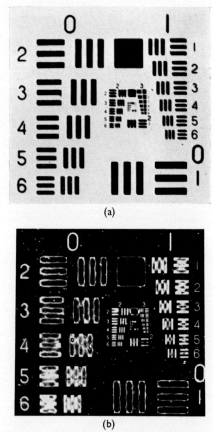

Fig. 8. Differentiation and contrast reversal by use of a blocking filter.

—also the timing marks. This was accomplished without knowledge of the signal or its position in the field of view. Excellent work has been carried out on geophysical data (Jackson, 1965) making use of separation by orientation.

D. Amplitude Filters

1. Contrast-enhancement

In the transform of a low-contrast scene the zero-order diffraction term (or d.c. term) is large and the higher frequencies contain relatively little energy.

If a small-amplitude filter is placed to cover the zero-order to allow less of this light to pass through to the image plane, the other frequencies are not changed. Hence we have changed the relative energy distribution in the spatial frequency spectrum. The image now has improved contrast, the improvement depending upon the amplitude transmittance of the filter.

2. *Differentiation—Contrast Reversal*

The type of amplitude filter described above could be used in a more general sense to change the balance of the high- and low-frequency contributions to the image. Since edges in an object represent high-frequency information it is clear that the edges of the object may be enhanced. Figure 8(a) shows a high-contrast bar target that comprised the original object. A filter that has an amplitude transmittance that has a maximum in the centre and then tapers linearly to zero at an intermediate frequency value is used to operate on the transform of the bar target object: the effect on the larger bars is to differentiate, i.e. only the edges are visible—but as a bright outline on an otherwise dark background. At higher frequencies a contrast reversal has taken place. The bars now appear bright on a dark background. Notice, however, that this reversal is a one-way process. It is not possible to start with Fig. 8(b) and obtain Fig. 8(a) by the same technique.

In a later section we shall discuss aberration balancing and illustrate further applications of amplitude filters both alone and in conjunction with a phase filter. Finally we may note that holographic filters are amplitude-only filters but do in fact operate on both the amplitude and phase of the transform.

III. IMAGING OF COMPLEX OBJECTS

Consider an object where, $A(x,y)$, the complex amplitude transmittance is given by

$$A(x, y) = b(x, y) e^{i\phi(x, y)},$$

where $b(x,y)$ is the amplitude and $\phi(x,y)$ is the phase. In the transform plane we write the complex amplitude of the Fourier transform of $A(x,y)$ in two parts, where the area immediately surrounding the zero frequency point is written separately from the remainder. Hence

$$U(\xi, \eta) = U_0(\xi, \eta) + U_1(\xi, \eta), \tag{8.5}$$

where

$$U_0(\xi, \eta) = C_1 \iint \exp\left\{\frac{-ik}{f}(\xi x + y\eta)\right\} dx\, dy,$$

$$U_1(\xi, \eta) = C_1 \iint [A(x, y) - 1] \exp\left\{\frac{-ik}{f}(\xi x + y\eta)\right\} dx\, dy. \tag{8.6}$$

The contribution U_0 can also be thought of as the light distribution in the transform plane when no object is present, and U_1 is the contribution from diffraction; f is the focal length of the transform lens.

The filter function that operates on this transform is a complex transmission function given by

$$F(\xi, \eta) = a\,e^{i\alpha}, \tag{8.7}$$

where a and α are both constants. Furthermore this filter only exists over the region of $U_0(\xi, \eta)$.

The complex amplitude distribution after the filter is thus

$$U'(\xi, \eta) = a\,e^{i\alpha}\,U_0(\xi, \eta) + U_1(\xi, \eta). \tag{8.8}$$

The resulting image complex amplitude distribution is then given by

$$\Psi(x', y') = \Psi_0(x', y') + \Psi_1(x', y'), \tag{8.9}$$

where

$$\Psi_0(x', y') = a\,e^{i\alpha}\,C_2 \iint U_0(\xi, \eta)\exp\left\{\frac{-ik}{f}(x'\,\xi + y'\,\eta)\right\}d\xi\,d\eta$$

$$\Psi_1(x', y') = C_2 \iint U_1(\xi, \eta)\exp\left\{\frac{-ik}{f}(x'\,\xi + y'\,\eta)\right\}d\xi\,d\eta. \tag{8.10}$$

For simplicity the focal length of the second lens is also written as f. Both the above integrals may be set with infinite limits. Hence,

$$\Psi_0(x', y') = Ca\,e^{i\alpha}$$
$$\Psi_1(x', y') = C[A(x', y') - 1]. \tag{8.11}$$

The resultant image intensity is then

$$I(x', y') = \Psi(x', y')\,\Psi^*(x', y')$$
$$= C^2 |a\,e^{i\alpha} + A(x', y') - 1|^2. \tag{8.12}$$

Substituting for $A(x', y')$ we have

$$I(x', y') = |C|^2 \{a^2 + b^2(x', y') + 1 - 2a\cos\alpha - 2b(x', y')\cos\phi(x', y')$$
$$+ 2ab(x', y')\cos(\alpha - \phi(x', y'))\}. \tag{8.13}$$

We shall now look at this resultant intensity under a variety of circumstances.

A. Phase Objects

A phase object (Section VI.C.1, Chapter 10) is one that contains no absorption and hence $b(x, y) = 1$; the object is then described by a complex transmission $e^{i\phi(x, y)}$. Thus Eq. (8.13) becomes

$$I(x', y') = |C|^2 \{a^2 + 2[1 - a\cos\alpha - \cos\phi(x', y') + a\cos(\alpha - \phi(x', y'))]\}. \tag{8.14}$$

The real problem that we are interested in here is how to make $\phi(x,y)$ visible as an interpretable intensity variation. Phase-contrast techniques are limited to situations where $\phi(x,y)$ is small. We shall deal with this approximation first and evaluate the various phase-contrast techniques.

1. Small-phase Approximation

The approximation for small phase differences is defined by taking the expansion of the exponential and keeping first order terms only,

$$e^{i\phi(x,\,y)} \simeq 1 + i\phi(x,y). \tag{8.15}$$

This is equivalent to the approximation that

$$\begin{aligned} \cos\phi(x,y) &\simeq 1, \\ \sin\phi(x,y) &\simeq \phi(x,y). \end{aligned} \tag{8.16}$$

Under this approximation Eq. (8.14) further reduces to

$$I(x',y') = |C|^2 \{a[a + 2\phi(x',y')\sin\alpha]\}. \tag{8.17}$$

B. Dark Ground

In the dark ground-method (Section VI.C.2, Chapter 10) a simple blocking filter is used to eliminate U_0 entirely. Hence in our notation $a = 0$. Under this condition Eq. (8.13) becomes

$$I(x',y') = |C|^2 \{b^2(x',y') + 1 - 2b(x',y')\cos\phi(x',y')\}. \tag{8.18}$$

Equation (8.18) describes the intensity distribution in the image of a complex object under dark-ground conditions. For a phase-only object, the resultant intensity is found from Eq. (8.14) or (8.18) to be

$$I(x',y') = 2|C|^2 [1 - \cos\phi(x',y')]; \tag{8.19}$$

the intensity maps the cosine of the phase of the object.

In the small-phase-difference object, $I(x',y')$ approaches a constant as seen from Eq. (8.17). However, a more careful reappraisal from Eq. (8.12) with the condition of Eq. (8.15) shows that the intensity varies as $\phi^2(x',y')$.

C. Phase-contrast Methods

1. Positive Phase-contrast (Bright Contrast)

The filter here is set so that $a = 1$ and $\alpha = \pi/2$.

a. *Small-phase approximation*. Under the small phase approximation we have from Eq. (8.17)

$$I(x',y') = |C|^2 [1 + 2\phi(x',y')]. \tag{8.20}$$

This shows Zernike's important result that the intensity distribution in the image is linearly related to the object phase.

   b. *General phase object.* From Eq. (8.13)

$$I(x', y') = |C|^2 [3 + 2(\sin \phi(x', y') - \cos \phi(x', y'))]; \qquad (8.21)$$

this is a more complicated and less useful relationship.

   c. *Complex object.* Now from Eqn. (8.13)

$$I(x', y') = C^2 [2 + b^2 - 2b(\cos \phi(x', y') - \sin \phi(x', y'))]; \qquad (8.22)$$

an even more difficult distribution to interpret. Clearly b has to be measured separately if any attempt to determine ϕ is to be made.

2. Negative Phase-contrast (Dark Contrast)

The filter here is constructed such that $a = 1$, $\alpha = 3\pi/2$.

   a. *Small-phase approximation.*

$$I(x', y') = |C|^2 [1 - 2\phi(x', y')]. \qquad (8.23)$$

   b. *General phase object.*

$$I(x', y') = |C|^2 [3 - 2(\sin \phi(x', y') + \cos \phi(x', y'))]. \qquad (8.24)$$

   c. *Complex object.*

$$I(x', y') = C^2 [2 + b^2 - 2b(\sin \phi(x', y') + \cos \phi(x', y'))]. \qquad (8.25)$$

3. Phase-contrast with Contrast Enhancement

The results obtained above for the small-phase approximation in both positive and negative phase-contrast are appropriate techniques for obtaining an image intensity linearly related to the object phase. However, the contrast is quite low since all the d.c. energy is allowed to pass through to the image space. Clearly contrast enhancement can be achieved if $a < 1$. Hence it is usual to make the filter partly absorbing. This gives the general result for positive and negative phase-contrast as follows.

Positive phase-contrast with absorption

$$I(x', y') = |C|^2 a[a + 2\phi(x', y')]. \qquad (8.26)$$

Negative phase-contrast with absorption

$$I(x', y') = |C|^2 a[a - 2\phi(x', y')]. \qquad (8.27)$$

4. An Example

For an example let us consider that the object consists of a one-dimensional phase variation given by

$$\phi(x) = p \cos qx$$

and that it exists only inside a given area, and outside that area $\phi(x) = 0$. The image intensity distribution, with positive phase-contrast and $a = 1$, has the form shown in Fig. 9(b). With $a < 1$ the image intensity can be optimized to approach the curve of Fig. 9(c). The equivalent curves for negative phase-contrast are shown in Figs. 9(d) and (e).

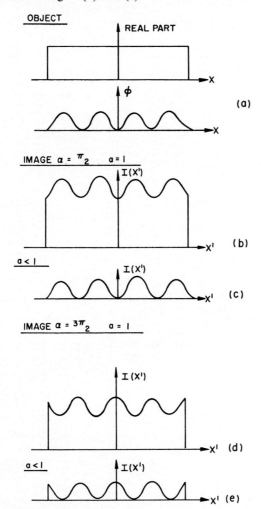

Fig. 9. The intensity distribution in the image of a phase object under various phase contrast arrangements.

D. Schlieren Systems—Knife-edge Test

The particular Schlieren system (Section VI.C.4, Chapter 10) that will be discussed here is the classical Foucault knife-edge test (see, e.g. Linfoot, 1955). The filter completely removes one half plane of the transform and the zero order. For simplicity we shall treat the problem one-dimensionally. For small phase differences in the object we have a distribution in the transform plane

$$U(\xi) = \int [1 + i\phi(x)] \exp\{-ikx\xi/f\} \, dx$$
$$= \delta(\xi/\lambda f) + i\phi(\xi/\lambda f), \qquad (8.28)$$

where $\phi(\xi)$ is the Fourier transform of $\phi(x)$. We now multiply this function by a step function representing the knife edge to give

$$U'(\xi) = S(\xi) [\delta(\xi) + i\phi(\xi)],$$

where

$$S(\xi) = \begin{cases} 1, & \xi \geqslant \xi_0 > 0 \\ 0, & \xi < \xi_0 \end{cases} . \qquad (8.29)$$

We note that $[S(\xi)][\delta(\xi)]$ is zero everywhere, as long as the zero order is being removed by the knife edge. Thus

$$U'(\xi) = i S(\xi) \phi(\xi). \qquad (8.30)$$

The complex amplitude distribution in the image plane is again given by transformation of $U'(\xi)$. Hence

$$\Psi(x') = i \int S(\xi) \phi(\xi) \exp\left\{\frac{-ik}{f} \xi x'\right\} d\xi. \qquad (8.31)$$

However

$$S(x') = \frac{1}{2}\left[\delta(x') + \frac{i}{\pi x'}\right].$$

Therefore

$$\Psi(x') = \frac{i}{2} [\phi(x') + iH\{\phi(x')\}],$$

where

$$H\{\phi(x')\} = \frac{1}{\pi} f \frac{\phi(x'') \, dx''}{x'' - x'}, \qquad (8.32)$$

and is the Hilbert transform of $\phi(x')$; f indicates the Cauchy principal value at $x'' = x'$. We may further note that

$$\phi(x'') = -\frac{1}{\pi} f \frac{\phi(x) \, dx}{x'' - x} \qquad (8.33)$$

and

$$\frac{d\phi(x')}{dx'} = \frac{1}{\pi} H\{\phi(x')\}. \qquad (8.34)$$

Then

$$\Psi(x') = \frac{i}{2}\left[\phi(x')_e^t + \frac{i}{\pi}\left(\frac{d\phi(x')}{dx'}\right)\right]. \tag{8.35}$$

If the phase object contains sharp discontinuities, then at the location of the edge

$$\frac{d\phi(x')}{dx'} \gg \phi(x')$$

and

$$\Psi(x') = \left(-\frac{1}{2\pi}\right)\frac{d\phi(x')}{dx'}. \tag{8.36}$$

IV. ABERRATION BALANCING

A. Basic Concepts

One of the prime motivations for the work in optical processing carried out by Maréchal and Croce (1953) was to correct aberrated images. Tsujiuchi illustrated the use of this technique in a series of papers (1960a, b, 1961) which are perhaps best read in his review paper (Tsujiuchi, 1963) which discusses a variety of aberrations. Clearly considerable technological advantage can be gained if photographs taken with an aberrated optical system can be corrected by subsequent coherent processing. Within some definable limits this can be accomplished. The major problem area—still as yet unsolved—is how to handle the photographic grain. Furthermore it must be stressed that the impulse responses, or the transfer function, of the aberrated system must be known.

Let the intensity impulse response of the incoherent photographic imaging process be $S(x, y)$. Then the image of a given object intensity distribution $I_0(r, s)$ is given by

$$I_{im}(x, y) = \int I_0(r, s) S(x - r, y - s) \, dr \, ds. \tag{8.37}$$

Equation (8.37) represents the usual convolution process. This intensity distribution is recorded on film on the linear portion of the H and D curve with a $\gamma = 2$.

This photographic record is now the input to the coherent optical processing system. The amplitude transmitted through the negative is then

$$A(x, y) = [I_{im}(x, y)]^{\gamma/2} \tag{8.38}$$

The complex amplitude distribution in the transform plane is then

$$U(\xi, \eta) = C \iint I_{im}(x, y) \exp\left\{\frac{-ik}{f}(x\xi + y\eta)\right\} d\xi \, d\eta, \tag{8.39}$$

and

$$U(\xi, \eta) = C[\tilde{I}_0(\xi/\lambda f, \eta/\lambda f)\, \tilde{S}(\xi/\lambda f, \eta/\lambda f)], \qquad (8.40)$$

which is the product of the object spectrum $\tilde{I}_0$ and the Fourier transform of the intensity pulse response (i.e. the transfer function τ). Hence we may write

$$U(\xi, \eta) = C[\tilde{I}_0(\xi/\lambda f, \eta/\lambda f)\, \tau(\xi/\lambda f, \eta/\lambda f)]. \qquad (8.41)$$

Let the filter function $F(\xi, \eta)$ be the reciprocal of τ, i.e.,

$$F(\xi, \eta) = \frac{1}{\tau(\xi/\lambda f, \eta/\lambda f)}. \qquad (8.42)$$

Thus immediately after the filter the complex amplitude is simply

$$U'(\xi, \eta) = C\tilde{I}_0(\xi/\lambda f, \eta/\lambda f). \qquad (8.43)$$

Upon retransformation to the image plane

$$\Psi(x', y') = C'\, I_0(x', y'). \qquad (8.44)$$

The convolution has been undone and the effect of the aberrated lens removed. The resultant intensity distribution in the image plane is finally

$$I(x', y') = |C'|^2\, I_0^2(x', y'). \qquad (8.45)$$

If this is recorded on film with $\gamma = \frac{1}{2}$ then the intensity transmittance of this final record is the original object intensity.

Another possible experimental technique exists that achieves the same result and avoids the necessity for controlling the gamma of the film. If the original photographic film is pre- or post-fogged and then developed, the amplitude transmittance contains a number of terms which, if the fogging is large enough, can be written as a binomial expansion and only the linear terms in $I_0(x, y)$ retained together with the constant. Upon transformation the term in $\tilde{I}_0(\xi/\lambda f, \eta/\lambda f)$ proceeds as before and the constant transforms to a delta function which proceeds through the filter unaltered. The problem with this approach is the additional noise introduced into the process by the fogging exposure.

We must be quite careful in the interpretation of this result. We have to determine the form of $\tau(\xi, \eta)$. Clearly if τ contains any zeros the value of $F(\xi, \eta)$ is required to be infinity—an impossible situation; furthermore small values of τ create equally serious problems. Hence for those spatial frequencies that were recorded, some processing can be carried out to get a flatter transfer function; both the contrast and phase of the recorded spatial frequencies can be changed. This process will become clearer if we discuss some typical examples.

B. Examples

1. Astigmatic Image

The aberration of astigmatism produces a transfer function in one direction that is Gaussian and constant in the perpendicular direction. We will assume here that the diffraction limited system would have an impulse response that

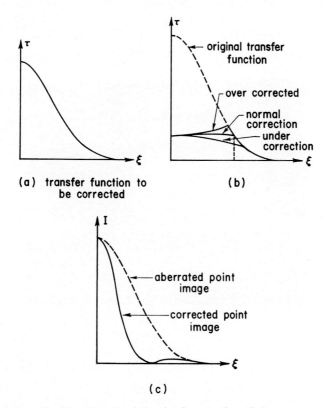

(a) transfer function to be corrected

(b)

original transfer function

over corrected

normal correction

under correction

(c)

aberrated point image

corrected point image

Fig. 10. Aberration balancing for an astigmatic image.

is essentially a delta function by comparison. The aberrated transfer function is shown in Fig. 10(a). The correction filter is essentially the inverse of this function except that some acceptable level of contrast has to be decided upon. Hence the corrected curve is flat out to some spatial frequency; then the actual transfer function curve is followed. Figure 10(b) shows the corrected transfer function curve together with curves for over- and under-correction (see Tsujiuchi, 1963).

The effect on the image of a point is shown in Fig. 10(c); the point intensity distribution in the corrected image is very much narrower.

2. *Defocused Image*

An approximate form for the defocused impulse response is a blur circle of uniform intensity. (In actual fact the impulse response is much more complicated than this.) The transfer function is then a first-order Bessel function divided by its argument. The effect of the defocusing is very nicely seen in the image of a converging bar target shown in Fig. 11. It will be noted that the contrast of all frequencies is considerably lower than the in-focus image. Certain frequencies (approximately represented by a particular vertical

Fig. 11. Out-of-focus image of a converging bar target.

plane through the converging bars) have a contrast that is zero. After the first zero position a region in which the contrast is reversed is visible, i.e. the black bars show up as white bars and vice versa. Figure 12(a) shows the transfer function that has to be corrected; this is, of course, a two-dimensional symmetric function. The negative value of τ is interpreted as a contrast reversal. Again an appropriate contrast level is picked and the first maximum region of the transfer function flattened out; the resulting transfer function would be that of Fig. 12(b). Now a phase filter is applied (c) to give a final transfer function (d). There are certain frequencies that have zero contrast; these are, of course, lost for ever. The corresponding point images are shown in Figs. 12(e)–(g).

The techniques can be carried out, but the difficulty of manufacturing the filter makes it a somewhat impractical technique in general. However, if a large amount of data has to be processed then the effort to make the filter is rewarded. Until the advent recently of the use of holographic filters, complex filters were manufactured in two portions—the amplitude portion by the

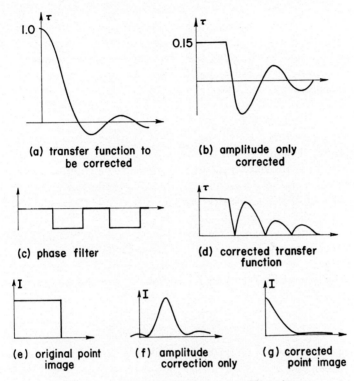

(a) transfer function to be corrected

(b) amplitude only corrected

(c) phase filter

(d) corrected transfer function

(e) original point image

(f) amplitude correction only

(g) corrected point image

Fig. 12. Aberration balancing for an out-of-focus image.

appropriate exposure on film and the phase portion by controlled evaporation of transparent material, e.g. magnesium fluoride onto a glass substrate.

3. *Motion-blurred Image*

The problem of linear object motion during photography can be treated in a similar way to the aberration just described. It is necessary to know the impulse response of the system. The image of a point object will be essentially a line in the direction of motion. This can be treated approximately as a rectangular function and hence the transfer function to be corrected is a sinc† function. The filter required to correct the motion blur in the original photograph is a complex filter comprising separate amplitude and phase portions,

† The term sinc means $(\sin x/x)$ (see, e.g., Jennison, 1961).

The procedure is almost identical to that for the defocused image except that the sinc function replaces the Bessel function and the problem is one- dimensional not two-dimensional. (For further discussions on motion blur see the NASA Seminar Proceedings).

In concluding this section it is of importance to note that the impulse response of an incoherent system is a real positive function since it is an intensity distribution. If it is also symmetric then its Fourier transform —the transfer function—is also real (but not necessarily positive). For an asymmetric impulse response, the transfer function is complex. Hence in general it is necessary to produce a complex filter with an arbitrary phase.

V. Holographic Filters

Holography (Chapter 9) has had an important impact on coherent optical processing; it has provided a method for preparing a complex filter directly, often without requiring detailed knowledge of the filter function itself. Hence many of the applications described in Section IV can be accomplished much more simply than by conventional filter fabrication. The basic process, however, remains the same. The idea of preparing holographic filters for optical processing was first seriously discussed by Vander Lugt (1964) in a paper on signal detection by complex spatial filtering. This idea has since been extended by a number of workers to aberration balancing and image-motion compensation.

A. Detection Filtering

Vander Lugt's idea for complex spatial filtering was to produce a filter such that the output consisted of a delta function in the image plane at the location of the object for which the filter was designed. The principle can be illustrated in the following way. (Figure 13 shows the basic system for producing the required filter.)

The desired form of an impulse response $S(x,y)$ is placed in the xy plane. The distribution in the transform plane is then $\tilde{S}(\xi/\lambda f, \eta/\lambda f)$ and a plane reference wave of the form $a_0 \exp(-2\pi i \alpha \xi)$ is added at an angle θ and $\alpha = \sin\theta/\lambda$. The resulting interference pattern is recorded as an intensity distribution in the ξ, η plane. Hence,

$$I(\xi, \eta) = |a_0 \exp(-2\pi i \alpha \xi) + \tilde{S}(\xi/\lambda f, \eta/\lambda f)|^2$$
$$= a_0^2 + |\tilde{S}(\xi/\lambda f, \eta/\lambda f)|^2 + a_0 \tilde{S}(\xi/\lambda f, \eta/\lambda f) \exp 2\pi i \alpha \xi$$
$$+ a_0 \tilde{S}^*(\xi/\lambda f, \eta/\lambda f) \exp -2\pi i \alpha \xi. \tag{8.46}$$

A record of this intensity distribution now constitutes the filter that will be used to process data. This system will be recognized as a holographic system that produces a hologram usually referred to as a generalized hologram (Leith and Upatneiks, 1962) or Fourier-transform hologram (Stroke, 1965).

Let the input to the usual coherent optical chain be $A(x,y)$; then the transmittance through the filter is

$$U'(\xi, \eta) = a_0{}^2 \tilde{A}(p,q) + |\tilde{S}(p,q)|^2 \tilde{A}(p,q) + a_0 \tilde{S}(p,q) \tilde{A}(p,q) \exp 2\pi i \alpha \xi$$
$$+ a_0 \tilde{S}^*(p,q) \tilde{A}(p,q) \exp - 2\pi i \alpha \xi, \qquad (8.47)$$

where p and q are written for $\xi/\lambda f$ and $\eta/\lambda f$ respectively.

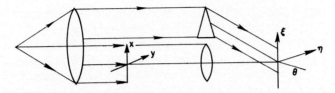

Fig. 13. Schematic diagram of an optical system for producing a holographic filter.

This distribution is now retransformed to the image plane to give a complex amplitude distribution $\Psi(x',y')$ shown in Eq. (8.48) where a multiplicative constant has been omitted:

$$\Psi(x',y') = a_0{}^2 A(x',y') + S(x',y') \circledast S^*(x',y') \circledast A(x',y')$$
$$+ a_0 S(x',y') \circledast A(x',y') \circledast \delta(x' + \alpha \lambda f, y') \qquad (8.48)$$
$$+ a_0 S^*(x',y') \circledast A(x',y') \circledast \delta(x' - \alpha \lambda f, y').$$

The first two terms in Eq. (8.48) continue on along the axis whereas the third and fourth terms are displaced upwards and downwards from the axis. The third term is the convolution of S with A. The fourth term is the cross correlation of S and A. Hence if the filter is made to correspond to some particular A then the output given by this fourth term is a delta function at the location of the particular A.

This technique has been applied to the detection of particular objects in a scene by making S correspond to the desired object. Then at the location of that object the filtered output consists of a bright spot of light while the remainder of the field is dark. Examples are the detection of geometric shapes, alphanumerics and isolated signal in random noise.

B. Application of Holographic Filters

Once it had been established that complex filters could readily be made holographically it was a natural development to extend the ideas to the problem of correcting aberrations. It is also important to realize the extra flexibility that holographic filters impart to the process of spatial filtering.

The filters had always been fabricated to operate on the optical transform of the input since the optical transform could readily be determined and the appropriate filter constructed. The filtering process could, in fact, be carried out in any plane with the appropriate filter. Clearly holographic filters can be built for use in the transform plane or any other plane with equal facility.

Leith *et al.* (1965) and Upatnieks *et al.* (1966) applied the idea of holographic filtering to the correction of spherical aberration. To produce the filter the light from a point source in the focal plane of the lens is propagated through the aberrated lens and interfered with a collimated reference wave. The recorded hologram is the filter or corrector plate. To use the filter it is replaced in the same position with respect to the lens as it had when made.

This basic idea was used by several workers to form images through various aberrating media (see, e.g. Goodman, 1968). Very often the hologram can be recorded so that the reference beam as well as the object beam pass through identical portions of the aberrating medium (e.g. turbulence). Under these circumstances the hologram is essentially unaffected by the presence of the aberration and the image formed from the hologram is unaberrated. The deconvolution process has further been exploited for image motion or other aberrations (Stroke, 1965; Stroke and Zech, 1967; Stroke, 1969).

VI. CONCLUSIONS

Only the fundamentals of optical data processing have been discussed in this chapter, together with a few simple examples. Many other results are described in the literature. Cutrona *et al.* (1960) have reported numerous techniques for performing a variety of tasks from the formation of Fourier and Laplace transforms of given inputs to the design of multi-channel cross-correlators and auto-correlators. Coherent optical processing also played a very vital part in synthetic-aperture radar (Cutrona *et al.*, 1966). Integration filters have been described by DeVelis and Reynolds (1968). Clearly most of these applications are outside the scope of this short survey. Mention should perhaps also be made of the interest in signal subtraction (Stroke, 1966; Bromley *et al.*, 1969) which can become an important technique in the subtraction of complex fields.

Many problems have yet to be solved before optical processing can be widely used in data handling. The major problem is noise, particularly that generated by photographic grain. The relief phase image associated with the density image in a photographic record can be removed by the use of index matching fluids, but a smaller (in magnitude) phase image is still present caused by small index variations inside the emulsion.

Another area of considerable study is the production of real-time filters or adaptive filters that would allow the optical processing system to be used for filtering operations where the filter could be generated, directly in the systems, on command.

Finally it is pertinent to comment on the comparison between optical processing and digital processing. Both methods have applications. Digital processing is perhaps best performed when a single input has to be processed by a number of different filters. However, it should be recognized that a two-dimensional input contains a large number of resolvable elements which have to be scanned and read-in to the computer usually through a memory. This can be a severe problem for large-format high-resolution photographic inputs. Finally, after processing, a "picture" has to be produced from the output of the computer. Excellent work has been done in this field notably by Harris (1964, 1966). Optical methods come into their own when large quantities of two-dimensional data have to be processed through the same filtering operation. Once the filter has been made the process is essentially instantaneous and two-dimensional. Both methods clearly have a role to play and optical techniques will probably be used for processes in their own right but also as peripheral equipment to computers to pre-process data to limit the amount of information that has to be handled digitally.

References

Abbe, E. (1873). *Arch. mikrosk. Anat.*, **9**, 413.
AGARD Conference Proceeding No. 50 (1970). (NATO Advisory Group for Aerospace Research and Development). "Opto-electronic Signal Processing Techniques."
Bromley, K., Monahan, M. A., Bryant, J. F. and Thompson, B. J. (1964). *Appl. Phys. Lett.* **14**, 67.
Cagnet, M., Françon, M. and Thrierr, J. C. (1962). "Atlas of Optical Phenomena," Prentice-Hall, New Jersey.
Chaudhuri, B. and Hargreaves, A. (1956). *Acta Crystallogr.*, **9**, 793.
Cheatham, T. P. and Kohlenberg, A. (1954). *I.R.E. Conv. Rec.* **IV**, 6.
Cutrona, L. J., Leith, E. N., Palermo, C. J. and Porcello, L. J. (1960). *I.R.E. Trans. Inf. Theory* IT-**6**, 386.
Cutrona, L. J., Leith, E. N., Porcello, L. J. and Vivan, W. E. (1966). *Proc. Instn elect. Engrs* **54**, 1026.
DeVelis, J. B. and Reynolds, G. O., (1967). "Theory and Applications of Holography," Addison Wesley Publishing Co., Reading, Mass., U.S.A.
Duffieux, P. M. (1946). "L'Integrale de Fourier et ses Applications a l'optique," Chez l'Auteur, Faculte de Sciences, Universite de Besançon.
Elias, P., Grey, D. S. and Robinson, D. Z. (1952). *J. opt. Soc. Am.* **42**, 127.
Elias, P. (1953). *J. opt. Soc. Am.* **43**, 229.
Françon, M. (1950). "le Contraste de Phase en Optique et en Microcopie" Editions de la Revue D'optique Theoretique et Instrumentale, Paris.
Goodman, J. W., Huntley, W. N., Jackson, D. W. and Lehmann, N. (1966). *Appl. Phys. Lett.*, **8**, 311.
Goodman, J. W. (1968). "Introduction to Fourier Optics," McGraw Hill, New York.

Harris, J. L. (1964). *J. opt. Soc. Am.*, **54**, 606.

Harris, J. L. (1966). *J. opt. Soc. Am.*, **56**, 569.

Jackson, P. L. (1965). *Appl. Optics*, **4**, 419.

Jennison, R. C. (1961). "Fourier Transform and Convolutions for the Experimentalist," Pergamon Press, Oxford.

Kozma, A. and Kelly, D. L. (1965). *Appl. Optics*, **4**, 387.

Leith, E. N. and Upatnieks, J. (1962). *J. opt. Soc. Am.*, **52**, 1123.

Leith, E. N., Upatnieks, J. and Vander Lugt, A. (1965). *J. Soc. Am.*, **55**, 595.

Linfoot, E. H. (1955). "Recent Advances in Optics," pp. 128–175, Clarendon Press, Oxford.

Lipson, H. and Thompson, B. J. (1959). *Bull. natn. Inst. Sci. India*, **14**.

Maréchal, A. and Croce, P. (1953). Compt. Rendu, 237, 706.

NASA Seminar (1969). "Evaluation of Motion-Degraded Images," Cambridge, Mass. (NASA SP-193) (Available from Superintendent of Documents, U.S. Government Printing Office, Washington, D.C. 20402).

O'Neill, E. L. (1956). *I.R.E. Trans. Inf. Theory* IT-**2**, 56.

O'Neill, E. L. (Ed.) (1962). "Communication and Information Theory Aspects of Modern Optics," Published by General Electric Company, Syracuse, New York.

O'Neill, E. L. (1963). "Introduction to Statistical Optics," Addison Wesley Publishing Co., Reading, Mass., U.S.A.

Pollack, D. K., Koester, C. J., Tippett, J. T. (Eds.) (1963). "Optical Processing of Information," Spartan Books, Inc., Baltimore.

Porter, A. B. (1906). *Phil. Mag.*, **11**, 154.

Rhodes, J. E. (1953). *Am. J. Phys.*, **21**, 337.

Shulman, A. R. (1970). "Optical Data Processing," John Wiley & Son, New York.

Stroke, G. W. (1965). *Appl. Phys. Lett.* **6**, 201.

Stroke, G. W. (1966). "An Introduction to Coherent Optics and Holography," p. 90. Academic Press, London and New York.

Stroke, G. W. and Zech, R. W., (1967). *Phys. Lett.* **25A**, 89.

Stroke, G. W. (1969). *Optica Acta*, **16**, 401.

Taylor, C. A. and Lipson, H. (1964). "Optical Transforms. Their Preparation and Application to X-Ray Diffraction Problems," Bell, London.

Thiry, H. (1964). *Appl. Optics*, **3**, 39.

Thompson, B. J. (1959). Ph.D. Thesis, Manchester University.

Tippett, J. T., Berkowitz, D. A., Clapp, L. C., Koester, C. J. and Vanderburgh, A. (Eds.) (1965). "Optical and Electro-Optical Information Processing," M.I.T. Press, Cambridge, Mass., U.S.A.

Tsujiuchi, J. (1960a). *Optica Acta*, **7**, 243.

Tsujiuchi, J. (1960b). *Optica Acta*, **7**, 385.

Tsujiuchi, J. (1961). *Optica Acta*, **8**, 161.

Tsujiuchi, J. (1963). "Progress in Optics," Vol. III (Ed. E. Wolf), North Holland Publishing Co.

Upatnieks, J., Vander Lugt, A. and Leith, E. N. (1966). *Appl. Optics*, **5**, 589.

Vander Lugt, A. (1964). *I.E.E.E. Trans. Inf. Theory*. IT-**10**, 2, 139.

CHAPTER 9

Holography

J. Shamir

Department of Electrical Engineering, Technion,
Israel Institute of Technology, Haifa, Israel

I. Introduction

In the framework of geometrical optics, an object can be seen in three dimensions if the intensity and direction of each light ray emerging from it are detected. Conventional photography records only the intensity, and therefore the information on the three-dimensional character of an object is lost in the photographing process.

Diffraction theory indicates that the directions of the various light rays that emerge from a light wave-front are determined by the phase distribution on a plane cutting this wavefront. "Holography" or "wavefront reconstruction" are terms used for a two-step process: (a) the amplitude and relative phase distributions of a wavefront are recorded at a plane intersecting the wave; and (b) the wavefront is reconstructed from this record. The recorded wavefront is called a "hologram", and it contains all the information needed to reconstruct the original wave-amplitude and phase (or intensity and direction). Such a reconstruction may produce an image containing all the information on the object, i.e., a true three-dimensional image.

Holography was invented in the late forties by Gabor (1949, 1951). However, the absence at that time of bright, coherent light sources hindered the development of the idea until the invention of the laser. Leith and Upatnieks (1962, 1963, 1964) took up the subject using lasers and a new method. Their work was so successful that holography bloomed and in a few years developed into a wide field of research and applications. Today, holography is so highly advanced that even the coherence restrictions can be relaxed, and holographic techniques are now used in such diverse areas as acoustics, microwave technology, data processing and non-coherent photography.

This chapter is devoted to the analysis of the holographic process and a review of its possible applications.

II. Holography from Different Points of View

A. The Hologram as a Diffraction Grating

Figure 1 represents two coherent plane waves, U_1 and U_2, impinging on a screen S with an angle θ between them. Using complex notation, the two waves can be represented by

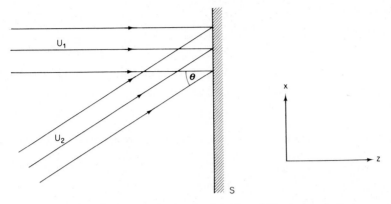

Fig. 1. Interference of two plane waves, U_1 and U_2, at a screen S.

$$\left.\begin{array}{l} U_1 = A_1 \exp\{ikz\} \\ U_2 = A_2 \exp\{ik(z\cos\theta + x\sin\theta)\} \end{array}\right\} \tag{9.1}$$

where A_1 and A_2 are the amplitudes, $k = 2\pi/\lambda$ is the wave number (λ is the wavelength), z and x are two of the spatial coordinates.

At the screen, interference occurs between the two waves, and the intensity distribution will be given by

$$\begin{aligned} I = |U_1 + U_2|^2 &= |U_1|^2 + |U_2|^2 + U_1^* U_2 + U_1 U_2^* \\ &= A_1^2 + A_2^2 + A_1 A_2 [\exp\{ik[z(\cos\theta - 1) + x\sin\theta]\} \\ &\quad + \exp\{-ik[2(\cos\theta - 1) + x\sin\theta]\}] \end{aligned}$$

or

$$I = A_1^2 + A_2^2 + 2A_1 A_2 \cos k[z(\cos\theta - 1) + x\sin\theta]. \tag{9.2}$$

Defining a spatial frequency by

$$\nu = k\sin\theta \tag{9.3}$$

and some phase factor

$$\phi(z) = kz(\cos\theta - 1), \tag{9.4}$$

we can rewrite Eq. (9.2) in the form

$$I = A_1^2 + A_2^2 + 2A_1 A_2 \cos[\nu x + \phi(z)]. \tag{9.5}$$

If the screen is replaced by a photographic plate, it records the intensity distribution I.

An exposed photographic plate may be described by its transmittance $T(x,y)$, where x and y are the plate coordinates. In general, a partially transparent object transmits a portion of the light impinging on it. If the impinging light intensity is I_0, the transmitted intensity I_T will be determined by the transmittance, T, of the object, according to the relation

$$I_T = TI_0.$$

When a conventional "negative" photographic emulsion is exposed to light, its density, D_n, increases. In the so-called linear region the relation

$$D_n = NE^\gamma$$

is approximately satisfied. Here γ is a parameter determined by the emulsion and processing, E is the exposure—the incident light intensity multiplied by the exposure time—and N is some normalization parameter. The transmittance of a negative transparency is given by

$$T_n = 1 - D_n.$$

To obtain a "positive" print from this transparency, another emulsion has to be exposed to a uniform light beam transmitted through the negative. The new emulsion will acquire a density

$$D_p = T_n{}^\gamma$$

and a transmittance

$$T_p = 1 - D_p = 1 - (1 - D_n)^\gamma = 1 - (1 - NE^\gamma)^\gamma.$$

In conventional photography we want to reproduce a beam intensity proportional to the original beam intensity; thus we would require $\gamma = 1$. A diapositive emulsion saves one step; in this case, exposure causes bleaching and the transmittance will be given by the relation

$$T = NE^\gamma. \tag{9.6}$$

Here again we would require $\gamma = 1$ for a conventional transparency. In holography we are not interested in the transmittance, T, but in a related parameter, the amplitude transfer function.

The amplitude transfer function operates on the complex amplitude and not on the intensity. If a wave of amplitude U_0 impinges on the plate, which has a transfer function $H(x,y)$, the outgoing wave will have an amplitude

$$U_T = H(x,y) U_0.$$

$H(x,y)$ may be complex, so that it can operate on the amplitude and phase of the impinging wave. To find the relation between T and H we may write

$$I_T = |U_T|^2 = |H(x,y) U_0|^2 = |H(x,y)|^2 |U_0|^2 = |H(x,y)|^2 I_0.$$

Thus, if H is real, we obtain

$$H(x,y) = T^{1/2}(x,y) \propto E^{\gamma/2} \propto I^{\gamma/2}. \tag{9.7}$$

Choosing $\gamma = 2$ (in contrast to conventional photography), we have

$$H(x, y) \propto I(x, y). \tag{9.8}$$

Substituting Eq. (9.5) into Eq. (9.8), we conclude that the photograph of the interference pattern produced by two plane waves can yield a sinusoidal grating as a first approximation. To simplify calculations, we put $A_1 = A_2$, and obtain

$$H(x, y) = N\{1 + \cos [\nu x + \phi(z)]\} \tag{9.9}$$

where N is again some normalization function (not the same as above) which restricts H to the interval $0 \leqslant H \leqslant 1$.

From Eq. (9.9) it is evident that it is not necessary to require a positive record. A negative one will do just as well, the difference between the two being only a phase shift by π. Applying the techniques of scalar diffraction theory (see Chapter 1, and, for example, Born and Wolf, 1959; Lipson and Lipson, 1969; Goodman, 1968), we can find how a plane wave will be altered by H. Suppose a plane wave

$$U_0 = A \exp ikz$$

impinges on the developed plate from the left, the emerging wave to the right will be

$$
\begin{aligned}
U &= U_0 H(x, y) \\
&= AN \exp ikz(1 + \cos \nu x) = AN \exp ikz[1 + \tfrac{1}{2}(\exp i\nu x + \exp -i\nu x)] \tag{9.10}
\end{aligned}
$$

where, for simplicity, we put $\phi(z) = 0$.

It is easy to show (see Chapter 1) that, at a sufficient distance from the plate, the wave will become the Fourier transform of U. (Using a lens which may be also the lens of our eye, we transfer the far field to the focal plane of the lens; thus, looking at the diffracted beam, we actually see this far field.)

$$\psi(u) = \int_{-\infty}^{\infty} U \exp \{-iux\} \, dx \tag{9.11}$$

in which $u = k \sin \alpha$—the spatial frequency of the diffracted beam—α is the angle between the propagation direction of a light ray and the z coordinate. (At the focal point of a lens u is the coordinate.)

Substituting Eq. (9.10) into Eq. (9.11), we obtain

$$\psi(u) = AN \exp ikz(\delta(u) + \tfrac{1}{2}[\delta(u - \nu) + \delta(u + \nu)]). \tag{9.12}$$

This represents three beams: one which is transmitted without diffraction—the zero order beam—and two diffracted beams in the $\alpha = \pm\theta$ direction. The wave with $\alpha = +\theta$ is a replica of the original wave, while the other one ($\alpha = -\theta$) is a conjugate wave. It is worth noting that this contrasts with the usual results of a rectangular grating, where higher-order diffraction beams are also present. In Fourier-transform theory, we say that these are the higher harmonics of the fundamental *spatial frequency* (Section III.C, Chapter 1). Similarly, higher harmonics will also be present when Eq. (9.7) is not exactly satisfied, as in a usual physical situation.

The meaning of Eq. (9.12) is that a recorded interference pattern of two plane waves is capable of reproducing these waves; it "remembers" the angle between the two waves. The light wave emerging from a physical object can be described as a superposition of a large number of elementary plane waves. If this wave interferes on a photographic plate with a coherent reference wave, each of the elementary plane waves will produce its own "grating". If the developed plate is illuminated by the reference wave, all these waves will be reconstructed according to Eq. (9.22). The superposition of the reconstructed elementary waves reproduces the original wave. In addition, there is a second wave, the "conjugate" wave, the meaning of which will be discussed in the next section.

The wave-front recorded in this way is the Leith-Upatnieks hologram (Leith and Upatnieks, 1962, 1963, 1964, 1967).

B. The Hologram as a Fresnel Zone Plate

A second way to analyse the holographic process is to consider an object as composed of a large number of point sources. The wave emerging from a point source of strength, A_0, is a spherical wave which has an amplitude

$$U_0 = A_0 \frac{1}{r} \exp i\mathbf{k} \cdot \mathbf{r} \tag{9.13}$$

at a distance r from the source. Here again the wave vector is $\mathbf{k}$. In a spherical wave $\mathbf{k}$ has everywhere the direction of $\mathbf{r}$; thus Eq. (9.13) can be written as

$$U_0 = A_0 \frac{1}{r} \exp ikr \tag{9.13'}$$

with

$$r = \sqrt{x^2 + y^2 + z^2}.$$

At a sufficient distance, z_0, from the source 0 (see Fig. 2) and if the dimensions of the screen are small compared to z_0, the quadratic approximation may be used:

$$\sqrt{z_0{}^2 + x^2 + y^2} \approx z_0 \left(1 + \frac{x^2 + y^2}{2z_0{}^2}\right).$$

In this approximation Eq. (9.13') becomes

$$U_0 = A_0 \frac{1}{z_0} \exp(ikz_0) \exp\left\{i\frac{k}{2z_0}(x^2 + y^2)\right\}. \tag{9.14}$$

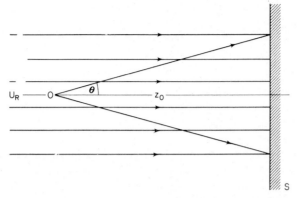

Fig. 2. Interference of a spherical wave with a coherent, plane wave, background.

Let this wave interfere with a coherent reference plane wave

$$A_R \exp(ikz_0). \tag{9.15}$$

(We neglect a possible constant phase.)

The intensity observed at a screen S (Fig. 2) will be:

$$I = |U_R + U_0|^2 = |U_R|^2 + |U_0|^2 + U_R^* U_0^* + U_R^* U_0^*$$

$$= A_R^2 + (A_0/z_0)^2 + \frac{A_R A_0}{z_0}\left[\exp\left\{i\frac{k}{2z_0}(x^2 + y^2)\right\} + \exp\left\{-i\frac{k}{2z_0}(x^2 + y^2)\right\}\right]$$

$$= A_R^2 + (A_0/z_0)^2 + 2\frac{A_R A_0}{z_0}\cos\frac{k}{2z_0}(x^2 + y^2). \tag{9.16}$$

Assuming Eq. (9.7) to hold for a photographic plate replacing the screen, we obtain a sinusoidal Fresnel zone plate. The transfer function of the plate has the form,

$$H(x, y) = A + B\left[\exp\left\{i\frac{k}{2z_0}(x^2 + y^2)\right\} + \exp\left\{-\frac{k}{2z_0}(x^2 + y^2)\right\}\right] \tag{9.17}$$

where the constants A and B satisfy the relation,

$$A \geqslant 2B.$$

The illumination of this plate by a plane wave, say U_R Eq. (9.15), results in three diffracted waves: the zero-order plane wave,

$$U_{d.c.} = A U_R = A A_R \exp ikz, \tag{9.18}$$

a diverging spherical wave,

$$U_v = B U_R \exp\left\{i\frac{k}{2z_0}(x^2 + y^2)\right\} = B A_R \exp\{ikz\}\exp\left\{i\frac{k}{2z_0}(x^2 + y^2)\right\}, \tag{9.19}$$

which seems to diverge from a point at a distance z_0 behind the plate, and a converging spherical wave

$$U_c = BU_R \exp\left\{-i\frac{k}{2z_0}(x^2 + y^2)\right\} = BA_R \exp\{ikz\} \exp\left\{-i\frac{k}{2z_0}(x^2 + y^2)\right\} \quad (9.20)$$

which converges towards a point at a distance z_0 from the plate. The virtual image-wave U_v is the exact reconstruction of the original wave, while the conjugate wave, U_c, produces the real image of the source, at a distance z_0 from the plate.

The first hologram ever made, the Gabor hologram (Gabor, 1949, 1951), was of this kind. If there are many point sources, each one will produce its own zone plate, which will be capable of reproducing its image. Comparing the Gabor hologram to the Leith–Upatnieks hologram, it has two drawbacks: (a) the three reconstructed waves are collinear, which makes the observation

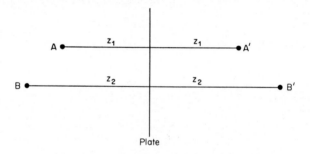

Fig. 3. Illustrating the pseudoscopic nature of the real image.

of the images difficult, and (b) only transparent, or small, objects can be recorded.

The zone-plate picture of the hologram may be generalized to include also the Leith–Upatnieks hologram. In this case, the zone plates will be off-axial, (or astigmatic) and it can be shown that again each elementary source will produce an interference pattern that reproduces it.

To end this section, the pseudoscopic nature of the real image is illustrated with the help of Fig. 3. Observing the hologram from the right, object A is closer than object B. However, in the real image space B' comes before A'.

C. Mathematical Treatment of Holography

Figure 4 is the schematic diagram of a hologram-recording system. It consists of a light source S_0 illuminating an object, a reference source S_R and a photosensitive plane E. Denoting the coordinates of the object by (x_1, y_1, z_1), we can define a scattering function $f(x_1, y_1, z_1)$ in such a way that the impinging wave U_1 will be scattered to produce a wavefront

$$U = f(x_1, y_1, z_1) U_1. \qquad (9.21)$$

The function f is equivalent to the usual transfer function, when transmitted waves are considered. A volume element $d\mathbf{r}_1$ at a point B on the object, can

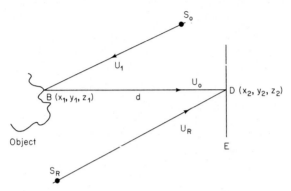

Fig. 4. Principle of holographic recording.

be regarded as a point source of strength $U d\mathbf{r}_1$. The light amplitude arriving at point $D(x_2, y_2, z_2)$ on E from the point B, is a portion of an elementary spherical wavelet

$$\Delta U_0 = U \frac{\Delta \mathbf{r}_1}{d} \exp -ikd$$

where d is the distance between B and D. The total amplitude arriving at D from the whole object constitutes the object beam

$$U_0(x_2, y_2, z_2) = \int_{\text{object}} U \frac{1}{d} \exp -ikd \, d\mathbf{r}_1.$$

$$U_0(x_2, y_2, z_2) = \int_{\text{object}} U_1 f(\mathbf{r}_1) \frac{1}{|\mathbf{r}_2 - \mathbf{r}_1|} \exp \{-ik|\mathbf{r}_2 - \mathbf{r}_1|\} \, d\mathbf{r}_1. \qquad (9.22)$$

If there is no emulsion, U_0 is the light wave that makes the object visible. In order to record the wave U_0, a reference wave, U_R, is provided. In most cases, U_R will be uniform spherical or plane wave, but it does not have to be specified here.

The two waves interfere to produce the intensity pattern,

$$I(x_2, y_2, z_2) = |U_R + U_0|^2 = |U_R|^2 + |U_0|^2 + U_R^* U_0 + U_R U_0^*. \qquad (9.23)$$

Assuming the validity of Eq. (9.7) for a photographic plate at E, we obtain a transfer function for the photo-emulsion:

$$H(x_2, y_2, z_2) \propto I(x_2, y_2, z_2). \tag{9.24}$$

If a uniform reference wave is used, the term $|U_R|^2$ causes a uniform darkening of the plate. This can be described as a d.c. bias level, which is modulated by the other terms. The zero-order diffraction term of the object, $|U_0|^2$, contains information on the amplitude of the object wave, but not on its phase relations. Therefore, it is included, by most authors, in the d.c. term. Following this way of reasoning, we can write the transfer function in the form

$$H = H_{d.c.} + U_R^* U_0 + U_R U_0^*. \tag{9.25}$$

The hologram can now be illuminated by a coherent wave and its diffraction observed. For the present, we shall deal only with the illumination by the original reference wave, U_R (Fig. 5).

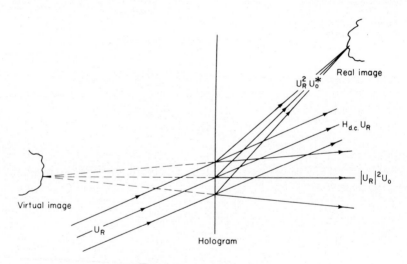

Fig. 5. Reconstruction process.

The wave emerging from the hologram will include three terms.

$$U_R H = U_{d.c.} + U_v + U_c \tag{9.26}$$

where $U_{d.c.} = U_R H_{d.c.}$ is the zero-order diffracted beam,

$$U_v = |U_R|^2 U_0$$

is a reconstruction of the original wave multiplied by a constant factor. Viewing U_v, is the equivalent of looking at the real object, and it is called the virtual image. The pseudoscopic real image is constructed by the conjugate beam,

$$U_c = U_R^2\, U_0^*.$$

The inclusion of the phases of U_R may cause distortion of the real image, in addition to it being pseudoscopic.

In this section, the basic principles of holography have been studied in three possible ways. More refined details, as the non-exact validity of Eq. (9.7), and the resolution restrictions of a photographic plate, will be dealt with in the next section. Many variations of the basic system will also be discussed later.

III. GENERAL CONSIDERATIONS

A. Emulsion Response

Until now, it has been assumed that Eq. (9.7) holds for the whole dynamic range of the photo-sensitive emulsion. If it were so, it would be advantageous to use a reference beam of the same average intensity as the object beam:

$$\left.\begin{array}{l} U_R = A \exp\{i\phi_R(x, y, z)\} \\ U_0 = A(x, y, z) \exp\{i\phi_0(x, y, z)\} \end{array}\right\} \tag{9.27}$$

with $\langle A(x,y,z)\rangle \approx A$ satisfied.

Instead of Eq. (9.25) we may now write

$$H \cong 2A^2[1 + \cos(\phi_R - \phi_0)]. \tag{9.28}$$

The maximum possible value of the fringe contrast,

$$\kappa = \frac{H_{\max} - H_{\min}}{H_{\max} + H_{\min}}, \tag{9.29}$$

is obtained when Eq. (9.28) is satisfied. This value is unity, and is equivalent to a 100% modulation depth in communication.

However, a physical recording material does not satisfy Eq. (9.7), and the $\log H - \log I$ curve has usually a form like the one illustrated in Fig. 6. In order to obtain linear response, a bias level has to be added in order to "throw" the modulation into the linear region. This is usually achieved by increasing the reference wave intensity. It was found (Kiemle and Röss, 1969; Friesem et al., 1967), that intensity ratios

$$2 \leqslant \frac{|U_R|^2}{\langle|U_0|^2\rangle} \leqslant 10 \tag{9.30}$$

11

are all satisfactory in obtaining good reconstructions. Equation (9.29) yields
for the contrast an interval:

$$\frac{2\sqrt{10}}{11} \leqslant \kappa \leqslant \frac{2\sqrt{2}}{3}.$$

Another way to obtain the bias is to provide a uniform background light
(which need not be coherent) or a pre-exposure of the plate (Shamir *et al.*,
1971). This method is advantageous when a strong laser is not available.

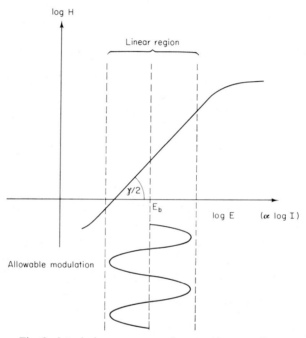

Fig. 6. A typical response curve for a positive recording.

B. *Emulsion Resolution*

Section III.A dealt with the depth of modulation in a hologram. If we
continue to use communication terminology we may also talk about the
modulation frequency and bandwidth. By frequency we mean the spatial
frequency defined by Eq. (9.5), and the bandwidth is related to the range
of frequencies recorded. A hologram may be compared to an electronic circuit
element, with a specified spectral response. The information capacity of the
hologram will be higher, the wider its frequency response.

Figure 7 illustrates a typical response curve of a photographic plate.

Somewhere in the neighbourhood of the spatial frequency ν_r lies the number given by the manufacturer as the resolution of the emulsion in lines per millimetre. The higher the resolution, the smaller the grain size; the smaller the grain, the higher the exposure needed. As a result, a good holographic plate should have higher resolution, which in turn requires a strong laser, or long exposure time. As an example, a 50 mW He–Ne laser, which is about the strongest laser of this kind, requires about 1 sec to make a good hologram.

Since a hologram actually records interference fringes, the whole holographic recording system has to be interferometrically stable for relatively long periods. Otherwise, the interference fringes are washed out and information is lost. Therefore, in each holographic system, a compromise has to be

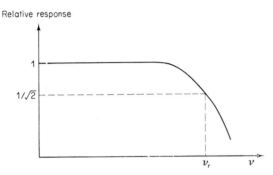

Fig. 7. Spatial frequency response curve of a typical photographic emulsion.

found between the required information content, and the allowable exposure time, which is limited by the stability of the set-up.

C. Coherence Requirements and Stability

A hologram is a record of a complex interference pattern. Such a pattern can be produced only by coherent light. In order to obtain a good hologram, each light ray impinging on the recording plate should traverse a path length that will vary less than the coherence length of the source. Abramson (1969, 1970) developed a graphic method, the Holo-Diagram, that may help in the alignment of a recording system. The method enables one to design the set-up in a way that will minimize the path differences.

The coherence length itself is not a very well defined notion. For our purpose, we may define it as the minimum path difference in a two-beam interferometer for which the interference pattern disappears. This somewhat complex definition is forced upon us, because most commercial lasers emit a beam with a complex mode structure. With such a beam, the interference pattern disappears for quite a short path difference; however, it may appear once more for a longer path difference. In a good holographic system the mode structure has to be reduced to a single axial mode. This is achieved by some mode selecting system, like the resonant reflector or intracavity etalon or thin-film mode selector (see for example, Magyar, 1969; McClung *et al.*, 1970; Brooks *et al.*, 1966; Smith, 1969).

Concluding this section, we shall sum up the main requirements of a holographic recording system, referring to Fig. 8.

Figure 8 is a schematic diagram of a simple holographic system for the recording of a three-dimensional object. The purpose of the lenses is to enlarge the beam, while the pinhole "spatial filter" serves to "clear up" the beam in

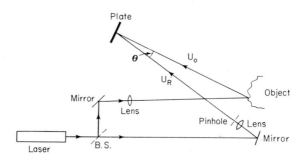

Fig. 8. A simple set-up for holographic recording.

order to produce a uniform, spherical reference wave. In order to provide a more uniform illumination of the object, a diffuser is sometimes placed in front of it.

A good hologram will be produced only if the following requirements are satisfied.

(a) The overall path-length of each ray reflected from the object to the recording plate should differ by less than the coherence length, from the overall reference-beam path-length.

(b) The stability of the system should be such that the path differences will vary no more than about a tenth of a wavelength during exposure. Some exceptions where this is not satisfied will be dealt with in the section on holographic interferometry.

(c) The wavelength of the laser should be stable, at least to a degree that will prevent the variation of the relative phases between the two beams, during exposure, by more than about $(2\pi)^{-1}$.

(d) The maximum angle, θ, between an arbitrary ray emerging from the object and one from the reference beam, should satisfy

$$k \sin \theta \leqslant v_r \qquad (9.31)$$

where v_r was defined in Fig. 7 and k is the wavenumber.

(e) The information has to be recorded in the linear region, for example by satisfying Eq. (9.30).

The light wavefront emerging from an object contains all the information needed for the reconstruction of a 3-D image. The above statement holds also when the light is transmitted through an optical system. A part of the information, namely the phase, is lost only when the wavefront is recorded by a square-law detector like a photographic emulsion. However, if the recording is carried out holographically, the phase information may be retained too. In the following, a variety of recording configurations will be described briefly.

A. Fresnel and Fraunhofer Holograms

Figure 8 describes one of the simplest forms of lensless holography. These are the Fresnel and Fraunhofer holograms. The term Fresnel hologram indicates that the plate is placed in the near field of the object. Since the first Gabor hologram (see Section II.B) was of this kind, sometimes these two names are used, erroneously, as synonyms. For similar reasons, the Leith–Upatnieks hologram (Section II.A) is sometimes identified with the Fraunhofer hologram—a holographic recording of the far-field pattern.

The Fresnel and Fraunhofer holograms constitute the main part of what is called "lensless 3-D photography". These configurations are extensively used for the recording of 3-D images and for holographic interferometry to be described later.

B. Fourier Transform Holograms

In the Fresnel and Fraunhofer holograms, the Fresnel and Fraunhofer diffraction patterns of the object are recorded respectively. The information on the object is contained also at a plane where the object beam is a spatial Fourier transform of the object (compare Section III.C, Chapter 1). If the hologram is recorded at this plane, it is called a Fourier transform hologram. Although such a hologram can be recorded without lenses (see for example Stroke, 1965), in most cases a lens is used to obtain the Fourier transform of an object (Fig. 9).

Regarding the object as a signal, its Fourier transform may be used for spectral analysis, filtering and correlation processes, and so on. This kind of hologram may thus serve as the basis for data processing. Optical data processing is now a wide field of research and applications, and therefore

Fourier-transform holograms are discussed in Chapter 8 which is devoted to this subject. We shall only note here the following property of Fourier holograms. Since they record the Fourier transform of the object, this transform can only be reconstructed. In order to reconstruct the image of the object, a second Fourier transformation has to be carried out. This can be done by inserting a lens in the reconstructed beam (Fig. 9b). There will be two reconstructed real images. (See also end of Section VII.C.)

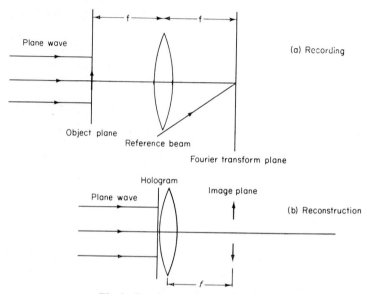

Fig. 9. Fourier transform hologram.

C. Image Plane Holography

For microscopic applications, where high resolution is needed, the object has to be placed close to the recording plane. A similar configuration is advantageous also when it is desired that the hologram reconstruction be viewed from a wide viewing angle. The main difficulty encountered when a 3-D object is placed near the photographic plate is the illumination of the object. The reference beam can be introduced through the back of the plate, so that there is no problem here. The object illumination problem was solved by McMahon and Caulfield (1970) using part of the reference beam transmitted through the plate. Another way around the difficulty is to construct the image of the object near the plate, using an imaging system (Rosen, 1967). In a special case of this technique the image of the object is focused on the plate (Kock et al., 1967). Interesting results were achieved in image-plane holography by Brandt (1969); if the focused image is used as the object beam and scattered light from the object as the reference beam, the coherence requirements of the source could be much relaxed.†

One of Brandt's techniques was to use a holographically generated zone plate as a double-focus lens; the positive focus constructed the image, while the negative focus supplied a diverging reference wave.

V. Volume Holograms

Until now, it has been assumed that holographic recording occurred only on the surface of the recording material. Planar gratings diffract a light beam with any arbitrary angle of incidence. Thus a simple plane hologram will reconstruct an image (possibly distorted) even if the illuminating beam does not come from the same direction as the original reference beam. In practice, however, photographic material has finite thickness, and therefore the interference fringes will be three-dimensional. In simple words, there will be dark layers across the emulsion, and they will block out any light coming from unfavourable directions. The angular region that allows transmission will be smaller, the thicker the recording material. The properties of volume holograms and their large information capacity were extensively studied by many authors (Leith *et al.*, 1966; Gabor and Stroke, 1968; Kogelnik, 1969). Here we shall limit ourselves to a few interesting applications.

A. Multiple Recording and Colour Holograms

Due to the direction sensitivity of thick emulsions, it is possible to record a number of holograms on the same plate. If for each recording the reference beam has a different direction, a specific reference beam will reconstruct only its own hologram. If the emulsion has high enough resolution, and the beams are sufficiently separated, high quality and independent reconstruction can be obtained from each hologram.

The number of holograms that can be recorded on a single plate is determined by the beam separation and maximum spatial frequency allowable. These, in turn, depend on emulsion thickness and resolution, respectively. A conventional holographic emulsion has a thickness between 7 μm and 25 μm and a maximum resolution of about 2,500 lines/mm. These numbers limit the number of holograms recordable on the same plate to less than ten. Photochromic materials, on the other hand, may exceed the above number by some orders of magnitude (Megla, 1966; Jackson; 1969, Baldwin, 1969). Friesem and Walker (1970) reported over a hundred separate recordings in a single photochromic plate.

One interesting application of multiple holograms is to record the same

† As the dispersion in this hologram is very low, it can be reconstructed even using white light.

object with light of different wavelengths. A coloured image may be obtained by reconstructing with all the colours together (Pennington and Lin, 1965; Collier and Pennington, 1967).

B. Reflection Holograms

It was mentioned earlier (see Section IV) that the reference beam may be introduced from the back surface of the plate (Fig. 10). If the emulsion is thick enough, standing waves will be produced in it by the two oppositely travelling waves. The developed emulsion will contain reflecting layers, equivalent to a three-dimensional grating. In a hologram of this kind, reconstruc-

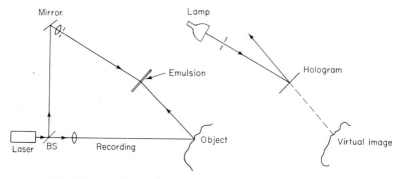

Fig. 10. Recording and reconstruction of a reflection hologram.

tion can be carried out only by light of wavelengths that satisfy the Bragg condition (Denisyuk, 1962, 1963; Stroke and Zech, 1966). From a different point of view, this set of reflecting layers can be described as an interference filter. Illuminated by white light, this filter transmits only a narrow spectral bandwidth which is capable of reconstructing the image. As a result, reflection holograms can be viewed by white light. However, because of emulsion shrinkage during development, the hologram cannot be reconstructed by the wavelength used to record it, unless this shrinkage is corrected (Nishida, 1970). In a conventionally developed plate a hologram recorded by red light will be reconstructed in green.

VI. SPECIAL HOLOGRAM TYPES

A. Phase Holograms

A hologram recorded in any manner with transmission characteristics of the form given by Eqs. (9.23) to (9.24) may be bleached to form a phase hologram (Cathey, 1965). The transfer function of a phase hologram has only phase variations. The bleaching process (Russo and Sottini, 1968; Burckhardt and Doherty, 1969) dissolves the metallic silver from the emulsion, leaving "holes" that introduce an effective phase variation in the wave transmitted.†

To analyse the phase hologram, we shall return to the elementary holographic grating of Section II.A, and rewrite here its transfer function:

$$H(x) = N[1 + \cos(\nu x + \phi(z))] \tag{9.32}$$

The bleaching process changes this into another transfer function,

$$\xi(x) = \exp\{i\alpha[H(x) + \rho]\} \tag{9.33}$$

where α is some proportionality factor, usually much less than unity, and ρ is a constant phase factor. Denoting $N\alpha = \beta$ Eq. (9.33) can be written as

$$\xi(x) = \eta \exp i\beta \cos \nu x \tag{9.34}$$

where η is some constant phase factor. The factors N and α being less than unity, β is also less than unity. Expanding Eq. (9.34) into series, we obtain

$$\xi(x) = \eta\left[1 + i\beta \cos(\nu x) - \frac{\beta^2}{2}\cos^2(\nu x)\ldots\right] \tag{9.35}$$

Rearranging Eq. (9.35) we obtain

$$\xi(x) = \eta\left[1 - \frac{\beta^2}{4} + i\beta \cos(\nu x) - \frac{\beta^2}{4}\cos(2\nu x)\ldots\right]. \tag{9.36}$$

Continuing in the same way that led to Eq. (9.12), we can find how a plane wave impinging on this phase grating will be diffracted. Using the notation of Eq. (9.12), we obtain the diffracted light to be

$$\psi(u) = \eta'\left\{\left(1 - \frac{\beta^2}{4}\right)\delta(u) + \frac{i\beta}{2}[\delta(u - \nu) + \delta(u + \nu)]\right.$$
$$\left. - \frac{\beta^2}{8}[\delta(u - 2\nu) + \delta(u + 2\nu)]\ldots\right\} \tag{9.37}$$

where η' is again a constant phase factor, different from η. The differences between Eqs. (9.37) and (9.12) are that there is no attenuation in the phase hologram, but there are also high-order diffraction terms. The phase factors appearing in Eq. (9.37) have no effect at all; thus they may be ignored. Due to the absence of attenuation, phase holograms may achieve high diffraction efficiencies. Burckhardt and Doherty (1969) report 45%. By properly adjusting β, the diffraction efficiency in the first order can be increased and the higher-order terms may be neglected.

† Bleaching processes are also available which change the metallic silver into transparent compounds having refractive index higher than that of the surrounding medium.

Also reflection holograms may be bleached, and it can be shown, theoretically, that diffraction efficiencies approaching 100% should be obtainable (Urbach and Meier, 1969).

B. Holography with Totally Reflected Waves

One way to use total reflection in holography is due to Stetson (1968). He introduced the reference beam from the back surface of the hologram at such an angle that it was totally reflected from the air-emulsion surface. This technique enables us to place a transparent object very closely to the plate, resulting in an ultra-high-resolution hologram. (0·8 μm resolution was obtained using He–Ne laser light of 0·63 μm wavelength.)

A more interesting technique is to record the evanescent wave which penetrates to a very small depth into the reflecting surface. Bryngdahl (1969) submerged the photographic plate in a liquid of refractive index higher than that of the emulsion. Light impinging on the emulsion at angles higher than the critical angle are totally reflected. However, some of the energy is still absorbed in the emulsion. The evanescent wave is also capable of interfering with another evanescent wave or with a conventional travelling wave. Using this interference, a hologram may be recorded. A hologram constructed by at least one evanescent wave, with the other evanescent or real, has some interesting features: first of all, it is a real planar hologram that is not affected by volume effects like emulsion shrinkage, internal reflections and scattering. Secondly, although a planar grating is formed, the Bragg condition must be satisfied because the Poynting vector lies in the hologram plane. This fact makes good reconstruction by white light possible. The reconstruction colour is a function of angle of incidence, because the evanescent wavelength of a specific vacuum wavelength is also a function of the angle of incidence. There is also an academic interest in this technique, which is the mere fact that such waves can be recorded.

C. Holography with Distorted Beams

In the derivation of Eq. (9.23), there was no restriction on the reference wave U_R. Usually a uniform, plane or spherical, wave is used for U_R, but this is not necessary. It is undesirable to introduce amplitude variation across the beam because then U_v (Eq. 9.26) will also contain amplitude variations. However, U_R may have any complex phase variation and, if the reconstruction is carried out with exactly the same wave, the virtual image should be perfect. This characteristic of holography can be used for encoding processes. Using complex U_R the reconstruction only can be carried out with the original U_R (Lanzl et al., 1968).

Another possibility is that the object beam undergoes distortion. If the distortion is stationary as those introduced by an optical system, the reconstruction may be corrected with the help of some compensating element (Leith and Upatnieks, 1966; Bryngdahl and Lohmann, 1969). However, this cannot be done when the distortion is varying with time, i.e., long-distance holography in turbulent media (Goodman *et al.*, 1969). Here it is advantageous to transmit the reference beam through the same optical path and then both beams undergo identical phase distortions. It is evident from Eq. (9.23) that these distortions cancel out in the recording.

VII. Applications of Holography

Holographic techniques are penetrating almost daily into new fields of application. It turns out that holography holds the first place in the uses of lasers. The widest application of holographic techniques is in the field of optical data processing, discussed in Chapter 8 of this book. In the present chapter we shall deal with some other aspects of the applications of holography. Three main fields will be discussed—microscopy, interferometry and optics technology.

A. 3-D Photography

The first, and natural application of holography—3-D photography—constitutes now only a very small portion of the huge field that developed from it. There is still some work being done in refining 3-D recording techniques, however; until some breakthrough occurs, it is unreasonable to believe in wide applications of the technique. For example, in order to record a live scene a high-power pulsed laser is needed (Brooks *et al.*, 1966; Zech and Siebert, 1968; McClung *et al.*, 1970). Such a system is very expensive and hazardous, definitely not applicable for everyday life. However, this technique may be used in research and technology for the 3-D recording of fast processes (Brooks *et al.*, 1965; Tanner, 1966). The advantage of holography over conventional techniques is in the 3-D recording, with no limitations by the focal depth of an optical system.† The process is recorded in three dimensions, and it can be reproduced and analysed, comfortably, using conventional recording techniques. Multiple recording (see Section V.A) can be applied as well if it is desired to record a number of stages in the development of the system.

Talking about 3-D photography, one would like to have also a 3-D television set. However, the resolution of a vidicon camera is orders of magnitude lower than needed for the transmission of holograms and this, of course, is not the

---

† The only limitation is the coherence length of the source.

only difficulty. In spite of these some work is being done also in this direction (Larsen, 1969).

B. Holographic Microscopy

The aim of Gabor (1949, 1951) in developing holography was to apply it in microscopy. The advantage of holographic microscopy over conventional microscopy is twofold: first, it will be shown below that, in principle, extremely high magnification should be attainable, and secondly, a 3-D record is obtained that is not limited by the very small focal depth of conventional microscopes. This is advantageous, especially when dynamic systems are studied, and the whole 3-D scene can be recorded in one photograph.

C. Magnification in Holograms

Magnification in holographic systems is treated in almost every general text on holography (Leith and Upatnieks, 1967; DeVelis and Reynolds, 1967; Goodman, 1968; Kiemle and Röss, 1969; Stroke, 1969). Since a detailed

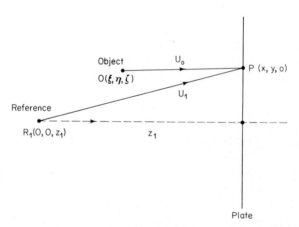

Fig. 11. Recording a Gabor hologram of a point object with a point reference source.

analysis of the subject is beyond the scope of this book we shall limit ourselves to a first-order treatment in a simple case. The discussion will be based on the imaging of a point object in the Gabor hologram (Section II), using a point source as a reference.

In Fig. 11, U_0 is the object beam and U_1 the reference beam. In our coordinate system an arbitrary point on the plate has the coordinates $(x, y, 0)$, the object point 0 has coordinates (ξ, η, ζ), and the reference source is at $R_1(0,0,z_1)$. The two spherical waves U_0 and U_1 have amplitudes at $P(x,y,0)$,

$$
\left.
\begin{aligned}
U_1 &= A_1 \frac{1}{z_1} \exp\left\{i \frac{k_1}{2z_1}(x^2 + y^2)\right\} \\[2mm]
U_0 &= A_0 \frac{1}{\zeta} \exp\left\{i \frac{k_1}{2\zeta}[(x - \xi)^2 + (y - \eta)^2]\right\}
\end{aligned}
\right\}
\tag{9.38}
$$

where we used the quadratic approximation of Eq. (9.14). The recording wave number is k_1. Exposure and suitable development of the plate yield a normalized transfer function (in first approximation),

$$
H(x, y) = |U_1 + U_0|^2 = |U_1|^2 + |U_0|^2 + U_1^* U_0 + U_1 U_0^*.
\tag{9.39}
$$

For further analysis we shall restrict ourselves to the virtual image term

$$
H_v(x, y) = U_1^* U_0.
$$

Illumination of the plate by a reconstructing wave U_2 produces a diffracted wave at the hologram plane,

$$
U_v = U_2 U_1^* U_0.
\tag{9.40}
$$

If $U_2 \equiv U_1$, the original wave, U_0, will be reconstructed. However, if $U_2 \neq U_1$, then also $U_v \neq U_0$. Let us try a reconstruction, using a point source of wavelength $\lambda_2 = 2\pi/k_2$, at a distance z_2 from the plate.

$$
U_2 = A_2 \frac{1}{z_2} \exp\left\{i \frac{k_2}{2z_1}(x^2 + y^2)\right\}.
\tag{9.41}
$$

Up to a constant (complex) factor, the virtual image wave will have a form

$$
\begin{aligned}
U_v &= \exp \frac{i}{2}\left[\frac{k_1}{\zeta}(x - \xi)^2 + (y - \eta)^2 + \left(\frac{k_2}{z_2} - \frac{k_1}{z_1}\right)(x^2 + y^2)\right] \\[2mm]
&= \exp \frac{i}{2}[\phi(x) + \psi(y)]
\end{aligned}
\tag{9.42}
$$

where we defined

$$
\phi(x) = \frac{k_1}{\zeta}(x - \xi)^2 + \left(\frac{k_2}{z_2} - \frac{k_1}{z_1}\right)x^2
\tag{9.43}
$$

with a similar expression for $\psi(y)$.

Developing Eq. (9.43) we obtain

$$
\phi(x) = \left(\frac{k_1}{\zeta} + \frac{k_2}{z_2} - \frac{k_1}{z_1}\right)x^2 - 2\frac{k_1}{\zeta}x\xi + \frac{k_1}{\zeta}\xi^2.
$$

If we put

$$a \equiv \frac{k_1}{\zeta} + \frac{k_2}{z_2} - \frac{k_1}{z_1} \left.\vphantom{\frac{k_1}{\zeta}}\right\} \tag{9.44}$$

$$b \equiv \frac{k_1}{\zeta}$$

Eq. (9.43) can be written as

$$\phi(x) = a\left[\left(x - \frac{b}{a}\xi\right)^2 + \left(\frac{b}{a} - \frac{b^2}{a^2}\right)\xi^2\right]. \tag{9.45}$$

Disregarding constant phase factors, we can write U_v (Eq. 9.42) in the form

$$U_v = \exp\frac{i}{2}\left[a\left(x - \frac{b}{a}\xi\right)^2 + a\left(y - \frac{b}{a}\eta\right)^2\right]. \tag{9.46}$$

This wave is exactly of the same form as U_0, but instead of diverging from the point $0(\xi, \eta, \zeta)$ it diverges from a new point, $0_1(\xi', \eta', \zeta')$, where

$$\xi' = \frac{b}{a}\xi; \quad \eta' = \frac{b}{a}\eta \quad \text{and} \quad \frac{k_2}{\zeta'} = a. \tag{9.47}$$

The magnification in the x and y direction can be defined as

$$M_{x,y} = \frac{d\xi'}{d\xi} = \frac{b}{a}, \tag{9.48}$$

and, in the z direction,

$$M_z = \frac{d\zeta'}{d\zeta}. \tag{9.49}$$

By substituting Eq. (9.44) we obtain

$$M_{x,y} = \left(1 + \frac{k_2\,\zeta}{k_1\,z_2} - \frac{\zeta}{z_1}\right)^{-1} \tag{9.50}$$

and

$$\frac{d\zeta'}{d\zeta} = \frac{k_2 k_1}{a^2\,\zeta^2} = \frac{k_2}{k_1}\left(\frac{a\zeta}{k_1}\right)^{-2} = \frac{k_2}{k_1}\left(1 + \frac{k_2\,\zeta}{k_1\,z_2} - \frac{\zeta}{z_1}\right)^{-2}. \tag{9.51}$$

Comparison with Eq. (9.50) yields

$$M_z = \frac{k_2}{k_1}M_{x,y}^2; \tag{9.52}$$

thus the magnifications in the lateral and longitudinal directions are different unless

$$\frac{k_2}{k_1}M_{x\,y} = 1. \tag{9.53}$$

This is a difficult condition to satisfy especially in 3-D objects, where ζ is not constant. Writing Eq. (9.50) in the form

$$M_{x,y} = \left[1 + \zeta\left(\frac{\lambda_1}{\lambda_2 z_2} - \frac{1}{z_1}\right)\right]^{-1} \tag{9.50'}$$

we may observe a few facts. Unless $M_{x,y} = 1$, a three-dimensional object will always be distorted since $M_{x,y}$ depends on the ζ coordinate of the object. Unit magnification in the (x,y) plane will be obtained when

$$\lambda_2 z_2 = \lambda_1 z_1. \tag{9.54}$$

In the special case of plane-wave reference and reconstructing beams $(z_1 = \infty$ and $z_2 = \infty)$, there is no magnification possible.

Gabor's idea (1949, 1951) was to obtain high magnification by change of wavelength. This can easily be demonstrated for a special case where $z_1 = z_2 = \zeta$. In this case we obtain

$$M_{x,y} = \frac{\lambda_2}{\lambda_1}; \tag{9.55}$$

in principle, making a hologram by X-rays $(\lambda_1 \sim 1 \text{ Å})$ and reconstructing by visible light $(\lambda_2 \sim 5{,}000 \text{ Å})$, a magnification of 5,000 is obtained, even without changing the optical configuration.

However, this result is not so good as it seems at first sight. We already saw that there are aberrations, even in the first approximation. Other kinds of aberrations will appear in a more exact treatment, including the usual aberrations of lenses.

At this point it is instructive to illustrate the similarity between holograms and thin lenses.

Substituting Eq. (9.44) into Eq. (9.47), we obtain

$$\frac{1}{\zeta'} = \frac{k_1}{k_2}\left(\frac{1}{\zeta} - \frac{1}{z_1}\right) + \frac{1}{z_2}. \tag{9.47a}$$

Using the same notation for a negative lens (the virtual image corresponds to a negative lens while the real image corresponds to a positive one) of focal length f, the well-known lens equation is

$$\frac{1}{\zeta} - \frac{1}{\zeta'} = -\frac{1}{f}. \tag{9.47b}$$

For $k_1 = k_2$, a comparison between Eqs. (9.47a) and (9.47b) yields for the hologram

$$\frac{1}{f} = \frac{1}{z_2} - \frac{1}{z_1}. \tag{9.47c}$$

This resembles the formula relating the focal length to the radii of curvature of the lens surfaces. The lateral magnification of a lens is known to be

$$M(l)_{x,y} = \frac{\zeta'}{\zeta}. \tag{9.48a}$$

Substituting Eqs. (9.47b) and (9.47c) into Eq. (9.48a) we obtain

$$M(l)_{x,y} = \left(1 + \frac{\zeta}{f}\right)^{-1} = \left[1 + \zeta\left(\frac{1}{z_2} - \frac{1}{z_1}\right)\right]^{-1}. \tag{9.50a}$$

Equation (9.50a) is the same as Eq. (9.50) with $k_1 = k_2$. The longitudinal magnification can be found by using Eq. (9.47b) to evaluate Eq. (9.49):

$$M_z = \left(\frac{\zeta'}{\zeta}\right)^2 = (M_{x,y})^2. \tag{9.52a}$$

and again, this corresponds to Eq. (9.52).

To end this section, an additional remark regarding Eq. (9.47) is worthwhile: ζ' may become negative if $\zeta > z$ and this will result in a real "virtual image". A real image of this kind can be used for microscopic study, which is difficult with an image formed beyond the hologram—the usual virtual image.

A special case occurs when $\zeta = z_1$, as in lensless Fourier-transform holograms (Stroke, 1969). From Eq. (9.47a) we obtain $\zeta' = z_2$, and if a plane wave is used for reconstruction, such that $z_2 = \infty$, ζ' is also ∞ and the "virtual image" can be obtained also as a real image, at the focal plane of a lens, together with the "real image".

D. The Problem of Resolution

The resolution in holographic systems is quite a complicated problem. Extensive discussions can be found in the literature cited in the previous section, and here we shall limit ourselves only to a few general remarks.

In conventional photography, the image is focused on the photographic plate. Two image points can be resolved on the photograph if they are separated by a distance exceeding the grain size (the resolution limit of the plate). In a hologram, on the other hand, the information on an image point is dispersed over the whole plate; thus the resolution of the emulsion has nothing to do with the resolution of the image. The resolution in a hologram will be determined by its numerical aperture as if it were the entrance pupil of an optical system (Champagne and Massey, 1969). Moreover, an exact calculation will show that the ultimate resolution will be only about half of the classical resolution limit. The basis of this is the fact that, after all, a hologram does not record *all* the information carried by the object wave. A hologram records the portions of the wave that are in phase with the reference wave, while one half of the information, i.e., that which is out of phase with the reference wave, is lost (Russell, 1969).†

Although the image resolution is not determined directly by plate resolution, it has some indirect effect. The plate resolution limits the recordable spatial bandwidth which, in turn, limits the angular view of the object.

The angle subtended by the plate at some point on the object is often larger than the allowable bandwidth and this will result in a decrease of the effective aperture causing a decrease in resolution. A way around this difficulty is to use a point reference source in the plane of the object and then the bandwidth requirements are limited by a large factor—especially if the object is much smaller than the plate as is the case in microscopy.

A final remark should be added to the resolution problem: holograms have to be recorded by coherent light. If coherent light impinges on a rough surface a speckled pattern appears limiting the resolution to about 20 l/mm. However, this can be overcome by using very smooth surfaces.

E. Applications of Holographic Microscopy

As should be evident from the previous section, holographic microscopy is not so promising as it seems at first sight. Although, in principle, very high magnification should be attainable there are, in general, large aberrations and also the resolution is not sufficient. However, holography is very important where three-dimensional information is needed and where focusing is difficult.

The first holographic microscope was already built in 1951 (Gabor and Goss, 1966), but it was difficult to use before the invention of the laser. Most promising is the use of holographic microscopy to dynamic particle fields where the whole volume can be observed together (Trolinger *et al.*, 1969; Feleppa, 1969).

Some additional comments on holographic microscopy will be given also in Section VIII.A, regarding changes of hologram scale and wavelength.

F. Holographic Interferometry

In the usual applications of holography, the motion of the object during exposure is a nuisance. However, if this motion has to be measured, holography may serve as an extremely sensitive and powerful tool.

G. Mathematical Treatment of Holographic Interferometry

We shall analyse holographic interferometry, referring to Fig. 12: $S_1(x_1)$ and $S_r(x_r)$ are points in the object illuminating source and reference source

---

† The disturbing appearance of the conjugate wave in the reconstruction is also a result of this "half information".

respectively. We use the notation (x) for simplicity, but it has to be understood that it stands for (x, y, z). Points on the object and photographic plate are denoted respectively by $0(\xi)$ and $P(x)$.

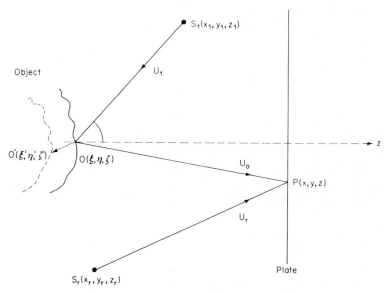

Fig. 12. Schematic diagram, explaining the principle of holographic interferometry. The object point $0(\xi, \eta, \zeta)$ is translated by the vector $\delta\xi$ to a point $0'(\xi', \eta', \zeta')$.

Defining the distances,

$$
\left.
\begin{aligned}
\mathbf{d}_1 &\equiv \mathbf{S}_1\,\mathbf{0} \\
\mathbf{d}_0 &\equiv \mathbf{OP} \\
\mathbf{d}_r &\equiv \mathbf{S}_r\mathbf{P}
\end{aligned}
\right\}
\tag{9.48}
$$

the wave illuminating the object will have an amplitude U_1 at 0, given by

$$
U_1 = A_1(d_1)\exp{-i\mathbf{k}_1 . \mathbf{d}_1}.
\tag{9.49}
$$

The object, having a scattering function $f(\xi)$, will produce a wave amplitude U_0 at the plate having the form

$$
U_0 = \int U_1 \frac{f(\xi)}{d_0} \exp\left(-(i)k d_0\right) d\xi.
\tag{9.50}
$$

If the object is stationary during exposure, U_0 is recorded holographically, producing a transfer function

$$
H \sim |U_r + U_0|^2.
\tag{9.51}
$$

Where U_r is the reference wave

$$
U_r = A_r \exp{-i\mathbf{k}_r . \mathbf{d}_r}.
\tag{9.52}
$$

Suppose the object moves now to an adjacent position, such that point $0(\xi)$ is translated to point $0'(\xi')$: then

$$\xi' = \xi + \delta\xi. \tag{9.53}$$

The wave impinging now on the plate will be

$$U_0' = \int U_1' \frac{f(\xi')}{d_0'} \exp(-ikd_0')\,d\xi'. \tag{9.54}$$

The dash on each quantity means that the coordinate ξ is replaced by ξ'.

If the previous plate is exposed once more, we shall obtain a double-exposed hologram with two transfer functions H and H' superposed.

$$H + H' = |U_r + U_0|^2 + |U_r + U_0'|^2. \tag{9.55}$$

Reconstruction in the usual manner by U_r will yield a superposition of two reconstructed waves. The virtual image-wave will have the form

$$U_v = U_0 + U_0'. \tag{9.56}$$

On the other hand, if the hologram is developed after the first exposure and replaced exactly in its previous position the zero-order diffracted part of U_0' will interfere with the reconstructed U_0, again producing the wave given by Eq. (9.56). In either case there will be interference between U_0 and U_0' the first corresponding to a recorded interference and the second being suitable for real-time observation of a changing U_0'.

To analyse the interference produced, we shall develop Eq. (9.56). First we observe that integration on $f(\xi')$ is the same as integration on $f(\xi)$ since it is the same object translated. We can thus write Eq. (9.54) in the form

$$U_0' = \int U_1(\xi + \delta\xi)\frac{f(\xi)}{d_0(\xi + \delta\xi)} \exp\{-ikd_0(\xi + \delta\xi)\}\,d\xi. \tag{9.54'}$$

Assuming a small displacement, i.e. $\xi \gg \delta\xi$, we may write in the denominator

$$d_0(\xi + \delta\xi) \sim d_0(\xi).$$

Substituting the relations

$$\mathbf{d}(\xi + \delta\xi) = \mathbf{d} + \delta\boldsymbol{\xi} \tag{9.57}$$

and

$$\mathbf{k}(\xi + \delta\xi) = \mathbf{k} + \delta\mathbf{k}$$

we obtain

$$\begin{aligned} U_1' &= A_1(d_1)\exp\{-i(\mathbf{k}_1 + \delta\mathbf{k}_1)\cdot(\mathbf{d}_1 + \delta\boldsymbol{\xi})\} \\ &= A_1(d_1)\exp\{-i\mathbf{k}_1\cdot\mathbf{d}_1\}\exp\{-i(\mathbf{k}_1\cdot\delta\boldsymbol{\xi} + \delta\mathbf{k}_1\cdot\mathbf{d}_1 + \delta\mathbf{k}_1\cdot\delta\boldsymbol{\xi})\} \end{aligned} \tag{9.58}$$

where we neglect the change of A_1 for the small distance $\delta\xi$.

From Eq. (9.58) we obtain

$$U_1' = U_1 \exp{-i\delta\psi} \tag{9.59}$$

where

$$\delta\psi = \mathbf{k}_1 . \delta\boldsymbol{\xi} + \delta\mathbf{k}_1 . (\mathbf{d}_1 + \delta\boldsymbol{\xi}).$$

Substituting Eqs. (9.59) and (9.57) into Eq. (9.54′) we obtain

$$U_0' = \int U_1 \exp i\delta\psi \frac{f(\xi)}{d_0} \exp\{-ik|\mathbf{d}_0 + \delta\boldsymbol{\xi}|\} \, d\xi. \tag{9.60}$$

In general this is a complicated integral and very difficult to evaluate. To proceed further we start simplifying the term

$$|\mathbf{d}_0 + \delta\boldsymbol{\xi}| = [(x - \xi - \delta\xi)^2 + (y - \eta - \delta\eta)^2 + (z - \zeta - \delta\zeta)^2]^{1/2} \tag{9.61}$$

by assuming that the cylindrical approximation

$$(z - \eta)^2 \gg (x - \xi)^2 + (y - \eta)^2$$

is applicable, and that the displacement is only in the z direction,† i.e.

$$\delta\xi = \delta\eta = 0.$$

Thus we have

$$|\mathbf{d}_0 + \delta\boldsymbol{\xi}| \simeq (z - \zeta - \delta\zeta) + \frac{(x - \xi)^2 + (y - \eta)^2}{2(z - \zeta - \delta\zeta)}. \tag{9.61a}$$

If we neglect $\delta\zeta$ in the denominator, we obtain

$$|\mathbf{d}_0 + \delta\boldsymbol{\xi}| \simeq (z - \zeta) + \frac{(x - \xi)^2 + (y - \eta)^2}{2(z - \zeta)} - \delta\zeta$$

or

$$|\mathbf{d}_0 + \delta\boldsymbol{\xi}| \simeq d_0 - \delta\zeta. \tag{9.62}$$

We also have

$$\mathbf{k}_1 . \delta\boldsymbol{\xi} = k_{1z} \delta\zeta$$

and

$$\delta\mathbf{k}_1 = \frac{d\mathbf{k}_1}{d\zeta} \delta\zeta$$

that, together with Eq. (9.62), transform Eq. (9.60) into

$$U_0' \simeq \int U_1 \frac{f(\xi)}{d_0} \exp\{-ikd_0\} \exp\{i(k - k_{1z} + (d\mathbf{k}_1/d\zeta).\mathbf{d}_1 + (d\mathbf{k}_1/d\zeta).)\} \delta\zeta \, \delta\boldsymbol{\xi} \, d\xi. \tag{9.60'}$$

---

† Generality is not lost by this assumption, since the z direction can be chosen arbitrarily.

If U_1 is a uniform plane wave, then $k_{1z} = k \cos\theta$ (Fig. 12), and $dk_1/d\zeta = 0$. The quantities k_{1z}, k and $\delta\zeta$ do not depend on ξ, so that they can be removed from the integration, and we finally obtain

$$U_0' \simeq \exp\{i(k_{1z} + k)\delta\zeta\} \int U_1 \frac{f(\xi)}{d_0} \exp\{-ikd_0\} \, d\xi$$

or

$$U_0' \simeq \exp\{i(k_{1z} + k)\,d\zeta\}\, U_0. \qquad (9.63)$$

Substitution into Eq. (9.56) yields

$$U_v = U_0[1 + \exp\{ik\delta\zeta(1 + \cos\theta)\}]. \qquad (9.64)$$

Observing the reconstruction, we can see an intensity distribution

$$I_v = |U_v|^2 = 2|U_0|^2\,(1 + \cos\varDelta) \qquad (9.65)$$

where we define

$$\varDelta = k\delta\zeta(1 + \cos\theta). \qquad (9.66)$$

I_v is exactly the image of the object, but it is superposed by dark lines, each connecting all points on the object satisfying

$$\varDelta = \pi(2n + 1) \qquad (9.67)$$

where n is an integer or zero.

More detailed treatment of the subject has been given by Stetson (1969), Brown et al. (1969) and Sollid (1969).

H. Strain Analysis and Non-Destructive Testing

Holographic interferometry produced a revolution in interferometric techniques. It introduced two remarkable advances. First, before the holographic era, all interferometric work had to be done with high-quality optical components; in a holographic system, any rough object can serve as a reflector (or transmitting media) without affecting the quality of the interference pattern. Secondly, in conventional interferometry, two processes could be compared only in real time, while holography is able to compare a system with its state in the past, or with the state of another system in the present or past.

These characteristics make holography a very powerful tool, applicable in research and technology. One such application is the measurement of strains induced in various objects. For example, this technique can be used to evaluate the behaviour of industrial products under different conditions. Due to the high sensitivity of the technique very small changes in conditions (pressure, temperature, or some applied forces) suffice to produce observable changes (Boone and Verbiest, 1969). This is also a requirement for non-destructive testing. Figure 13 is a reproduction from the article by Grant and Brown (1969) showing interference fringes produced in tyres due to small changes in pressure, temperature and creep. Faults in the tyre are easily visible in the interferometric recording.

Another application of this technique is the comparison of many similar objects. For example, in a production line a master object may be used to

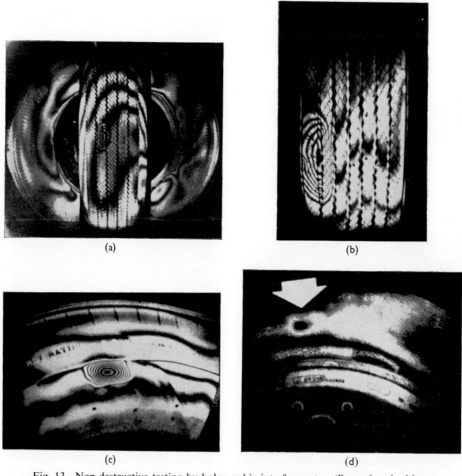

(a)   (b)

(c)   (d)

Fig. 13. Non-destructive testing by holographic interferometry. (Reproduced with permission from Grant and Brown, 1969.) (a) Double exposure, creep, 50 psi; shoulder separation between liner and first ply; separation between first and second plies in tread region (8·25 × 14, 4-ply tyre). (b) Real time, creep, 30 psi; large separation at tread edge between outer ply and tread (8·25 × 14, two-ply, tubeless tyre). (c) Double exposure, heat, 30 psi; medium separation between outer ply and sidewall (8·25 × 14, two-ply, tubeless tyre). (d) Real time, pressure change, 100 psi; separation at turnup of first wire bead; through 17 layers of ply and about $\frac{3}{4}$-in. deep; defect about 1 in. × $2\frac{1}{2}$ in. (25 × 6·75 18PR aircraft tyre).

record a hologram and each product can be compared with it interferometric-ally. Sometimes the master may even be an idealized object produced, for example, by computer (Pastor, 1969).

To end this section, another holographic method for estimating strains is worth mentioning.

In order to explain the method we shall return to Fig. 4, representing the principle of holographic recording. On the one hand, illuminating the de-veloped hologram with the reference beam U_R results in the reconstruction of the object beam U_0; on the other hand, a reconstruction may also be performed by illuminating with the object beam U_0, and then the reference beam U_R will be reconstructed. However, if U_0 is not exactly the same as during the recording process, the reconstruction of U_R will not be perfect. The intensity of the reconstructed beam may be used as a measure for the correlation between the original object beam and the reconstructing beam. Marom (1970) devised a way to use this technique in real-time strain measurements.

I. Vibration Analysis

Until now stationary or quasi-stationary systems have been discussed; it was assumed that during exposure the system is stationary. To analyse non-stationary systems, we shall return to Eq. (9.63) and assume that $\delta\zeta$ is a function of time, or using Eq. (9.66) we may rewrite (9.63) as

$$U_0'(t) = U_0 \exp\{i\Delta(t)\}. \tag{9.68}$$

The intensity to be recorded by the plate is a function of time:

$$I(t) = |U_r + U_0 \exp\{i\Delta(t)\}|^2,$$

or

$$I(t) = |U_r|^2 + |U_0|^2 + U_r^* U_0 \exp\{i\Delta(t)\} + U_r U_0^* \exp\{-i\Delta(t)\}. \tag{9.69}$$

The transfer function H of the plate will be determined by the exposure time, T.

$$H \simeq \int_0^T I(t)\,dt$$

$$= (|U_r|^2 + |U_0|^2)\,T + U_r^* U_0 \int_0^T \exp\{i\Delta(t)\}\,dt + U_r U_0^* \int_0^T \exp\{-i\Delta(t)\}\,dt. \tag{9.70}$$

Illumination by the reference beam, U_r, produces a virtual image,

$$U_v = U_0 \int_0^T \exp\{i\Delta(t)\}\,dt, \tag{9.71}$$

which is the wave from a stationary object, multiplied by the integral

$$M = \int\limits_{0}^{T} \exp\{i\Delta(t)\}\,dt. \tag{9.72}$$

A stationary object can be defined by requiring that $\Delta(t)$ does not change more than about $\pi/4$, during the time T. If $\Delta(t)$ changes much Eq. (9.72) can be replaced by the average of the argument:

$$M \simeq T\langle\exp\{i\Delta(t)\}\rangle. \tag{9.72'}$$

A random motion (noise), or linear displacement, yields $M = 0$, which means that no reconstruction is possible. However, for a periodic function M may have values different from zero. For example, if an object point, $P(\xi, \eta, \zeta)$,

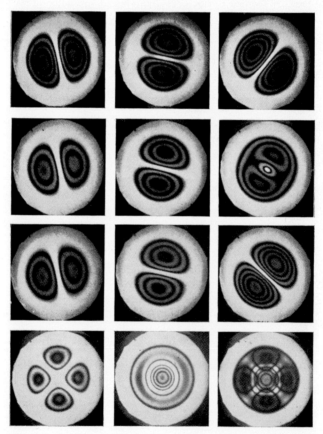

Fig. 14. Reconstruction from time average holograms of vibrating objects. (Reproduced, with permission, from Molin and Stetson, 1969.)

vibrates sinusoidally with an amplitude $a(\xi, \eta, \zeta)$, Eq. (9.72') will yield the zero-order Bessel function

$$M \simeq |J_0(4\pi a/\lambda)|^2 \qquad (9.73)$$

where λ is the wavelength (Powell and Stetson, 1965; Stetson, 1968). Equation (9.73) has zeros, and these are the interference fringes produced in "time-average holography". Such fringes are reproduced in Figs. 14 and 15.

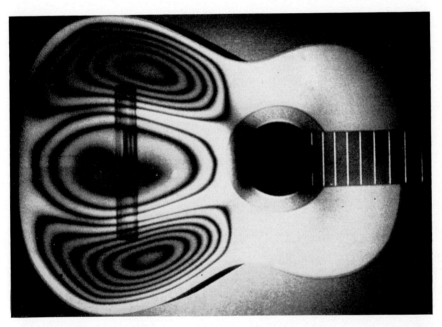

Fig. 15. Time average hologram of a vibrating guitar. (Reproduced with the permission of the Institute of Optical Research, Stockholm.)

Time-average holography is only one holographic method for the observation of vibrating objects. A second method is to record a hologram of the stationary object, replacing it, and observing live interference fringes. This is very useful in real-time analysis of vibrating objects. The third method is the stroboscopic recording. If the vibration frequency of the object is known, it may be synchronized to a stroboscopic illumination in such a way that the object will be recorded only at two points on its vibration cycle. The record produced by the stroboscopic method is equivalent to a double exposed hologram, yielding more defined fringes than the time-average hologram (Watrasiewicz, 1968; Zaidel' et al., 1969).

J. Flow Visualization (Interferometry of Phase Objects)

Equation (9.65) was obtained for an object that moved after the first exposure of the hologram. A similar relation can be obtained for a phase object that is altered after the first exposure.

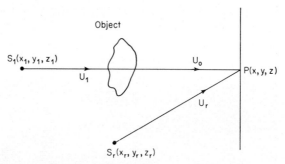

Fig. 16. Holography of phase objects.

Figure 16 is a schematic diagram for the holographic recording of phase objects. A phase object can be described by a transfer function

$$f(\xi, \eta, \zeta) = \exp\{i\phi(\xi, \eta, \zeta)\}, \tag{9.74}$$

which changes the illuminating beam, U_1, into

$$U_0 = U_1 f(\xi, \eta, \zeta) = U_1 \exp\{i\phi(\xi, \eta, \zeta)\}. \tag{9.75}$$

A simple recording and reconstruction of U_0 will yield an intensity distribution

$$|U_0|^2 = |U_1 \exp\{i\phi(\xi, \eta, \zeta)\}|^2 = |U_1|^2. \tag{9.76}$$

Thus, the object would be invisible. However, if before that, or after that, U_1 was recorded separately the double-exposed hologram will reconstruct a virtual image wave $U_1 + U_0$, with an intensity distribution,

$$\begin{aligned} I(\xi, \eta, \zeta) &= |U_1|^2 |1 + \exp\{i\phi(\xi, \eta, \zeta)\}|^2 \\ &= 2|U_1|^2 |1 + \cos\phi(\xi, \eta, \zeta)|. \end{aligned} \tag{9.77}$$

The object will be described by interference fringes running along constant phase lines. The same method can be used to visualize changes of phase in the existing object.

Phase-object holography is a very powerful tool for flow analysis. Flow patterns and shock waves can be observed dynamically or recorded in a double-exposed hologram. In each case a 3-D picture may be reconstructed (Tanner, 1965, 1966). High-quality optics becomes unnecessary; and a process can be recorded through any transparent material (Heflinger et al., 1966).

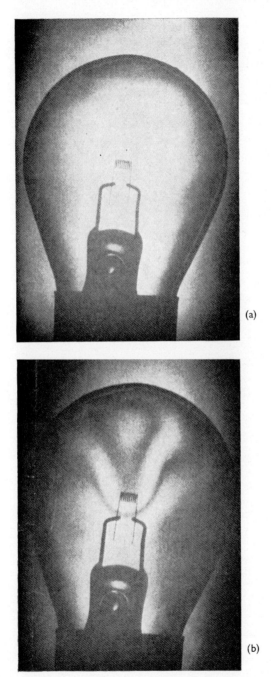

(a)

(b)

Fig. 17. Reconstruction from (a) conventional hologram; (b) double-exposed hologram, of a glowing lamp, showing convection currents. (Reproduced, with permission, from Heflinger *et al.*, 1966.)

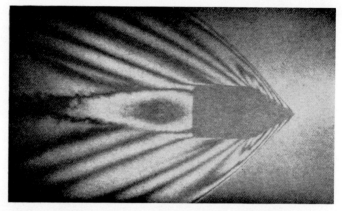

Fig. 18. Shock wave pattern of a flying bullet, reconstructed from a double-exposed hologram. (Reproduced, with permission, from Heflinger *et al.*, 1966.)

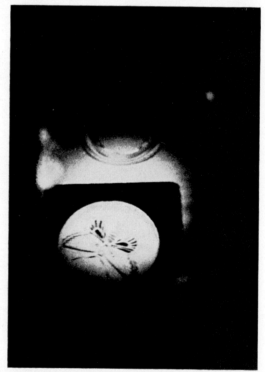

Fig. 19. Reconstruction from hologram of a birefringent phase object recorded with circularly polarized light.

A useful method in phase-object holography is to supply a background fringe pattern (for example with the help of a gas prism). The phase object will bend these lines, and this makes the measurements easier (Jahoda, 1969).

A special kind of phase object is a birefringent object. Its characteristics can be recorded using linearly polarized light (Rogers, 1966; De and Sévigny, 1967). However, the birefringent characteristics of an object can best be analysed using a single exposure with circularly polarized light. Reconstruction

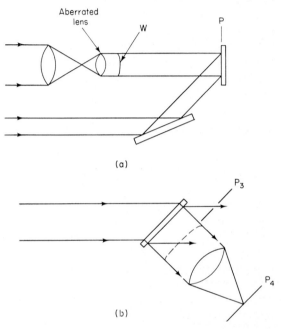

Fig. 20. Holographic correction of lens aberrations. (Reproduced, with permission, from Upatnieks *et al.*, 1966.)

of such a hologram can show, for example, a 3-D view of photoelastic fringes (Fig. 19) (Politch *et al.*, 1970).

K. *Holography in Optical Technology*

Holographic interferometry can be used in the optical workshop to test and manufacture optical components. Optical surfaces may be compared to idealized surfaces, and even to computer-generated functions (Pastor, 1969). While holographic interferometry can help in the conventional manufacture of optical components, general holographic techniques may be applied to new methods of producing them. The building blocks of holography—the diffraction grating and Fresnel zone plate—are natural candidates for such production. And, indeed, high-quality gratings and zone plates have been obtained by simple holographic techniques (George and Matthews, 1966; McMahon and Franklin, 1969; Labeyrie, 1969).

A hologram recorded by the light leaving an optical component, such as a lens, is capable of reproducing the wave emerging from the component together with its aberrations. Since the conjugate reconstructed wave is inverted, it may serve as a correction for the lens itself. Upatnieks *et al.* (1966) developed a method to produce phase-correction plates for lenses. Figure 20 illustrates the basic principle of the procedure. Using this method, various plates can be produced to correct the aberrations of optical systems.

Another interesting application of holographic techniques is to measure the coherence length of light sources. A holographic record of a long object will reconstruct only those parts of the object that are within the coherence length of the source (Staselko *et al.*, 1969).

VIII. Extension of Holographic Techniques to Other Fields

Holography is not limited to light waves and, in principle, any wave motion may produce holograms. Gabor's idea—to record a hologram by electron waves and reconstruct by light—is still the dream of many workers in microscopy. Although much work is being done to evaluate the expected characteristics of electron and X-ray holograms (Stroke, 1968; Winthrop and Worthington, 1965; Stroke, 1969), they are likely to remain a dream for a long time to come. The main reason for this is the absence of coherent electron-wave and X-ray sources. However, incoherent "semi-holographic" techniques, to be discussed later, may be applied to these regions of the spectrum as well.

Since lasers are available in the ultra-violet and infra-red, holography is possible in these regions although there are still difficulties with the recording materials (Simpson and Deeds, 1970). However, at present the most interesting extension of holography is in microwaves and acoustics. There is no coherence problem in these regions; thus there is nothing to hinder the flourishing of microwave and acoustic holography.

A. Acoustic Holography

Acoustic holography is a typical example of a peculiar phenomenon in science; i.e., a great advance must take place in a particular field of research in order to realize what could have been done for a long time in a quite different field. Acoustic holography started only after optical holography reached a high stage of development, although acoustic technology was adequate for this purpose long before that. From a historical point of view, no less interesting is the fact that optical holography was translated into acoustics "word by word"; the main point in optical holography is the recording of phase with the help of a reference beam; it took some time to realize that acoustic detectors are sensitive to amplitude and phase, and thus a reference beam is superfluous, (Metherell, 1969).

Actually imaging with sound is not new. Sonar devices have been known for a long time and they produce images on a screen similar to radar. However, as in optical photography, phase information is lost and 3-D imaging is not possible. The importance of imaging with sound stems from the fact that acoustic waves are able to penetrate deep into media that are opaque for electromagnetic waves. Sound waves are reflected from irregularities in the medium and this makes these waves a useful probe to detect these irregularities. The most useful procedure in acoustic holography would be to make a hologram with acoustic waves and reconstruct by light to produce a 3-D image of irregularities and scattering objects immersed in the medium.

A 3-D picture of the layers in the earth crust should prove an extremely valuable tool in search for oil, water and minerals. For oceanography such a tool is no less valuable, while in medicine it may even replace X-rays (Thurston, 1969; Weiss and Holyoke, 1969). Other potential applications of acoustic holography are in nondestructive testing and military uses.

B. Recording Techniques

The most straight-forward translation of optical holography is the use of some sonosensitive plate which will record the interference pattern of an object wave with a coherent reference wave (Greguss, 1966, 1967). Another version of this technique is the replacement of the sonosensitive plate by the surface of a liquid (Marom et al., 1967) or a flexible membrane. The surface wave pattern on the liquid may serve as a reflection phase hologram, when illuminated from above by laser light. Surface wave holography makes it possible to view, in real time, objects "illuminated" by ultra-sonic waves.

Using these techniques an acoustic reference wave has to be provided in the same way as in optical holography. If the acoustic wave is recorded electronically the reference phase may be provided internally to the detector (Massey, 1967). The hologram using this method may be produced by scanning the acoustic field (with the help of a microphone or piezoelectric transducer) mixing the signal with the reference and recording the output pattern. The recording can be done by photographing a synchronous light bulb (which may be attached to the scanning detector) or by displaying the intensity pattern on a CRT screen and photographing this pattern. A stationary detector array may also be used instead of a scanning detector. In this case the whole field may be sensed together and, since it is possible to sense the amplitude and phase relations, there is no need for a reference at all. The reference can be omitted even in scanning devices when the scanning velocity is known (Metherell, 1968).

An interesting recent recording technique is due to Metherell *et al.* (1969) —temporal reference acoustical holography. This method applies holographic interferometry and dispenses altogether with the acoustic reference wave. The acoustic detector is a membrane or a liquid surface. A doubly-exposed optical hologram is recorded of the surface with a time interval of half an acoustic cycle between the two exposures. To show that this record is a hologram of the acoustic field we shall analyse a simple case.

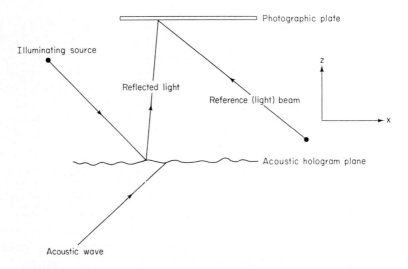

Fig. 21. Recording scheme for a "temporal reference acoustical hologram".

Let us assume that an oblique acoustic plane wave is incident on the recording surface (Fig. 21 describes the configuration). At a certain time, the surface deformation in the z direction can be given in the form

$$z_1 = A(1 + \cos \nu x) \qquad (9.78)$$

where A is a constant and ν is the acoustic spatial frequency. Half an acoustic cycle later, the deformation will be

$$z_2 = A(1 - \cos \nu x) \qquad (9.79)$$

the difference between the two exposure moments being

$$\Delta z = z_1 - z_2 = 2A \cos \nu x.$$

If we substitute this into Eq. (9.65), we obtain the intensity distribution in the reconstructed wave from this hologram,

$$I = k[1 + \cos(2A \cos \nu x)],$$

where k is some constant. If $2A$ is small enough this may be photographed to produce a grating of spatial frequency ν which is adequate to reconstruct the acoustic wave. Using a high-frequency acoustic wave (4·8 MHz) with a high-power double-pulsed laser, this technique is insensitive to object motion and environmental noise.

C. The Reconstruction

In general, the usual procedure in acoustic holography is the recording of an acoustic wavefront and reconstructing by a light wave. The result should be an optical wavefront identical in shape to the original acoustic wave. There are two main points to be considered here: (a) the reconstructing wavelength will, usually, differ from the recording wavelength by at least two or three orders of magnitude; (b) the scale of the hologram itself will usually be changed between recording and reconstruction.

To analyse the situation, we shall proceed in a similar way to Section VII.C. The transfer function of a recorded hologram will be given by (taking only the virtual-image term)

$$H_\nu(x,y) = U_1^* U_0 = c \exp \frac{i}{2} \left[\frac{k_1}{\zeta}(x - \xi)^2 + \frac{k_1}{\zeta}(y - \eta)^2 - \frac{k_1}{z_1}(x^2 + y^2) \right] \quad (9.82)$$

where c is some constant and the notation of Section VII.C was used.

If the hologram is demagnified by a factor P, the spatial frequency will increase. Denoting the new plate coordinates by x', we can write

$$Px' = x, \qquad Py' = y, \qquad (9.83)$$

so that the new transfer function will be

$$H_\nu'(x',y') = c \exp \frac{i}{2} \left[\frac{k_1}{\zeta}(Px' - \xi)^2 + \frac{k_1}{\zeta}(Py' - \eta) - \frac{k_1}{z_1}((Px')^2 + (Py')^2) \right].$$
$$(9.84)$$

Illuminating by a reconstructing beam U_2 (Eq. 9.41) and omitting the primes (since (x',y') are now the coordinates of the plate) we can proceed as from Eq. (9.42) to Eq. (9.47); however, the parameters a and b will now be defined by the relations

$$\left. \begin{aligned} a &= \frac{k_1 P^2}{\zeta} - \frac{k_1 P^2}{z_1} + \frac{k_2}{z_2} \\ b &= \frac{k_1 P}{\zeta}. \end{aligned} \right\} \qquad (9.85)$$

The magnification in the lateral direction is

$$M_{x,y} = \left[P + \zeta \left(\frac{k_2}{k_1} \frac{1}{P} \frac{1}{z_2} - \frac{P}{z_1} \right) \right]^{-1}. \tag{9.86}$$

The longitudinal coordinate will be transformed by

$$\zeta' = \frac{k_2}{a} = \left[\frac{k_2 P^2}{k_2} \left(\frac{1}{\zeta} - \frac{1}{z_1} \right) + \frac{1}{z_2} \right]^{-1}, \tag{9.87}$$

and the longitudinal magnification will again be given by

$$M_z = \frac{k_2}{k_1} M_{(x,y)}^2. \tag{9.88}$$

If a nondistorted image is to be obtained the following relations should be satisfied: $M_{x,y}$ should not depend on ζ, and thus from Eq. (9.86) we obtain

$$\frac{k_2}{k_1} = \frac{z_2}{z_1} P^2. \tag{9.89}$$

Then

$$M_{(x,y)} = P^{-1}, \tag{9.90}$$

and, since usually $P \gg 1$, a highly demagnified image will be reconstructed. To obtain a true image, we would also require $M_{(x,y)} = M_z$; thus from Eqs. (9.88) and (9.90) we have

$$\frac{k_2}{k_1} M_{x,y} = 1$$

or

$$\frac{k_2}{k_1} = \frac{\lambda_1}{\lambda_2} = P. \tag{9.91}$$

Substituting Eq. (9.91) back into Eq. (9.89) we obtain the relation between the reference and reconstructing wave curvatures,

$$\frac{z_1}{z_2} = P, \tag{9.92}$$

as a necessary condition for undistorted reconstruction. A special case is

$$z_1 = z_2 = \infty.$$

These results may be summed up as follows. A hologram recorded by acoustic waves, may be reconstructed by light waves. In general, the reconstructed image will be highly distorted, but this does not mean that it cannot be accurately analysed. To eliminate distortions Eqs. (9.91) and (9.92) have to be satisfied; but then the image will be very small and it will have to be observed through a microscope. This is possible because the image is formed very close to the plate (we can see this by substituting Eq. (9.92) into Eq. (9.84) and obtain $\zeta' = \zeta/P$).

Some more refined discussion on image formation in acoustic holography is given in the literature (for example, Metherell, 1969; Mueller *et al.*, 1969).

D. Microwave Holography

Microwave sources are coherent like acoustic sources. Phase and amplitude detection is also possible; thus the recording techniques are similar to those used for acoustic holograms (Larson, 1969). The most common recording technique is some way of scanning the hologram plane and mixing the signal with an internal reference. The output is displayed on an oscilloscope screen and recorded on a photographic plate. As in acoustic holography the reconstruction is carried out using laser light. Therefore the considerations of Section VIII.C apply also for microwave holography.

Perhaps the most interesting application of microwave holography is in the field of radar data processing. One of these is the synthetic-aperture side-looking radar (Cutrona *et al.*, 1966). Here a relatively small antenna, carried by a low-flying airplane, scans the "hologram plane". However, the scan is done only along a horizontal line and so the "hologram" has only one dimension. The second dimension corresponds to the perpendicular range from the line of flight. Since the hologram is essentially one-dimensional the reconstruction also has to be done in one dimension, i.e., with the help of cylindrical optics. Applying the technique, good reconstructions were obtained resembling aerial photographs.

E. Incoherent Holography

In principle, holography needs coherent sources. However, every source of radiation has a limited coherence and this limited coherence can be utilized in holographic processes. If, for instance, an interferometric arrangement is used, where the optical path length of the object beam equals that of the reference beam, a hologram may be obtained (Peters, 1966).

A self-luminous object may be recorded holographically as well. If the light emerging from the object is divided and then recombined in some interferometric system in such a way that each point on the object provides its own reference point, a hologram may be obtained (Lohmann, 1965). Another way is to use the second-order coherence of the object and correlate intensity fluctuations (Davydova and Denisyuk, 1969).

All these techniques are tedious and may be applied to more or less plane objects. They can be used with light but are almost impossible with X-rays, for example. An entirely different approach may be derived by looking at a hologram from another point of view; a conventional photograph records the intensity distribution of the object wave at one point, the pinhole of a pinhole camera, which is the basis for all conventional photography. A stereoscopic view of an object can be obtained by using two observation points. This is adequate to obtain a depth perception but it is not a true 3-D image. More information about the 3-D nature of the object will be obtained from more than two observation points. A hologram represents an infinite number of observation points distributed continuously over a portion of a plane. This is the reason why a hologram is capable of reproducing a true 3-D image.

It turns out that the human eye does not need an infinite number of observation points; a limited number of these are quite adequate to "see" a good 3-D image.

Since we have only two eyes we can observe only two pictures simultaneously and that is why stereoscopic pairs and not triplets or multiplets are used. Holographic techniques may be used to remedy this drawback. The principle of the technique is as follows. An object is photographed in the conventional way from a large number of observation points. Transparencies of each photograph are recorded holographically, side by side on a single plate. Reconstruction from this hologram produces an image of the object in such a way that each portion of the hologram reproduces the image from a slightly different viewing point, very similar to a true hologram. In this way quite high-quality 3-D images can be obtained.

The multiple photograph can be produced by scanning with a single camera (Hildebrand and Haines, 1969; De Bitetto, 1969). It can be done by using the fly's eye lens—a multiple, two-dimensional lens array (Pole, 1967). In electron beam and X-ray photography the technique of successive recording positions can also be applied (Groh and Kock, 1970) or a multiple-pinhole camera may be used (Stroke et al., 1969). Sometimes these techniques are advantageous even when coherent sources are available. For example, it was found that enormous power densities are required in the application of acoustic holography to medical diagnostics. These power densities are harmful for a living body but they can be much reduced if the so-called stereo-holographic methods described above are used.

IX. Conclusions

The invention of the laser stimulated the development of holography and in a relatively short time it has penetrated into various fields of research and application. Although the basic requirement of holography is a highly coherent radiation source, there is a tendency to use non-coherent sources as well.

The purpose of this chapter was to explain the basic physical principles underlying the holographic process and its applications. In most cases only simplified models have been used in order to avoid the sinking of the physical principles into the sea of mathematical formulae. More rigorous and extensive treatment can be found in the literature cited.

REFERENCES

Abramson, N. (1969). *Appl. Opt.*, **8**, 1235–1240.
Abramson, N. (1970). *Appl. Opt.*, **9**, 97–101.
Baldwin, G. D. (1969). *Appl. Opt.*, **8**, 1439–1446.
Boone, P. and Verbiest, R. (1969). *Opt. Act.*, **16**, 555–567.
Born, M. and Wolf, E. (1959). "Principles of Optics," Pergamon Press, Oxford.
Brandt, G. B. (1969). *Appl. Opt.*, **8**, 1421–1429.
Brooks, R. E., Heflinger, L. O., Wuerker, R. F. and Briones, R. A. (1965). *Appl. Phys. Lett.*, **7**, 92–94.
Brooks, R. E., Heflinger, L. O. and Wuerker, R. F. (1966). *IEEE, J.*, **QE-2**, 275–279.
Brown, G. M., Grant, R. M. and Stroke, G. W. (1969). *J. acoust. Soc. Am.*, **45**, 1166–1179.
Bryngdahl, O. (1969). *J. opt. Soc. Am.*, **59**, 1645–1650.
Bryngdahl, O. and Lohmann, A. (1969). *J. opt. Soc. Am.*, **59**, 1245–1246.
Burckhardt, C. B. and Doherty, E. T. (1969). *Appl. Opt.*, **8**, 2479–2482.
Cathey, W. T., Jr. (1965). *J. opt. Soc. Am.*, **55**, 457.
Champagne, E. B. and Massey, N. G. (1969). *Appl. Opt.*, **8**, 1879–1885.
Chau, H. H. M. (1969). *Appl. Opt.*, **8**, 1209–1211.
Collier, R. J. and Pennington, K. S. (1967). *Appl. Opt.*, **6**, 1091–1095.
Cutrona, L. J., Leith, E. N., Porcello, L. J. and Vivan, W. E. (1966). *Proc. IEEE*, **54**, 1026–1032.
Davydova, I. N. and Denisyuk, Yu. N. (1969). *Opt. Spectr.*, **26**, 223–225.
De Bitetto, D. J. (1969). *Appl. Opt.*, **8**, 1740–1741.
De, M. and Sévigny, L. (1967). *J. opt. Soc. Am.*, **57**, 110–111; *Appl. Phys. Lett.*, **10**, 78–79.
Denisyuk, Yu. N. (1962). *Sov. Phys. Doklady*, **7**, 543–545.
Denisyuk, Yu. N. (1963). *Opt. Spectrosc.*, **15**, 279–284.
DeVelis, J. B. and Reynolds, G. O. (1967). "Theory and Applications of Holography," Addison Wesley Publ. Co. London and New York.
Feleppa, E. J. (1969), *Physics Today*, **22**, No. 7, 25–32.
Friesem, A. A., Kozma, A. and Adam, G. F. (1967). *Appl. Opt.*, **6**, 851.
Friesem, A. A. and Walker, J. L. (1970). *Appl. Opt.*, **9**, 201–214.
Gabor, D. (1949). *Proc. R. Soc.*, **197**, 454–487.
Gabor, D. (1951). *Proc. phys. Soc.*, **64**, 449–469.
Gabor, D. and Goss, W. P. (1966). *J. opt. Soc. Am.*, **56**, 849–858.
Gabor, D. and Stroke, G. W. (1968). *Proc. R. Soc. A*, **304**, 275–284.
George, N. and Matthews, J. W. (1966). *Appl. Phys. Lett.*, **9**, 212–215.
Goodman, J. W. (1968). "Introduction to Fourier Optics," McGraw-Hill Book Co. San Francisco.
Goodman, J. W., Jackson, D. W., Lehmann, M. and Knotts, J. (1969). *Appl. Opt.*, **8**, 1581–1586.
Grant, R. M. and Brown, G. M. (1969). *Materials Evaluation*, **27**, No. 4, 79–84.
Greguss, P. (1966). *J. Phot. Sci.*, **14**, 329–332.

Greguss, P. (1967). *J. acoust. Soc. Am.*, **42**, 1186.

Groh, G. and Kock, M. (1970). *Appl. Opt.*, **9**, 775–777.

Heflinger, L. O., Wuerker, R. F. and Brooks, R. E. (1966). *J. appl. Phys.*, **37**, 642–649.

Hildebrand, B. P. and Haines, K. A. (1969). *J. opt. Soc. Am.*, **59**, 1–6.

Jackson, G. (1969). *Optica Acta*, **16**, 1–16.

Jahoda, F. C. (1969). *Appl. Phys. Lett.*, **14**, 341–343.

Kiemle, H. and Röss, D. (1969). "Einführung in die Technik der Holographie," Akademische Verlagsgesellschaft, Frankfurt am Main.

Kock, W. E., Rosen, L. and Stroke, G. W. (1967). *Proc. IEEE*, **55**, 80–81.

Kogelnik, H. (1969), *Bell Syst. Tech. J.*, **48**, 2909–2947.

Labeyrie, A. (1969). *In* "Optical Instruments and Techniques," Oriel Press, London.

Labeyrie, A. and Flamand, J. (1969). *Opt. Comm.*, **1**, 5–8.

Lanzl, F., Mager, H. J. and Weidelich, W. (1968). *Phys. Lett.*, **27A**, 35–36.

Larsen, A. L. (1969). *Bell Syst. Tech. J.*, **48**, 2507–2527.

Leith, E. N. and Upatnieks, J. (1962). *J. opt. Soc. Am.*, **52**, 1123–1130.

Leith, E. N. and Upatnieks, J. (1963). *J. opt. Soc. Am.*, **53**, 1377–1381.

Leith, E. N. and Upatnieks, J. (1964). *J. opt. Soc. Am.*, **54**, 1295–1301.

Leith, E. N. and Upatnieks, J. (1966). *J. opt. Soc. Am.*, **56**, 523.

Leith, E. N. and Upatnieks, J. (1967). *In* "Progress in Optics," Vol. VI, 3–52 (E. Wolf, Ed.), North-Holland Publ. Co. Amsterdam.

Leith, E. N., Kozma, A., Upatnieks, J., Marks, J. and Massey, N. (1966). *Appl. Opt.*, **5**, 1303–1311.

Lipson, S. G. and Lipson, H. (1969). "Optical Physics," Cambridge University Press, Cambridge.

Lohmann, A. W. (1965). *J. opt. Soc. Am.*, **56**, 1555–1566.

Magyar, G. (1969). *Opt. Tech.*, **1**, 231.

Marom, E. (1970). *Appl. Opt.*, **9**, 1385–1391.

Marom, E., Boutin, H. and Mueller, R. K. (1967). *J. acoust. Soc. Am.*, **42**, 1169.

Massey, G. A. (1967). *Proc. IEEE*, **55**, 1115–1117.

McClung, F. J., Jacobson, A. D. and Close, D. H. (1970). *Appl. Opt.*, **9**, 103.

McMahon, D. H. and Caulfield, H. J. (1970). *Appl. Opt.*, **9**, 91–96.

McMahon, D. H. and Franklin, A. R. (1969). *Appl. Opt.*, **8**, 1927–1929.

Megla, G. K. (1966). *Appl. Opt.*, **5**, 945–960.

Metherell, A. F. (1968). *Appl. Phys. Lett.*, **13**, 340–343.

Metherell, A. F. (1969). "Acoustic Holography," Vol. 1, Plenum Press. New York.

Metherell, A. F., Spinak, S. and Pisa, E. J. (1969). *Appl. Opt.*, **8**, 1543–1550.

Molin, N.-E. and Stetson, K. A. (1969). *J. Sci. Instr. Ser. 2*, **2**, 609–612.

Mueller, R. K., Marom, E. and Fritzler, D. (1969). *Appl. Opt.*, **8**, 1537–1542.

Nishida, N. (1970). *Appl. Opt.*, **9**, 238–240.

Pastor, J. (1969). *Appl. Opt.*, **8**, 525–531.

Pennington, K. S. and Lin, L. H. (1965). *Appl. Phys. Lett.*, **7**, 56–57.

Peters, P. J. (1966). *Appl. Phys. Lett.*, **8**, 209–210.

Pole, R. V. (1967). *Appl. Phys. Lett.*, **10**, 20–22.

Politch, J., Shamir, J. and Ben Uri, J. (1970). *Appl. Phys. Lett.*, **16**, 496–498.

Powell, R. L. and Stetson, K. A. (1965). *J. opt. Soc. Am.*, **55**, 1593–1595; 1694–1695.

Rogers, G. L. (1966). *J. opt. Soc. Am.*, **56**, 831.

Rosen, L. (1967). *Proc. IEEE*, **55**, 79–80.

Russell, B. R. (1969). *Appl. Opt.*, **8**, 971–973.

Russo, V. and Sottini, S. (1968). *Appl. Opt.*, **7**, 202.

Shamir, J., Politch, J. and Ben Uri, J. (1971). *Am. J. Phys.* **39**, 840–841.

Simpson, W. A. and Deeds, W. E. (1970). *Appl. Opt.*, **9**, 499–501.

Smith, P. W. (1969). *Bell Syst. Tech. J.*, **48**, 1405.

Sollid, J. E. (1964). *Appl. Opt.*, **8**, 1587–1595.

Staselko, D. I., Denisyuk, Yu. N. and Smirnov, A. G. (1969). *Opt. Spectr.*, **26**, 255–229.

Stetson, K. (1968). *Laser Focus*, **4** (November), 30–31; Symposium on the Engineering Uses of Holography, 17th–20th Sept., Glasgow.

Stetson, K. (1969). *Optik*, **29**, 386–400.

Stroke, G. W. (1965). *Intl. Cong. on X-ray Optics and Microanalysis*, Orsay (September), 30–46.

Stroke, G. W. (1969). "Introduction to Coherent Optics and Holography," 2nd ed., Academic Press, Inc. (London).

Stroke, G. W., Hayat, G. S., Hoover, R. B. and Underwood, J. H. (1969). *Opt. Comm.*, **1**, 138–140.

Stroke, G. W. and Zech, R. G. (1966). *Appl. Phys. Lett.*, **9**, 215–217.

Tanner, L. H. (1965). *J. Sci. Instr.*, **42**, 834–837.

Tanner, L. H. (1966). *J. Sci. Instr.*, **43**, 353–358.

Thurston, F. L. (1969). *J. acoust. Soc. Am.*, **45**, 895–899.

Trolinger, J. D., Belz, R. A., and Farmer W. M., (1969). *Appl. Opt.* **8**, 957–961.

Upatnieks, J., Vander Lugt, A. and Leith, E. N. (1966). *Appl. Opt.*, **5**, 589–593.

Urbach, J. C. and Meier, R. W. (1969). *Appl. Opt.*, **8**, 2269–2281.

Watrasiewicz, B. M. (1968). *Opt. Tech.*, **1**, 20–23.

Weiss, L. and Holyoke, E. D. (1969). *Surgery*, **128**, 953–962.

Winthrop, J. W. and Worthington, C. R. (1965). *Phys. Lett.*, **15**, 124–226.

Zaidel', A. N., Malkhasyan, L. G., Markova, G. V. and Ostrovskii, Yu. I. (1969). *Sov. Phys.–Technical Phys.*, **13**, 1470–1475.

Zech, R. G. and Siebert, L. D. (1968). *Appl. Phys. Lett.*, **13**, 417–418.

CHAPTER 10

Optical Transforms in Teaching

S. G. LIPSON

Physics Department, Technion, Israel Institute of Technology,
Haifa, Israel

I. Introduction

There hardly exists a branch of physics which does not somehow use Fourier Transforms. This book is largely concerned with their use in optics and crystallography, but almost any physicist will be able to point out examples of the use of Fourier Transforms in his own subject, whether it be experimental or theoretical. From the point of view of the teacher, one can see the great advantages of a clear exposition of the properties of transforms, and in particular of the relationship between the characteristics of the function and those of its transform; it is the purpose of this chapter to discuss some of the ways in which this relationship can be made vivid by the use of optical transforms. A measure of the success of the method is that some of the demonstrations will appear to be trivial; some basic property of Fourier transforms has then become instinctive.

The rival to optical transforms in the classroom is the electronic display. On the one hand we prepare a mask to represent the function and form its transform by an optical diffraction system; on the other hand we generate the function as a time-dependent voltage and analyse it by means of a swept-frequency filter. The optical method is not only simpler and requires less complicated equipment; it also allows the transforming of two-dimensional functions and is not limited by causality (see Section IV.B).

This chapter is not intended as an introduction to Fourier theory. It will be assumed that the reader is already familiar with the basic facts of Fourier transforms and that he can find their mathematical proofs elsewhere (Lipson and Lipson, 1969). Moreover, in Chapter 1, the relationship between Fraun-hofer diffraction and the Fourier transform has been dealt with. Here we are concerned with the way in which optical transforms can be used to illustrate the transform-function relationship.

Firstly we shall describe an experimental method for producing easily visible diffraction patterns for classroom demonstration. The inclusion of practical details does not indicate the results of exhaustive experiments to find the ideal conditions, but simply that we have found the system described to give satisfactory results, and we wish to save the time of the teacher who is interested in trying these methods. Following this we shall discuss some of the basic properties of Fourier transforms and how they can be demonstrated with this apparatus. Then follow illustrations of a few applications of these transforms to subjects outside optics. A similar programme then deals with image formation and its applications.

II. OPTICAL TRANSFORMS

A. The Demonstration of Diffraction Patterns

Once it was necessary to use a spectrometer, and at the end of the lecture the students queued to look down the microscope at the diffraction patterns of a few objects. The optical diffractometer, described in Chapter 1, has a similar disadvantage. With the advent and ubiquity of cheap He–Ne lasers, we can now project diffraction patterns on to a screen, and fulfil all the requirements of a satisfying demonstration.

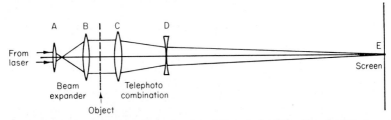

Fig. 1. Apparatus used for the demonstration of Fraunhofer diffraction. In order to expand the beam diameter from 1 mm to 6 mm, A has a focal length of 5 cm and B 30 cm; the distance $AB = 35$ cm. The telephoto combination consists of C with 20 cm focal length and D with -5 cm focal length. C and D are separated by just over 15 cm. The effective focal length is approximately $4 \times DE$.

The following points are to be considered in designing the demonstration equipment.

(a) The objects have to be fairly large. It is difficult, without special equipment, to make accurate masks with a total size of less than 5 mm and detail on a scale smaller than 0·2 mm.

(b) The average cheap laser has a beam width of about 1 mm. We must therefore expand the beam by a factor of about five, or preferably more.

(c) At the other end of the equipment, we want to produce a diffraction pattern that is visible to a whole class. This means an external diameter of about 200 mm, or even more. Usually the screen can be no further than 10 m from the apparatus. Now, for detail of 0·2 mm to produce a pattern 200 mm in diameter requires a focal length of about 40 m. To get this focal length into a 10 m space is a classic example of the requirements for a telephoto combination (and we should not forget to point this out to the students, too).

The equipment we have used for such demonstrations is shown in Fig. 1, the legend containing details of typical dimensions. Most of our masks have been made from old film, or by photographing large-scale drawings. In this latter process, variations in emulsion thickness do not seem to produce serious phase errors, and by using the negative as an object, very intense zero orders can be avoided. When using old film, it can be cut cleanly with a sharp razor blade and a steady hand; and in the absence of a pantograph specially designed for the purpose (Taylor and Lipson, 1964), arrays of holes can be made accurately with the help of a drill-press having a work-holder controlled in two dimensions by lead-screws. A small drill or needle inserted to a pre-determined depth makes adequate holes. In addition to such black-and-white, or real, objects we should also like to be able to demonstrate the effects of phase-changes. In general it is difficult to do this, but there are exceptions. A microscope slide or piece of mica can be used to introduce an unknown phase-change, and this can be varied slowly by tilting the sheet. Alternatively, a thin prism can be used to give a phase gradient; one half of a Fresnel biprism is excellent for the purpose. The various techniques for making masks are described by Taylor and Lipson (1964).

B. The Reciprocal Relationship

In the following sections we shall describe optical demonstrations of some of the most basic properties of Fourier transforms. The object is in general a complex function of two dimensions, x and y. When the mask is cut out of film, the function always has a real value of 0 or 1; when parts of the "1" are covered with glass or mica, they have a value $\exp(i\phi)$. Although we are thus limited to functions with $|f(x,y)| = 1$ or 0, the range of possibilities is so large that very little more could be asked (Section II.E.5).

On the screen we see the transform $a(k_x, k_y)$. The actual relationship between the screen dimensions (u, v) and (k_x, k_y) is

$$(u, v) = \frac{F\lambda}{2\pi}(k_x, k_y)$$

where F is the focal length of the telephoto combination and λ is the wavelength of the light. We assume, of course, that the angle of diffraction is small.

If we compare the diffraction patterns of two objects, similar except for a difference in scale, their diffraction patterns are again similar. If the scale of the object is given by parameter b, and that of the transform by p, we have

$$pb = \text{constant}.$$

As the scale of the object increases, that of the transform decreases.

We can illustrate this by the concrete example of a slit of width b. The transform has amplitude

$$a(k) = b\frac{\sin\frac{1}{2}kb}{\frac{1}{2}kb}.$$

The scale parameter p can well be the distance between the first zeros on each side of the centre. In that case

$$pb = 2\pi.$$

Two relevant demonstrations are: (a) the transform of a variable slit, (b) the transform of a rectangular aperture; the centre peak of this is similar in shape to the aperture, but rotated by 90°.

C. Symmetry Properties

Diffraction patterns are often centrosymmetric even when the object has no apparent symmetry. This is in fact a property of real objects only, and appears algebraically as the result:

$$\text{if} \quad f(\mathbf{r}) = f^*(\mathbf{r})$$
$$\text{then} \quad a(\mathbf{k}) = a^*(-\mathbf{k}).$$

Since we observe only the intensity, $|a(\mathbf{k})|^2$, the pattern appears to be centrosymmetric. The phase difference between $a(\mathbf{k})$ and $a(-\mathbf{k})$ is occasionally quite important, as in the example of Section IV.B.

It is very easy to demonstrate the symmetry, simply by using any deliberately unsymmetrical real object. Then cover one half with a microscope slide and the symmetry disappears, unless by mischance the phase change is exactly $2n\pi$.

D. Fourier Inversion

The Fourier inversion theorem shows that the Fourier transform of the Fourier transform of a function is equal to a constant times the original function. In symbols, if $a(\mathbf{k})$ is the transform of $f(\mathbf{r})$, then $f(\mathbf{k})$ is the transform of $a(\mathbf{r})$. This relationship is very useful in practice, and is of course the basis of the diffraction theory of optical instruments (Section III, Chapter 1). The easiest mathematical illustration is that of a pair of narrow slits (Section II.E.3), which has a sinusoidal transform. And the transform of a sinusoidal diffraction grating is a single pair of orders, the $\pm$ first orders (Fig. 2). Despite the simplicity of the idea to be demonstrated, it is very difficult to carry out experimentally; this is because of the phase problem (compare Section I.D, Chapter 1)—the impossibility of recording phases directly.

Ideally we should like to transform an object, photograph or reproduce its diffraction pattern, and use this reproduction as a second object. The transform of the second object should be the original one (Fig. 2b). The reasons for practical difficulty are as follows. Firstly a finite object always gives rise to an infinite transform. Secondly, observation of the diffraction pattern does not tell us the phases. Thirdly, photographic recording records intensity and not amplitude.

Let us see how these problems affect the above example. We cut out two slits and get a sinusoidal diffraction pattern (Fig. 2a). This is infinite in extent, but we can resign ourselves to cutting it off at some point and putting up with the resulting error. This will be smallest if the fringes are fine, i.e. if the original slits were widely spaced. So we photograph the fringes, taking care to use a film with linear response to intensity. Now we use the photograph as obstacle;

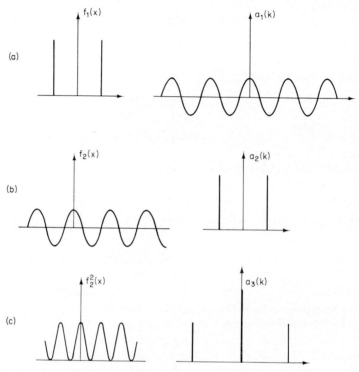

Fig. 2. The Fourier Inversion Theorem: (a) two delta-functions and their transform; (b) transform of a sinusoidal function; (c) transform of the square of the sinusoidal function in (b).

we find that the diffraction pattern contains strong ± first orders, and also a stronger zero order (Fig. 2c). Furthermore, any non-linearity of the film will be seen as weak second and higher orders. The reason for the strong zero order is that the photography has ignored the fact that alternate fringes in the pattern had opposite phases. And, in addition, the first orders are found to be twice as far apart as they ought to be, because photography has squared the amplitude and it is only by good fortune that squaring a sine still gives a sine, albeit of twice the frequency.

We have to plan the experiment rather carefully. First, the object has to be quite large, so that cutting off the wings of the transform will not seriously affect the second stage. Secondly, we must make sure that the whole of the pattern is real and positive, so that there is no phase information to neglect. Thirdly, we should fulfil one of the following conditions: either the pattern is of the "1 or 0" type, or the relative variations in intensity from place to place are not large. In either case the fact that the film records intensity and not amplitude makes little difference.

The most satisfactory demonstration turned out to use a wire gauze as object. Its period was 0·2 mm, but this is not critical. To make the second object we observed the intensities of the spots in the diffraction pattern and

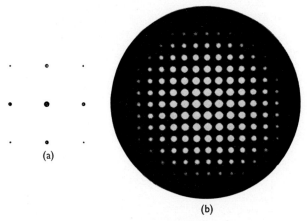

(a)

(b)

Fig. 3. A mask (a) to simulate the nine central orders of the diffraction pattern (Fig. 22a) of a piece of wire gauze, and its diffraction pattern (b).

made a set of pinholes in film to match them. Because of the inherent properties of wire gauze, the $(0,0)$, $(0,\pm1)$, $(\pm1,0)$, $(\pm1,\pm1)$ and all other diagonal orders are all positive. If orders higher than the first are neglected, we have satisfied the conditions for a true reproduction (Fig. 3). It is important that the pinholes be small, and they have area proportional to the amplitude (square root of the intensity) of the spots; otherwise only a small area of the final image appears and it is not a very accurate reproduction.

It will be appreciated that the above problems are identical with those facing the holographist, who is also trying to employ the inversion theorem in two separate operations. The complete solution is contained in the hologram, (Chapter 9); the above discussion has merely highlighted the problems. Of course, the ideal demonstration of the inversion theorem is the image-formation experiment in Section VI but by trying to separate the stages we can see just how much information is contained in the transform, and how all of it is required for reconstruction.

E. The Transforms of Some Useful Functions

Most of the basic ideas of Fourier theory can be derived in terms of the transforms of a small number of useful functions. These are:
(a) the square pulse, or "top-hat" function;
(b) the Dirac delta-function;
(c) a pair of delta-functions;
(d) an infinite array of delta-functions;
(e) Gaussian function.

By the use of convolution, multiplication, addition and the inversion theorem, we can generate from these a good approximation to almost any function we might need. In Section VI we shall deal with the operations; in this section we shall deal with the transforms of the elementary functions.

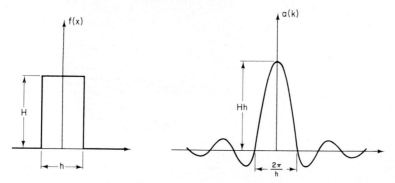

Fig. 4. The square pulse and its transform.

1. The Square Pulse

A square pulse of height H and width h has transform:

$$a(k) = Hh\frac{\sin\frac{1}{2}kh}{\frac{1}{2}kh}.$$

The form of this function $(\sin x/x)$ is very familiar to almost every physicist and has the following important characteristics (Fig. 4).
(a) The scale—say the distance between the first zeros—is inversely proportional to h; the function spreads out as pulse gets shorter;
(b) the amplitude of the function is proportional to the *area* under the pulse; in particular, the value at the origin, $k = 0$, is equal to hH. When we discuss energy considerations we shall return to this point (Section II.I).

2. *The Dirac Delta-function*

If we use a variable slit to demonstrate the dependence of the scale of the diffraction pattern on the slit width, we can immediately see what happens when the slit becomes extremely narrow. The first zeros move off the screen and the transform becomes almost uniform before it is too weak to observe. Clearly, as $h \rightarrow 0$,

$$a(k) \rightarrow Hh,$$

a constant value, independent of k. Now the Dirac delta-function, denoted by δ, is the limit of the square pulse as it gets narrower ($h \rightarrow 0$), but gets higher too ($H \rightarrow \infty$) in such a way that the area Hh remains constant. It is an impulse. For a unit delta-function, Hh takes the value unity. The transform of the delta function is a constant, equal to the strength Hh of the delta function.

3. *A Pair of Delta-functions*

A pair of delta-functions is the transform appropriate to the ideal Young's-slits experiment. If

$$f(x) = \delta(x - \tfrac{1}{2}b) + \delta(x + \tfrac{1}{2}b)$$

the transform is

$$a(k) = 2\cos\tfrac{1}{2}kb.$$

The optical transform is thus a series of sinusoidal fringes of period $4\pi/b$. The intensity distribution seen on the screen is

$$|a(k)|^2 = 4\cos^2\tfrac{1}{2}kb = 2(1 + \cos kb)$$

which indicates that the fringes have a period of $2\pi/b$.

We have assumed above that the two delta functions have the same phase. If we put a piece of glass over one of the slits used to demonstrate the transform, and tilt it so as slowly to alter the phase difference, we observe the fringes to move sideways. One should show, of course, that simple translation of the slits produces no observable effect; it simply alters phases. We can represent this situation by

$$f(x) = \delta(x - \tfrac{1}{2}b) + \exp i\phi . \delta(x + \tfrac{1}{2}b)$$
$$a(k) = 2\cos\tfrac{1}{2}(kb + \phi).$$

The fringes are shifted sideways proportionally to the phase difference.

4. *An Infinite Array of Delta-Functions*

Fourier originally proposed his theory as a method of analysis of a general periodic function; and although its extension as the Fourier Transform has wider applicability, the original form is still very important. Periodic functions can be incorporated by using the convolution theorem (Section II.F.3) together with the transform of an infinite lattice or array of delta-functions. In one dimension the series is written

$$f(x) = \sum_{n=-\infty}^{\infty} \delta(x-nb).$$

Its transform is

$$a(k) = \int_{-\infty}^{\infty} f(x)\exp(ikx)\,dx$$

$$= \sum_{n} \exp(iknb).$$

This infinite geometric series has the sum $(1 - 2i\pi m/b)^{-1}$ which diverges periodically when $kb = 2\pi m$. Either by imagination, or by rigorous mathematics, this can be seen to be equal to

$$a(k) = \sum_{m} \delta(k - 2\pi m/b)$$

which is also a regularly spaced set of delta functions, with period reciprocally related to b.

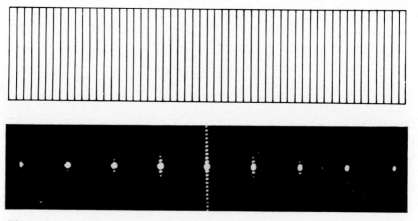

Fig. 5. A grating of narrow lines (the negative of the drawing, a) and its transform (b).

In two dimensions, we find that the transform of a periodic array of delta-functions leads again to a new periodic array, the reciprocal lattice (Section I.E, Chapter 1). If the direct lattice has unit translations **a** and **b** inclined at an angle γ, then the reciprocal lattice has units

$$a^* = 2\pi/a\sin\gamma \quad \text{and} \quad b^* = 2\pi/b\sin\gamma.$$

The axis **a*** is directed at right angles to **b**, and **b*** to **a**. Diffraction gratings, of which we shall have more to say later (Section II.G), are easily made by photographing large-scale drawings, and the example of an array of delta-functions corresponds to a grating with very fine lines. This will be seen to give a large number of orders, all approximately equal in intensity (Fig. 5). A two-dimensional array of holes is easily simulated by gauze, which can be distorted so as to give an angle between the axes, and tilted to remove the equality between a and b.

5. The Gaussian Function

For many reasons the transform of a gaussian function is very important, although its general properties can be guessed from a comparison with the square pulse. The gaussian function is (Fig. 5)

$$f(x) = (2\pi\sigma^2)^{-1/2}\exp{-(x^2/2\sigma^2)}.$$

The normalizing factor $(2\pi\sigma^2)^{-1/2}$ ensures that there is unit area under the curve, and σ is the width parameter.

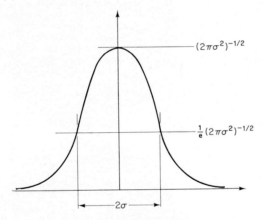

Fig. 6. The Gaussian function.

On the basis of a general similarity to a square pulse of width 2σ and of unit area, we should expect that

(a) that the transform has width-parameter $2\sigma^{-1}$;
(b) that the value at $k = 0$ is unity (the area).

In fact it is easy to show that the transform of $f(x)$ is also a gaussian

$$a(k) = \exp{(-\tfrac{1}{2}k^2\sigma^2)}$$

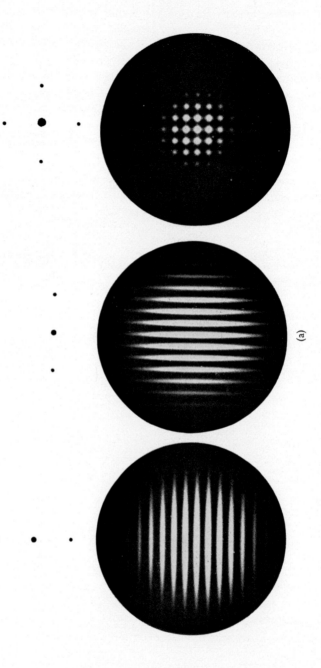

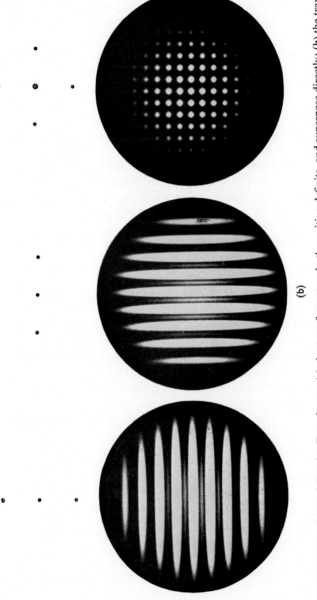

(b)

Fig. 7. Addition of Fourier Transforms: (a) the transforms are both positive definite, and superpose directly; (b) the transforms contain both negative and positive parts.

with width-parameter σ^{-1}. Amongst the reasons for the importance of this function are the following.

(a) It can be shown that the product of the widths of the function and of its transform is smaller for this function than for any other. This is important in quantum-mechanical applications;

(b) there are no side-bands either in the function or its transform. This property is unique, and is the origin of (a); it is important in establishing the concept that side-bands in a transform result from discontinuities in the function or its derivatives, and also, practically in the technique of apodization of lenses (Section V.E.3) (Born and Wolf, 1966).

To demonstrate the latter property, it is possible to make a blurred slit either photographically or by vacuum evaporation. In the latter method, the mask, a wire, must be supported some distance in front of the glass substrate; by preparing a number of slits of different degrees of sharpness, gradual disappearance of the side-bands can be observed.

F. The Operations of Addition, Multiplication and Convolution

1. Addition of Real Functions

It may seem trivial to discuss the transform of the sum of two functions, since it is simply the sum of the transforms; and it would be trivial if it were not for the complex nature of transforms in general. If we observe initially the transforms of the two functions separately, and then the transform of their sum, it will in general appear that the addition is not straightforward. For the intensity of the sum is only equal to the sum of the intensities if the phases of the two functions are the same. And what we observe is the intensity. It is worth demonstrating two extreme cases.

(a) The set of three equivalent holes, the centre one having more than twice the area of the outer ones, has an all-positive real transform:

$$f(x) = \delta(x - b) + 2\delta(x) + \delta(x + b)$$
$$a(k_x) = 2(1 + \cos 2kb).$$

Similarly the same function expressed as a function of y has an all-positive real transform. The sum of the two, which is a cross with the centre hole at least four times the area of the outer ones, is then simply the arithmetical superposition of the two functions of k_x and k_y (Fig. 7a).

(b) We can use a similar example to show that arithmetical superposition does not always hold. If the centre holes of the triads are *smaller* than twice the areas of the outer ones, some regions of each transform have negative sign and cancel positive parts of the other transform (Fig. 7b).

2. Addition of Complex Transforms

In both the above examples the transforms were real. When the transforms are complex, addition is less straightforward. For example, we can consider the example of a triangle of holes, to which is added a single hole at its centre. The single hole adds a constant vector to the whole of the transforms and knowing this can enable some information about the phases in the transform to be deduced by comparison between the transforms of the triangle and the triangle-plus-extra-hole. By adding the extra hole in two known places, complete information about the phases can be deduced from the three transforms obtained.

3. The Operation of Convolution

The two operations of convolution and multiplication are complementary in Fourier transforms; that is to say that the transform of the product of two

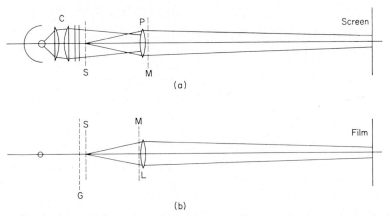

Fig. 8. Convolution illustrated by an out-of-focus optical instrument: (a) a projector; (b) a camera. S is the slide and M the second mask. In (a), notice that the condenser C is not focused on the projection lens P. In (b), the illumination system consists of a light source and ground-glass screen G.

functions is the convolution of their individual transforms. Because of the inversion theorem (Section II.D), it follows that the transform of the convolution of two functions is the product of their transforms. The mathematical proof of this statement is given in every exposition of Fourier transforms, but this still leaves the concept of a convolution to be appreciated.

The operation of convolution is the property of any out-of-focus instrument. Consider, for example, a projector (Fig. 8)—which can be used for demonstration—which focuses on the screen a sharp image of a mask (a slide) containing a single hole, or delta function. If the projector is defocused, the image becomes

364 S. G. LIPSON

a disk of light,† the disk resulting from the shape of the projection lens. By placing a second mask in front of this lens, the image becomes a scaled reproduction of this mask, centred on the original sharp image of the slide. Now replace the delta-function in the slide-holder by a more complicated slide —for example, a number of holes. Then each hole appears on the screen as a reproduction of the mask in front of the projection lens. This image is the *convolution* of the two masks. The process can obviously be extended to more complicated continuous functions (Fig. 9). If the slide-holder mask gives a sharp image $f_1(\mathbf{r})$, and the projection-lens mask a light distribution $f_2(\mathbf{r})$ when the slide is a single hole, then the joint image is clearly

$$F(\mathbf{r}) = \int f_1(\mathbf{r}') f_2(\mathbf{r} - \mathbf{r}') \, d\mathbf{r}'.$$

This is the convolution of $f_1(\mathbf{r})$ and $f_2(\mathbf{r})$.

To illustrate the use of the convolution function, the following demonstrations have proved suitable.

   (a) Object = a group of holes, representing say a benzene ring.
      Mask = two holes (giving a pair of rings).
   (b) Object = a row of small holes (periodic array of delta-functions).
      Mask = a large hole, preferably non-circular (illustrating formation of any periodic function by convolution).
   (c) Object = an equilateral triangle of holes.
      Mask = the same (illustrating the self-convolution function).

4. *Multiplication and Convolution*

Once the operation of convolution has been appreciated, we can return to the transforms of convolutions and products. Consider, as an example, the transform of a diffraction grating with wide lines. This function (Fig. 10) is the convolution of the periodic array of delta-functions with a single wide pulse, whose width is that of the individual lines. The transform of this convolution is the product of the transforms of the periodic array and the single line. Because the single line is necessarily narrower than the spacing between the delta-functions in the array, the central maximum of the $\sin\frac{1}{2}kb/\frac{1}{2}kb$ function always contains at least the zero and first orders. This type of grating is made most easily by photographing a drawing.

† To observe these effects without complications arising from the filament of the projector lamp, it is advisable to defocus the condenser lens until the projection lens is uniformly illuminated in the absence of a slide.

Fig. 9. Convolution of two-dimensional functions illustrated by means of an out-of-focus camera. The camera lens has been masked by three apertures (b), (d) and (f) and the slide (a) photographed through them. (c), (e) and (g) are the results. (From Lipson and Lipson, 1969.)

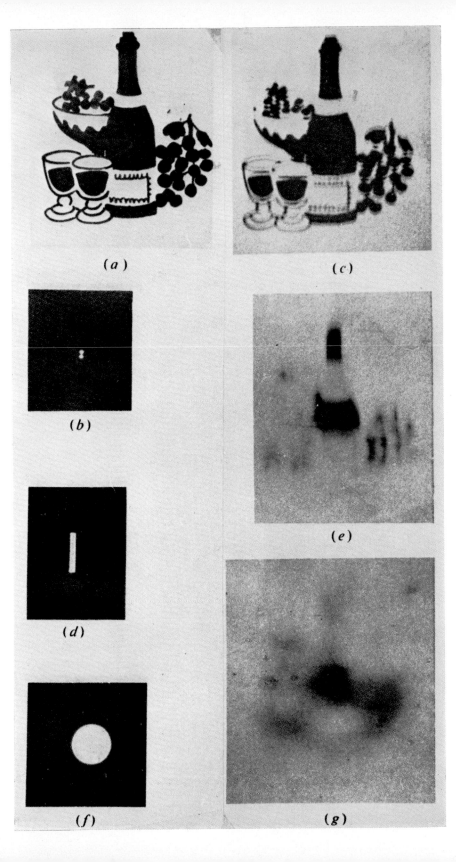

(a)

(b)

(c)

(d)

(e)

(f)

(g)

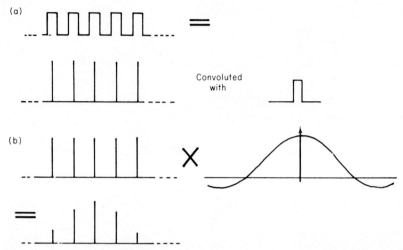

Fig. 10. Transform of a grating of wide lines, derived by means of a convolution: (a) The grating, and its breakdown into a convolution between the functions discussed in Sections II.E.1 and II.E.4; (b) the transform, as a product.

A second example, built out of the same elements, is that of a finite grating. In this case the infinite array is multiplied by the square pulse, which must now be longer than the period of the array. The transform is the convolution of the grating orders with the $\sin\frac{1}{2}kb/\frac{1}{2}kb$ function, which effectively broadens

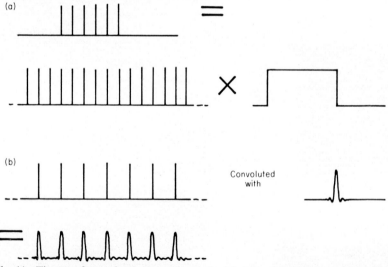

Fig. 11. The transform of a finite grating derived by describing it as a product; (a) is the grating and (b) its transform.

each order (Fig. 11). It is possible, with a little patience, to construct an animated demonstration of this point, by progressively cutting out more and more lines from a long grating, and showing how the grating orders get wider as a result (Lipson and Lipson, 1969, Appendix IV).

Such demonstrations necessarily give rise to questions about some of the curious properties of convolution functions. For example, what happens (in the first demonstration) when the line-width becomes equal to the spacing, and the grating no longer exists? Or, why is the spectrum of the limited grating dependent on the length in a stepwise manner?—it does not matter if the square-pulse contains 3·1 or 3·9 lines, but it does matter if it contains 3·9 or 4·1!

5. *Crystals*

One of the most important convolutions in physics is the crystal. We first construct a three-dimensional lattice of delta-functions (the axes do not need to be orthogonal). Then we convolute this lattice with the contents of the unit

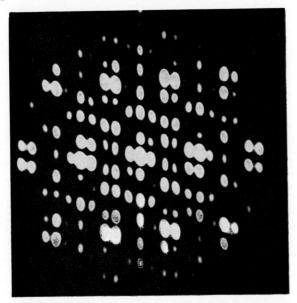

Fig. 12. The transform of four identical groups of holes, lying at the corners of a square. (From Taylor and Lipson, 1964.)

cell. Clearly we are involved here with two transforms; the first is that of the lattice, and is called the *reciprocal lattice* (Section I.E, Chapter 1) and the second is that of the unit-cell contents and is called the *structure factor* (Section I.D, Chapter 1). The transform of the crystal (and this is what is measured in crystal diffraction experiments) is the product of these two transforms, and is called the *weighted reciprocal lattice*. We can demonstrate the formation of a weighted reciprocal lattice by optical transforms in two dimensions. Two dimensions are realistic enough, because in practice any diffraction experiment on a three-dimensional crystal will separate out a particular set of points for observation, as a result of the necessity to conserve energy between ingoing and outgoing beams (Section III.A). If the wavelength is small compared with the lattice spacing, as it usually is in electron diffraction, this set of points lies on a plane of the three-dimensional reciprocal lattice, and so we are back again to two dimensions. In the case of X-rays or neutrons, where the wavelength is comparable with the spacing, the situation is more complicated, since the chosen set of points lies on a sphere (Section I.E, Chapter 1).

It is easy, although a little tedious, to make a mask showing a simple group of holes lying on a lattice. In fact, even four groups lying at the corners of a square is enough to show fairly convincingly that the principle is sound (Fig. 12).

G. Periodic Errors

On the basis of what we have so far discussed, we can see that the transform of any periodic structure is a set of sharp (delta-function) orders, whose relative amplitudes are determined by the structure of the unit which is being repeated. Exact periodicity is, however, a mathematical fiction, and in practice we have to put up with less than perfection. Crystals, for example, contain dislocations; and diffraction gratings contain all sorts of errors, although the careful manufacturer will reduce their magnitudes so as to be almost unobservable. In this section we shall deal with the influence of periodic errors in a regular structure. This study has many applications. One is to the understanding of ghost images in diffraction gratings. A second is to the X-ray diffraction patterns of ordered alloys, in which an atomic superlattice is formed. By neutron-diffraction studies, the superlattice resulting from magnetic ordering can be discovered. On the other hand, a periodic signal carried on a higher-frequency carrier wave is another example which can be described as a periodic error or perturbation on a periodic structure.

Our basic example is a diffraction grating containing a periodic error. This is a one-dimensional structure. The error will usually have the periodicity of the lead-screw in the dividing engine used to rule the grating.

The result of such errors is to introduce ghost orders. The relationship of ghost to main orders can be summarized as follows (Born and Wolf, 1966; Lipson and Lipson, 1969).

(a) The ghost orders are separated from the main orders by an angle inversely proportional to the period of the error. Since this is usually much larger (10–1000 times) than the period of the grating itself, the ghosts are very close to the main orders.

(b) If there is a periodic error in line strength, the grating can be represented as the product of a perfect grating and an error function. The transform is a

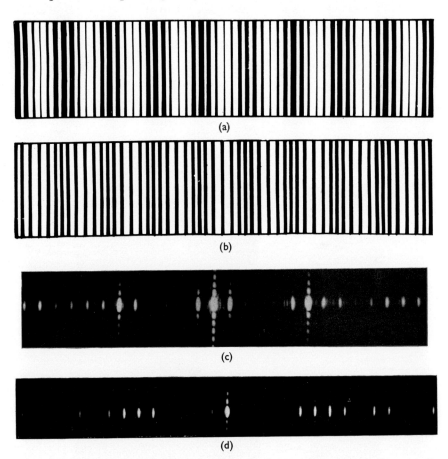

(a)

(b)

(c)

(d)

Fig. 13. Diffraction gratings with periodic errors: (a) periodic error in line-strength (amplitude-modulation); (b) periodic error in position (phase-modulation); (c) the diffraction pattern of (a); (d) the diffraction pattern of (b). In both examples, slight errors in execution of the drawings lead to background effects and weak ghosts in addition to those expected from the theory. The Fourier transform is very sensitive to such errors.

convolution, and the ghosts occur around each order in proportion to the intensity of the order.

(c) If there is a periodic error in line position, the ratio between ghost intensity and order intensity is not a constant, but depends on the line shape.

However, there are no ghosts around the zero order, and the above ratio generally increases towards higher orders.

Once again, these effects can admirably be demonstrated by the use of gratings drawn out on a large scale and photographed down. We have used gratings with about 50 lines, containing errors with a period of 6 lines (Fig. 13).

H. Random Distributions and the Correlation Function

The Fourier transform of a function is a non-local property of the function; that is, the value of the transform at any particular value of k depends on the value of the function at all values of x, via the definition of $a(k)$ as an integral. For example $a(0)$, the zero order, is the mean value of $f(x)$. Another

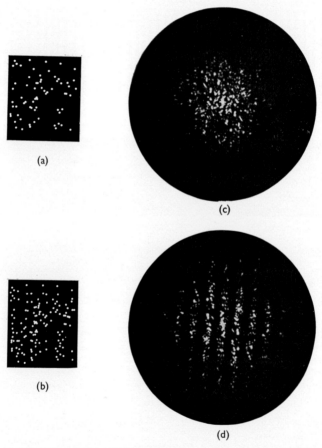

(a)

(c)

(b)

(d)

Fig. 14. Random arrays, (a) and (b), and their diffraction patterns (c) and (d) respectively. That (b) is made up from pairs of correlated points is clear from its diffraction pattern (d).

useful relationship involving the whole of the function $f(x)$ is the Wiener-Khinchin theorem, which states that the intensity $|a(k)|^2$ of the Fourier transform is the transform of the self-correlation function of $f(x)$. The self-correlation function is defined as

$$c(x) = \int f(x') f^*(x' - x) \, dx'$$

and the proof of the Wiener-Khinchin theorem follows directly from the correlation theorem.

This relationship is important from two points of view. Firstly the only information available from a single optical transform or other diffraction pattern is the intensity distribution $|a(k)|^2$. Thus the self-correlation function $c(x)$ in fact contains all the knowledge of the function $f(x)$ available from its optical transform. The deduction of $f(x)$ from $c(x)$ needs extra information (such as phases) or inspired guesswork, and is an art in itself. The second importance of $c(x)$ is that sometimes it is, in its own right, an important feature of $f(x)$. For example, suppose that we build up $f(x)$ out of an apparently random sequence of identical functions. A complete specification of the entire sequence would be too detailed for any practical use; whereas the self-correlation function, which specifies the distribution of spacings in the sequence, is more directly useful. Figure 14 shows two two-dimensional arrays of holes and their transforms. It is clear that the first contains no correlation (apart from that occurring accidentally in a finite sample) whereas the second contains pairs of holes; hence the transform is multiplied by a sinusoidal fringe pattern. The information in the transform is very clear; that in the original distribution is, to say the least of it, obscure.

I. Some Further Properties of Fourier Transforms

In this section we shall mention a few additional properties of Fourier transforms which can add to our understanding of physical systems.

1. Energy and Power Distributions

It is clear, since the Fourier Transform is really only a mathematical operation, that the total power in the original function must be equal to that in the transform. Thus, on physical grounds, we see that

$$\int_{-\infty}^{\infty} |a(k)|^2 \, dk = \int_{-\infty}^{\infty} |f(x)|^2 \, dx.$$

This can also be proved analytically. As an illustration, consider the transform of a square pulse, of height H and width h, which we saw in Section II.E.1

to be $hH \sin \frac{1}{2}kh / \frac{1}{2}kh$. The power in the transform is proportional to the square of the amplitude times the scale:

$$P_k \propto (hH)^2 \cdot \left(\frac{1}{h}\right) = hH^2.$$

The power in the function is likewise

$$P_x \propto H^2 \cdot h = hH^2.$$

This can, of course be demonstrated with any function. What is interesting, however, is that only the *total* powers integrated over all k are equal. Now in an optical system, only part of the transform, corresponding to $|k| < 2\pi/\lambda$ is observable; thus the equality must be true for a restricted range. We have in one dimension

$$\frac{2\pi}{\lambda} \int_{-\pi/2}^{\pi/2} |a(\theta)|^2 \cos \theta \, d\theta = \int_{-\infty}^{\infty} |f(x)|^2 \, dx.$$

2. *Transform and Function Alike*

We have already met two functions which are identical (except for a scale-change) to their transforms. These are the infinite array of delta-functions and the gaussian. Are they the only functions with this property, and is it of any significance?

It turns out that the Gaussian is just the first of an infinite of functions possessing this property. These are the functions

$$f_n(x) = H_n(x) \exp -x^2$$

where $H_n(x)$ is the nth Hermite polynomial (see Schiff, 1968; Koppelman, 1969). The functions $f_n(x)$ are well known as the eigenfunctions of the simple harmonic oscillator in quantum mechanics. Their importance in optics is that they represent the various modes of a confocal optical resonator. (This is a system consisting of two spherical mirrors with their foci coincident; in the most popular form, the radii of the mirrors are equal and equal to their separation.) The amplitude distribution in this optical resonator, because of the focusing and the repeated return of the light through the same regions, is continuously changing back and forth between $f(x)$ and its Fourier transform. It is thus reasonable—and can be proved to be so—that these functions represent the sections of a stable mode. The two-dimensional distribution $f(x,y) = H_n(x) H_m(y) \exp -(x^2 + y^2)$ describes the various modes of a laser. One can see, by constructing models similar to the functions $f_n(x)$ out of the units in Section II.E, how this similarity property arises (Fig. 15).

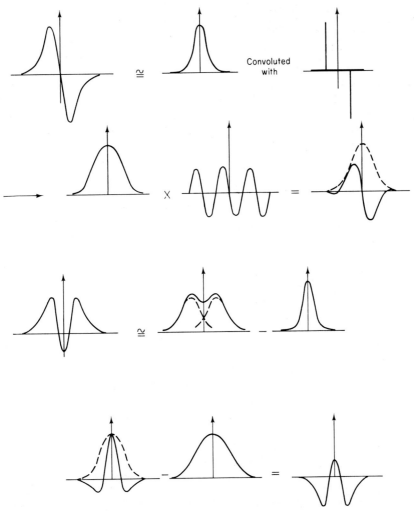

Fig. 15. Illustrating the Fourier Transforms of $f_n(x) = H_n(x) \exp{-x^2}$. In both cases the function has been approximated by functions whose transforms are discussed in Section II.E, combined using the operations in Section II.F. (a) is $f_1(x)$; (b) is $f_2(x)$.

III. SOME APPLICATIONS OF OPTICAL TRANSFORMS

Optical transforms have been widely used in crystallography as a convenient and direct physical way of solving the Fourier-analysis problems involved. We shall mention here a number of aspects of Section II which directly illustrate characteristics of X-ray, neutron and electron-diffraction photographs and techniques used in their interpretation. In addition to this, we shall

13

discuss some aspects of other branches of physics, such as electronics and radio astronomy, where optical analogues can also play their part.

A. Two and Three Dimensions

As mentioned in Section II.D.5, a real material should be represented by a three-dimensional function $f(x, y, z)$, whereas we have so far only dealt with two-dimensional functions. However, a diffraction pattern can at most be two-dimensional, since it is a function of angle, and can be recorded on a single plane or other surface. The "selection" which gives rise to this transformation is the requirement of conservation of energy during the diffraction process (called an "elastic" process).†

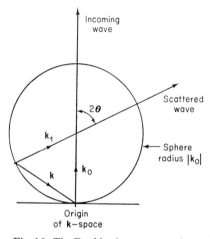

Fig. 16. The Ewald-sphere construction.

Consider the diffraction of an X-ray beam. The incoming beam has wave-vector $\mathbf{k}_0$ and photon energy $\hbar c k_0$ (c is the velocity of light and $\hbar$ is Planck's constant divided by 2π). The outgoing diffracted beam has wave-vector $\mathbf{k}_1$ and photon energy $\hbar c k_1$. To conserve energy, it is clear that

$$\hbar c k_0 = \hbar c k_1, \quad \text{or} \quad k_0 = k_1. \tag{10.1}$$

Now the change in wave-vector $(\mathbf{k}_1 - \mathbf{k}_0)$ results from diffraction, and the amplitude of the beam at $\mathbf{k}_1$ is just the amplitude of the Fourier transform of the material (in three dimensions) at wave-vector $\mathbf{k}$ where

$$\mathbf{k} = \mathbf{k}_1 - \mathbf{k}_0. \tag{10.2}$$

† There is also a class of *inelastic* diffraction phenomena (Bacon, 1962) in which energy is exchanged with the thermal motion of the crystal. To discuss this here would take us too far out of our way.

Thus the only parts of the Fourier transform that can result in observable diffraction are those satisfying Eqs. (1) and (2).

Figure 16 shows that the solution of these two equations defines a sphere in k-space, of radius k_0 $(=2\pi/\lambda)$, and we can only observe $a(\mathbf{k})$ when $\mathbf{k}$ lies on this sphere. It is called the *Ewald sphere*, or *sphere of reflexion* (Section I.E, Ch.1).

In this chapter we shall not use the properties of this sphere; we have only discussed its origin in order to justify the use of two-dimensional transforms as a real analogue to crystal diffraction. Details of the geometrical properties of the Ewald sphere are dealt with by Henry *et al.* (1960).

B. Transforms of Liquids and Solids

The difference between the liquid and solid states of the same material lies in the form of the ordering; and as we have seen, the Fourier transform particularly emphasizes periodicities in a structure. Around any particular atom or molecule (we shall say atom, for brevity) there lie other atoms in a more-or-less ordered fashion. In a liquid this order extends at best a few atoms away, and then the correlation of atomic positions with our original atom is negligible. On the other hand, in a solid crystal the order continues indefinitely and the atoms lie on a lattice. In practice, the material may be polycrystalline, in which case the order continues maybe for hundreds of thousands of atoms before it is destroyed at a grain boundary; or it may be amorphous, in which case there are only tens of atoms before the regularity is destroyed. Technically, the liquid shows *short-range order* and the solid *long-range order*.

1. Transform of a Crystal

In Section II.F.5, we have seen that a crystal is a convolution. When the crystal is finite (a crystalline in a polycrystalline material, for example) we can describe the crystal by:

(a) the lattice, which is a mathematical structure infinite in extent;
(b) the shape, which makes the crystal finite;
(c) the unit cell contents, which are repeated at every lattice point.

The complete crystallite is thus† [(a) multiplied by (b)] convoluted with (c). Its transform is thus: transform of (c) multiplied by [transform of (a) convoluted with that of (b)]. The transform of (c) is the structure-factor (Section II.F.5); the factor in square brackets is the reciprocal lattice with each point broadened depending on the size of the crystallite. For large crystals, the broadening may be obscured by other effects, such as the diameter of the incident beam of radiation, but for small crystals it can be observable.

---

† There is a subtle difference between this and (a) convoluted with (c) multiplied by (b), which might appear at first sight to be an equivalent alternative.

2. Transform of a Liquid or Amorphous Solid

Liquid or amorphous solids are characterized by short-range order. We can best illustrate the effect on the transform by means of the self-correlation function (Section II.I). Clearly this has a delta-function peak at the origin (we shall use one dimension for clarity)—see Fig. 17—because each atom is exactly correlated with itself. The position of the second atom is still well-defined, but

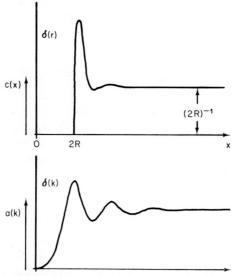

Fig. 17. A one-dimensional liquid: (a) its self-correlation function, and (b) the transform of (a). A major contribution to (b) is simply the cut-off at $x = 2R$, the diameter of the atoms, which results from the impossibility of placing two atoms closer than $2R$. Both $c(x)$ and its transform are symmetrical about their origins.

not exactly; hence it gives a narrow peak. As we go further away, the peaks get wider and smaller, and eventually merge into the background. In a liquid this happens after only three or four atoms. The spectral density of the transform, $|a(k)|^2$ is the transform of this self-correlation function, and is also shown in Fig. 17. Clearly in real life, where we have a two-dimensional transform, the peaks become rings (Furukawa, 1962).

C. The Effect of Temperature on a Solid

The effect of temperature is to give kinetic energy to all the molecules in a crystal. Whereas the crystal does not lose its long-range order—it does not melt—it does lose some of its short-range order. In seeing the effect of this motion on the Fourier transform it is convenient once again to use the self-correlation function.

In the stationary crystal, the intermolecular distances are known exactly to be periodic and so the correlation function is a periodic array of delta-functions. In the hot crystal, each of these delta functions is spread out, usually in an anisotropic manner, by the uncertainty in exact position of the molecules.

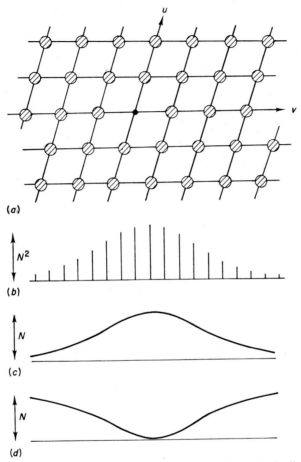

Fig. 18. (a) The self-correlation function of a structure with randomly displaced atoms. All peaks have a strength N; (b) one-dimensional representation of the transform of (a) assuming that the origin peak were as broad as the rest; (c) transform of the origin peak alone; (d) the difference between the transform of a δ-function at the origin and the transform of a broadened peak. (From Lipson and Lipson, 1969.)

But long range order is preserved—so that the correlation between two remote molecules is spread out by exactly the same amount as that between two close atoms. Thus the correlation function *remains periodic* and the peaks in the diffraction pattern remain sharp, although they are weakened (Fig. 18). In addition there is a certain amount of background introduced by correlation between the motion of atoms at close distances (and in particular the necessary correlation at zero distance, as in Section III.B.2), but it is the exact periodicity of $c(\mathbf{r})$ at large $\mathbf{r}$ which preserves the sharpness of the diffraction pattern (Lonsdale, 1949).

D. Superlattices

Superlattices occur mainly in alloys, where the existence of two components can lead to a secondary ordering of a longer period. Consider an alloy AB with equal numbers of A and B atoms. This could form a crystal with A and B atoms arranged randomly; on the other hand the A and B atoms could be arranged alternately, giving a subsidiary periodicity of twice the crystal lattice spacing. This is called the *super-lattice*. Or what happens in general is that the subsidiary ordering is sometimes not complete, but only has a short range. The situation is analogous to the diffraction grating with periodic errors (Section II.E). The errors give rise to ghosts orders. The above example would give us ghosts situated at half reciprocal-lattice vectors, and if the ordering is only short-range, the ghosts will not be as sharp as the main orders (Barrett and Massalski, 1966).

E. Neutron Diffraction and Magnetic Structure

Unlike X-rays, neutrons are sensitive to the magnetic moment of the atoms which scatter them. Thus it is possible to discover magnetic superlattices by means of neutron diffraction. For example, our alloy AB of Section III.D could have been an alloy of spin-up and spin-down atoms. If these are arranged on a lattice, as they would be in an anti-ferromagnetic material, neutron diffraction will produce ghost orders resulting from the magnetic superlattice. Just because X-rays are insensitive to magnetic ordering, it is possible, by comparing X-ray and neutron diffraction photographs, to separate magnetic and atomic superlattices (Bacon, 1966).

IV. Optical Transforms and Linear Networks

Many of the properties of linear electronic circuits, such as filters, linear amplifiers, and attenuators, can usefully be expressed in terms of their frequency response. This description is useful because the input signal to such a circuit can usually be characterized by its frequency range much more easily than its features described as a function of time. For example, human speech can easily be described as occupying the frequency range 10–10,000 Hz, whereas to describe it in any other way is much more difficult. However the response of the circuit to a given signal is also a matter of importance and it is in this sort of question that Fourier transforms are involved, and that optical transforms may be of some use as illustrations of general features.

A. Frequency and Impulse Responses

Suppose that a given circuit has a frequency response $f(\omega)$. That is to say, if we apply an input signal consisting of a pure tone of frequency ω, the output is also a pure tone of frequency ω, with an amplitude $f(\omega)$. For a filter, $|f(\omega)|$ will be less than, or equal to, unity; for a linear amplifier it may be greater than unity. Note also that the requirement of a *linear* system is already in evidence; if the response is not linear, the input sine-wave will not necessarily result in a pure sine-wave output, and harmonics will be generated. The properties of the system cannot then be described by the function $f(\omega)$ alone.

The response $f(\omega)$ is in general a complex number, and we shall see in Section IV.B. why it has always to be complex. The amplitude $|f(\omega)|$ describes the factor by which the input amplitude is increased; the phase of $f(\omega)$ gives the amount by which the phase of input signal is changed.

Suppose now that the input signal is not a pure sine wave. We can Fourier-analyse it into a linear combination of pure sine-waves and then investigate how each of these is changed by the circuit. Finally, because the circuit is linear, we add all the results together to get the output signal. Mathematically, the input is $u(t)$. This has Fourier spectrum $b(\omega)$. The component $b(\omega)$ is changed to $b(\omega)\,f(\omega)$ by the circuit, and the output signal $v(t)$ is the transform of $b(\omega)\,f(\omega)$:

$$b(\omega) = \int u(t)\exp(-i\omega t)\,dt$$

$$b(t) = \int b(\omega)\,f(\omega)\exp(i\omega t)\,d\omega.$$

The response to one particular input signal, a delta-function pulse, is an important characteristic of the circuit. This is called the *impulse response*. Clearly, when $u(t) = \delta(t)$, $b(\omega) = 1$ (transform of a delta function) and thus the impulse response

$$v_I(t) = \int f(\omega)\exp(i\omega t)\,d\omega.$$

The impulse response is the Fourier transform of the frequency response.

Optical transforms can be used to illustrate this relationship and to exemplify some of its consequences. Consider, for example, a low-pass filter $f(\omega) = 1$ ($|\omega| < \omega_0$), $f(\omega) = 0$ ($|\omega| > \omega_0$). This is equivalent to a slit of width $2\omega_0$ (from $-\omega_0$ to $+\omega_0$) and has a transform $2\omega_0 \sin \omega_0 t / \omega_0 t$. This is the response to an impulse; it has turned into a pulse of width $2\omega_0^{-1}$, together with fringe oscillations.

B. Causality and Symmetry

A moment's thought makes clear that there must be a flaw in the foregoing argument. For we have started with an impulse at $t = 0$, and ended with a

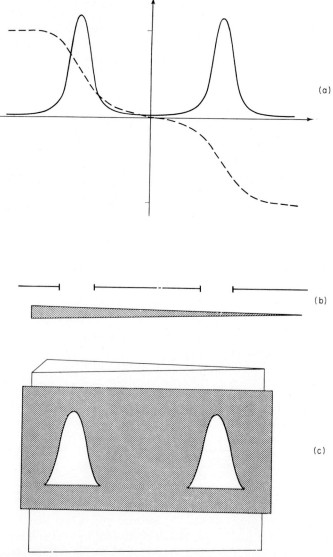

(a)

(b)

(c)

Fig. 19. (a) Complex transmission function of an L-C-R circuit; (b) a crude optical model of the function (dimensions are discussed in Section IV.B); (c) an improved version of (b) for the one-dimensional transform.

signal which starts ideally at $t = -\infty$, and is already appreciable when $t > -\omega_0^{-1}$. This is impossible physically, that the output should precede the input; it denies the *principle of causality*. It is required of any circuit that the output should be zero at least until the input begins, in this case until $t = 0$.

This requirement restricts the possible forms of $f(\omega)$. For example, the symmetry properties mentioned in Section II.C lead to the conclusion that if $f(\omega)$ is real, $v_I(-t) = v_I(+t)$. But $v_I(t < 0) = 0$ (by causality) and therefore either

$$v_I(t) = \delta(t) \quad \text{or} \quad v_I(t) = 0.$$

Both these solutions are trivial.

We conclude that $f(\omega)$ for a practical circuit cannot be entirely real, that the circuit must shift the phases.

Another conclusion arises from the symmetry properties. Since $u(t)$ and $v(t)$ are input and output signals, they must both be real, since an imaginary voltage is meaningless (u and v are actual voltages, not complex amplitudes of sine waves). It follows that $|f(\omega)|^2$ is symmetric, being the transform of a real function $v_I(t)$. Expressed completely $f(\omega) = f^*(-\omega)$; the circuit cannot distinguish between positive and negative frequencies. This is hardly surprising, but comforting to know that it is an inherent property of circuits in general.

We can make an optical model similar to a real circuit without too much difficulty. For example, a resonant L-C-R circuit has $f(\omega)$ as shown in Fig. 19, and this can be simulated by the model shown in Fig. 19(b). The diffraction pattern of this model is found to be mainly on one side of the zero; the non-causal part arises from the crudity of the model. We use a pair of slits backed by a thin prism (Section II.A) arranged to give 180° phase change between the two edges of each slit. For example, if the prism gives a deviation of $\frac{1}{2}°$, the required slit width is $\frac{1}{2}\lambda/\frac{1}{2}° \doteq 0.035$ mm. The separation between the inner edges of the slits must, in addition, correspond to zero phase difference, and so must be $0.070n$ mm, where n is any integer. It is obvious that the deviation of the prism has just served to shift the central peak of the transform to one side of the zero, but this is exactly the reason that a causal filter has to give rise to a phase shift.

C. A Note on Making Better Simulation Models

In two-dimensional models we have been limited to making "0 or 1" type of objects. In one-dimensional models, this is no longer a limitation, since we can use the second, y, dimension to replace the amplitude of $f(\omega)$. Since only one dimension is needed for these circuit illustrations, the method is applicable here. It is important to point out that the correct transform is obtained only along the k_x axis ($k_y = 0$). There we have a two-dimensional transform

$$a(k_x, 0) = \int\int f(x, y) \exp(ixk_x) \, dx \, dy$$
$$= \int [f(x, y) \, dy] \exp(ixk_x) \, dx.$$

But when $f(x,y)$ is of the "1 or 0" type, the integral $\int f(x,y) dy$ is just the y dimension of the opening at coordinate x, and so the method is justified. A better attempt at the simulation of the L–C–R circuit can then be made by using the model in Fig. 19(c).

V. Optical Transforms Applied to Antennae

An antenna is essentially a distributed coherent transmitter or receiver and is thus similar in principle to a diffracting aperture. Optical-transform analogies are not often directly applicable to the solution of antenna problems, but in many cases can indicate the existence of solutions in a particular direction. The reason for this indirectness is that one does not often know the exact form of the equivalent diffracting obstacle. In addition, the scale involved is very often of the order of a few wavelengths, in which case accurate optical simulation becomes something of a problem.

Consider, for example, the properties of a half-wave dipole antenna. This has a length of $\lambda/2$, and a width which is much smaller. As a general statement, one can deduce that the diffraction pattern is isotropic in the median plane of the aerial, and has a very broad maximum in the longitudinal plane, with the first zero at 90°, i.e. along the axis of the aerial. To deduce the exact distribution in the longitudinal plane, one would need to replace the axial current as a function of the position by $f(x)$. This itself requires knowledge of the radiated field in order to calculate it. The solution of such problems by optical methods is in practice limited to cases where the dimensions are much larger than the wavelength.

A. Resolving Power

Antennae are constructed on a large scale for two reasons—firstly to increase sensitivity, and secondly to increase directionality (in the case of a transmitting aerial) or angular resolution (for a receiving aerial). Clearly an aerial of outside dimension D can have an angular resolution no smaller than about λ/D, and analogy with optical systems can suggest the best way of achieving such resolution.

1. Dish Aerials

A dish aerial is used for maximum sensitivity. Provided that D, the diameter, is much greater than λ, the diffraction pattern of a circular aperture

$J_1(kr)/kr$, shows us that the central peak has angular radius $1 \cdot 2\lambda/D$ and this is the angular resolution obtainable according to the Rayleigh criterion with such an aerial. The side lobes (the second and higher maxima) are relatively small, indicating that there will not be much interference arising from sources located at larger angles than $1 \cdot 2\lambda/D$.

2. A Two-element Aerial

In cases where resolution is of prime importance, we shall see in Section VI.D.2 that the same space D can be utilized better by constructing two small

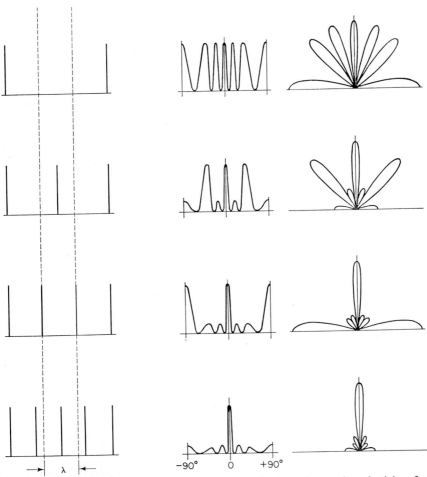

Fig. 20. Directionality of antennae. A comparison between the angular selectivity of a two-element aerial (a) and aerials in which the intervening space is filled by a regular array (b–d). The angular selectivity, measured at large distances, is the Fourier transform of the aerial strength as a function of position.

aerials, separated by D. Clearly sensitivity is lost, but now the optical transform of two points separated by D, $\cos\frac{1}{2}kD$, shows the limit of resolution to be $\lambda/2D$, an improvement by a factor 2·4. In this example the side-lobes are of equal strength to the central one, and much interference from sources at larger angles can be expected.

3. *The Diffraction-grating Approach*

To remove the effect of side-lobes, one can start to fill in the space between the two outer aerials with a regular sequence of aerials. The theory of such an array follows the principles of finite diffraction gratings, for which we know that the angular resolution of an order is $\frac{1}{2}\lambda/D$, where D is the total length of the grating. We do not gain in resolution by including more elements. But the angle of the first order is λ/b, where b is the separation between the elements, and clearly this gets larger as the separation is reduced (Fig. 20). We need to include elements until $\lambda > b$, in which case even the first order does not exist (it is at an imaginary angle, since $\sin\lambda/b > 1$). Then we have the following situation. First, in a radiating aerial, all the energy is channelled into the zero order (except for a relatively small amount in the subsidiary peaks resulting from the finiteness of the aerial) which has an angular width $\frac{1}{2}\lambda/D$. Secondly, in a receiving aerial, the angular resolution is $\frac{1}{2}\lambda/D$ and, because there are no orders of diffraction higher than the zero order, there is not much interference from signals outside the angle of resolution. In addition, the sensitivity of such an array is almost as good as a filled aerial of the same dimensions.

B. *Rotation*

When aerial arrays are constructed, they are often very large and the mechanical problems of rotation can be quite serious. It is possible, however, to simulate rotation of the aerial by introducing phase differences between the signals to or from the elements. This alters the direction of the orders, and in particular the zero order. For the series representing a grating with phase differences between the elements is

$$f(x) = \sum_n \delta(x - nb)\exp in\phi_0$$

and

$$a(k) = \sum_m \delta(k - \phi_0/b - 2\pi m/b)$$

which has its zero order at position $k = \phi_0/b$. The technique involves introducing the delay $n\phi_0$ into the nth element. It is not trivial, but is easier than rotating a large aerial which in the case of radio astronomy aerials, may be hundreds of meters in dimensions (Smith, 1962).

In microwave antennae, a similar technique is used for rotation. Here we have an array of waveguide terminations, and the delay is produced digitally by ferromagnetic cores which alter the inductance of a section of waveguide. In this example the advantage of the phase technique over physical rotation is that of the speed available.

C. Correlation-Function Techniques

It was realized when planning very large radio-astronomy antennae that there is no need for all the aerial to exist at the same time. By measuring, as a function of $\mathbf{r}$, the correlation function between the signals received at a pair of antennae separated by a variable vector $\mathbf{r}$, a picture of an incoherent source can be built up. In this case one has to pay for a reduction in hardware by an increase in the time involved in making an observation, since correlation-measurements involve integration over a period, and many such integrations have to be made.

The principle of the method follows from Section II.H. The self-convolution function of the received signals

$$c(\mathbf{r}) = \int f(\mathbf{r}') f^*(\mathbf{r}' - \mathbf{r}) \, d\mathbf{r}'$$

is measured. Since the kernel of the integral does not depend on $\mathbf{r}'$, it is sufficient to measure the time-average $\langle f(0) f^*(\mathbf{r}) \rangle$, generally within a fairly narrow frequency range (corresponding to forming an image in a single colour). Now the Fourier transform of $c(\mathbf{r})$ is $|a(\mathbf{k})|^2$, which is the intensity distribution of radio signals in the sky, lying in the frequency band of the signals. The reason for choosing a narrow frequency band is that we need to know λ in order to convert $\mathbf{k}$ into an angular variable. The instrument designed for radio astronomy observations working on this principle is the Mills Cross (Mills *et al.*, 1958).

VI. IMAGE FORMATION

In Section III, Chapter 1 image formation has been discussed on the basis of a double diffraction process, a direct application of the Fourier inversion theorem (Section II.D). In this section we shall describe experiments to illustrate some of the basic features of image reconstruction and the effect of various spatial filters on the image.

What must be emphasized at this stage is that when we form an image directly, we overcome the phase problem by using the diffraction pattern itself as the object for the second transform. Since we do not need to photogaph the diffraction pattern, we do not have the problem of storing the phases.

Thus we can use any object for the demonstration, and are not restricted as we were in Section II.B to objects having real positive transform.

In order to make a satisfactory demonstration of the image formation and spatial processes we need to bear the following points in mind.

(a) Since we want to "handle" the first diffraction pattern in an observable manner, it must be quite large, although need not be on the scale we needed for direct demonstration. We need, in fact, the finest detail in the diffraction pattern to be of the same magnitude as convenient dimensions for the spatial filters, such as a phase plate (Section III.C, Chapter 8), or a dark spot to

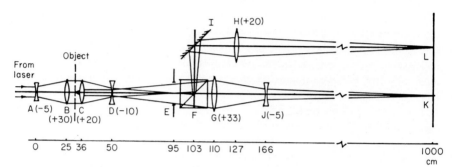

Fig. 21. The apparatus used to illustrate image-formation and spatial filtering. The lower scale shows distances measured from A in cm. The focal length in cm is given in brackets next to each lens. E is an iris diaphragm which can be set between 1 and 30 mm diam, F is a beam-splitter, if possible 30 mm square. G needs to be of good quality, since 30 mm aperture is used. The mirror I should be front-silvered so as to avoid multiple reflexions. The various spatial filters are inserted in the plane of E. The image, after filtering, is seen on the screen at K, and the filtered diffraction pattern at L.

simulate dark-ground illumination (Section II.C, Chapter 8). Taking this latter dimension as about 0·2 mm, and the beam diameter as 5 mm, we find that the focal length of the first diffracting lens must be about

$$f = \frac{0 \cdot 5}{\lambda} \times 0 \cdot 2 \text{ mm} = 1 \cdot 6 \text{ m}.$$

Once again, this can be achieved without trouble by using a telephoto combination; but we have had quite satisfactory results with a single 1 m focal-length lens.

(b) Immediately after the diffraction-pattern (spatial-filter, Section II, Chapter 8) plane, we need to extract part of the light to form a direct image of the pattern together with any modifications that have been made to it. This is done once again by means of a beam splitter. In this case it needs to be quite a large one, as the diffraction pattern may have interesting features 10–20 mm from its centre.

(c) The second diffraction stage is carried out by a second lens. We require a large final image (say 50–100 mm diam.) and there an overall magnification of about 10. Thus the entrance principal plane of the lens must be one-tenth of the distance from the object towards the image—about 500 mm from it. But the lens itself must be situated *after* the focal plane of the first diffracting stage. Thus a telephoto combination is almost certainly required here.

Based on these considerations, we arrive at an apparatus described by Fig. 21.

A. Resolving Power of a Microscope

The first application of the Abbe theory was to demonstrate the limited resolving power of a microscope (Section I.B, Chapter 1). Abbe, of course, was interested in the fundamental limitation N.A. $\leqslant 1$ (or μ for an immersion objective), but we can demonstrate the same ideas visually by limiting the numerical aperture to values of the order of 0.001. This is in fact very appropriate to an electron microscope (Sections I.A, Chapter 1 and VII.A, Chapter 10). Two types of object are valuable; the first is periodic, and our 0.2–0.5 mm gauze serves admirably; the second is not periodic, and contains detail of various degrees. Almost any black-and white drawing of high contrast, reduced to outside dimensions of about 5 mm, is satisfactory.

In the diffraction-pattern plane, we can see the well-known diffraction pattern of the gauze (Fig. 22a), and in the same plane we insert an iris diaphragm. The experiment consists of noticing how the sharpness of the final image (Fig. 22b) disappears as we cut out more and more of the diffraction pattern. When all that is left are the $(0,0)$ $(0,\pm1)$ and $(\pm1,0)$ order—assuming that the iris diaphragm is centred—the image is sinusoidal (Fig. 22c); it still has the right periodicity, but no further periodic detail. Any irregularities in the gauze, however, remain, since this is transmitted by the transform in between the orders. When the iris cuts out all but the $(0,0)$ order, the periodicity disappears; but the irregularities remain, getting less and less distinct as the diameter of the iris is further reduced (Fig. 22d).

Similarly, the non-periodic object gives an image which becomes progressively more and more blurred as the iris diameter becomes smaller. We have simply replaced the Fourier series (the optical transform of the periodic wire gauze) by a Fourier transform (that of the non-periodic object).

B. False Detail

Look carefully at the image of the gauze by the $(0,0)$ $(0,\pm1)$ and $(\pm1,0)$ orders (Fig. 22b). Not only is the image blurred, but there are bright spots at the crossover between the wires. Because we are not using all the transform, we have run into the danger of series-termination errors (Section II.C.3, Chapter 8), which produce spurious detail in the image. This was another of the points Abbe made in his original theory. In this example it is easy to see

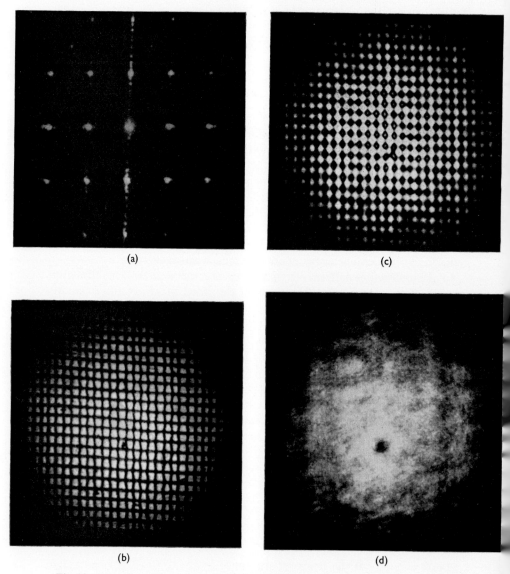

(a)

(c)

(b)

(d)

Fig. 22. Image formation using a wire gauze: (a) the diffraction pattern; (b) image using all orders of diffraction in (a); (c) image formed from the zero-order and four {0,1} orders; (d) image formed from the zero order only.

where they come from. Treating the problem one-dimensionally, the square-wave object (wire thickness equal to the spacing between the wires) gives order amplitudes:

$$0: \quad 1 \qquad \pm 1: \quad 0{\cdot}64 \qquad \pm 2: \quad 0 \qquad \pm 3: \quad -0{\cdot}21.$$

If we terminate the series after the first order (or second), we observe the sum:

$$f(x) = 1 + 0{\cdot}64(\exp -ix + \exp +ix)$$
$$= 1 + 1{\cdot}28 \cos x.$$

(x) has small negative maxima (at $x = (2n - 1)\pi$) which give weak spots just in the region which ought to be dark—and would be if we had included all the orders.

We can in fact emphasize such effects by choosing combinations of spots, but this is hardly appropriate—although considerable damage can be done to an image in this way. See, for example, Heidenreich (1964), pp. 324–331.

C. Spatial Filtering

An iris diaphragm is only one of a large variety of complex apertures that can be inserted into the diffraction pattern plane. The general name for these apertures is *spatial filters* (see also Section II, Chapter 8), the name being rooted in the analogy between the image-forming system and an electrical filtering circuit and about which we shall have more to say later (Section VII.B). The various filters which we shall discuss below are first the phase-contrast filter (corresponding to phase-contrast, dark-ground and Schlieren photography), then the filters appropriate to apodization, and finally spatial filters in general, with reference to some aspects of pattern recognition work.

1. Objects for Phase-contrast Demonstrations

A convenient phase-object (Section III.C, Chapter 8) is a light fingerprint on a clean glass slide; otherwise one can make holes in a piece of celluloid, or use an uneven piece of mica. Another object, particularly appropriate to Schlieren microscopy, is the hot air above a candle flame.

2. Dark Ground and Phase-contrast Microscopy

It is considerably easier to make a good demonstration of dark-ground microscopy than of phase-contrast. We simply put a small drop of ink (in Section VI we decided it would be about 0·2 mm diameter) on to a clean microscope slide, and adjust its position until it cuts out the zero order. To do this, remove the object, and adjust the spot to give a completely dark screen (or as dark as the quality of your lenses will allow!). A phase-object inserted now will reveal its phase variations in the image.

To make a $\pi/2$ phase plate is difficult. With luck, a hole in a piece of uniform celluloid ought to work, since the value of $\pi/2$ is not critical; but we have had no luck in this direction.

3. Babinet's Theorem

Babinet's theorem says that the diffraction patterns of two complementary screens are similar, complementary screens being defined as screens with clear and opaque parts such that the clear parts of one screen exactly match the opaque parts of the other. The limitations of this theorem are discussed by Lipson and Lipson (1969); within these limitations the theorem can be simply illustrated by using photographs of complementary drawings as diffracting objects. If an object is mainly clear, it has a strong zero order; on cutting this out (Section VI.C.2, Chapter 10) the image is found to be the complementary pattern.

4. Schlieren Technique

In the Schlieren technique we use a razor blade to eliminate one side of the diffraction pattern. Once again, phase variations are seen as contrast in the image, although in this case it is phase gradients perpendicular to the knife edge which give the effect. The image is weaker than in the previous cases, since we have cut out about half the light.

The candle flame can be used as object, emphasizing the hydrodynamic applications of this technique. It is also instructive to defocus the apparatus slightly, with no object in position, and to see the lens aberrations appearing on the screen. This is the Fresnel knife-edge test.

5. Apodization

Apodization—varying the transmission of different parts of a lens surface —can serve two purposes; one is to increase the resolution of an instrument somewhat, at the sacrifice of getting a direct image; the other is to remove termination errors (Section VI.B) at the sacrifice of resolution.

In the second use mentioned above, the spatial filter is a hole with fuzzy edges, best made by evaporation (Section II.E.5). If we return now to the experiment in which we formed an image of the gauze (Section VI.B); and if the diameter of the hole is such as to include just the $(0,0)$, $(0,\pm1)$ and $(\pm1,0)$ orders, the example which gave noticeable false detail; then we shall find that the false detail has disappeared. Clearly, in this case, the first orders have been weakened without affecting the zero order, and the first-order amplitude of $0\cdot64$ has been reduced below $0\cdot5$ at which value the false detail will vanish.

Apodization in the first sense, and its application to enhanced resolution, is applicable only to incoherent image-formation. We shall return to this point in Section VI.D.

6. *Pattern Recognition*

Chapter 8 discusses spatial filtering in general. In principle spatial filters can be made in much wider variety than electronic filters for the following reasons: they can be two-dimensional, they have no need to obey causality and they can be relatively detailed.

One of the uses they have been put to is that of pattern recognition, in which one seeks to find the position of a certain feature existing in the presence of a large noisy background (we can liken this to the use of a phase-sensitive detector in electronics by means of which a sinusoidal signal of given frequency and phase can be recognized when buried in noise).

The optical process works by taking the optical transform of the quotient of the transform of the signal (object) and that of the pattern $f(x)$ to be recognized. Clearly the noisy part of the signal will give a result which has no recognizable features. However, the pattern $f(x - x_0)$ occurring around an origin x_0, will contribute

$$a(k) \exp -ikx_0$$

to the transform, where $a(k)$ is the transform of $f(x)$. Multiplying this by the filter $a^{-1}(k)$ gives contribution $\exp -ikx_0$ to the transmitted amplitude, and the transform of this is $\delta(x_0)$—i.e. a delta function (bright spot) at the origin of the pattern in the original signal. If the function $f(x)$ occurs around a number of origins, a bright spot will appear at each one in the final image plane. Clearly the production of such $a^{-1}(k)$ filters is not simple, but has produced some quite remarkable results, and has been applied to the recognition of patterns such as fingerprints.

D. *Incoherent Image Formation*

Much practical image formation is carried out in incoherent light, or at least in partially-coherent light. In general this leads to slightly better resolution, although it is difficult to make any general statement to this effect (Thompson, 1969). The basic differences between coherent and incoherent illumination can be understood by considering a "gedankenexperiment" on the image-forming equipment described above; there are fundamental reasons why it should work, but it would be difficult to carry out in a convincing way.

1. *Resolution*

It is easiest to visualize partially coherent illumination (Section II.C, Chapter 2) as a superposition of plane-parallel coherent waves with random phases, incident on the object at various angles. The higher the degree of coherence, the smaller the range of angles required; the coherence distance is the wavelength divided by the range. We can consider the image formed, coherently, by each plane-parallel wave and then superimpose the resultant intensities to get the final image observed.

Consider, for example, the image of a sinusoidal diffraction grating giving zero and ± first orders. For coherent image-formation the aperture must be large enough to allow all three orders to pass. However, when the illumination is incoherent, the aperture need pass only two orders, since there will always exist directions of propagation of the component plane-waves for which one of the pairs $(0,+1)$ or $(0,-1)$ lies within the aperture. Thus all the information in the diffraction pattern will reach the image with a smaller aperture than in the coherent case. One can raise the objection that it would be possible to take a smaller off-centre aperture in the coherent case, and to use only the 0 and +1 orders. But then any information contained in the -1 order would be eliminated and one does not know *a priori* what this information is worth.

2. *The Michelson Stellar Interferometer, and Apodization*

Consider now the resolution of the images of two neighbouring incoherent points. Each one produces an image which is the transform of the aperture in the diffraction-pattern plane; if the aperture is a hole of radius R, this is the function $J_1(kR)/kR$ (J_1 is the first-order Bessel function). It has its first zero when

$$kR = 1 \cdot 2\pi.$$

If, however, the aperture were a slit, of width $2R$, the first zero would occur when

$$kR = \pi;$$

the circle gives a slightly more spread-out diffraction pattern. Now according to Rayleigh, two incoherent images can be considered to be resolved when the central maximum of one coincides with the first zero of the second. To find the ideal aperture to give highest resolution we must therefore look for one which has its first zero at the smallest angle (smallest value of k). Clearly the slit gives a slight better resolution than the circle. Of course it is obvious that a larger value of R would immediately solve the problem, but we mean to look for an aperture which lies within the physical bounds of the circle or one-dimensional slit. An immediate answer is a pair of narrow slits. If the separation is $2R$, the maximum allowable, the first minimum occurs at

$$kR = \tfrac{1}{2}\pi$$

and the resolution has been improved by a factor of two (Fig. 23). In the circular system, the circular disc is replaced by an annular ring. Of course the price paid for this increase in resolution includes a decrease in intensity and the fact that the image does not look much like the object. This is the basis of the Michelson Stellar interferometer (Lipson and Lipson, 1969) and also of one of the uses of apodization (Section VI, C.5), in which the centre of the lens of an image-forming instrument is blacked out—a rather strange idea to improve resolution.

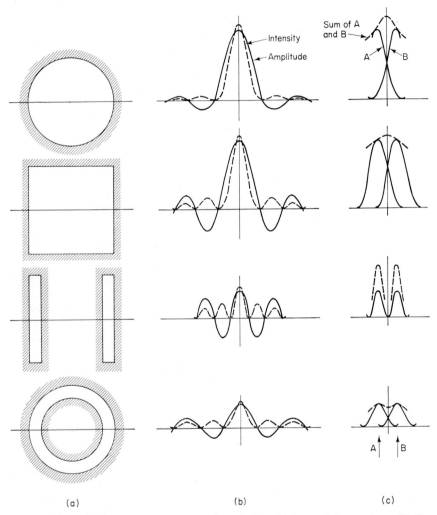

Fig. 23. Apodization for enhancement of resolution: (a) form of the aperture; (b) the amplitude —— and intensity – – – distribution in the focal plane, illumination being a single point source at infinity; (c) the intensity distribution —— resulting from two point sources at infinity, whose gaussian images would be at the positions indicated by the arrows A and B, when imaged through the apertures in (a). The broken lines show the sum of the intensities. The images are resolved only when using the lower two apertures.

VII. Some Applications of Image Formation Principles

The only non-optical example of image-formation is the electron microscope (Section I.A, Chapter 1), and we shall discuss below some of the applications of the ideas in the previous section to the working of these instruments. The mathematical principle of image-formation is, however, applicable to electronic filtering, and it is in fact by this analogy that the ideas of spatial filtering came to be born. This is also relevant to our discussion.

A. Electron Microscope

As in optical image formation, the principle of the electron microscope can be described in terms of double diffraction; first we form an electron diffraction pattern, and then the diffraction pattern of this pattern, which gives us an image of the original object (Fig. 24). The model described in Section VI is perhaps even more appropriate to electron microscopy than to optical microscopy, since it is a paraxial system (we noted that N.A. < 0.001) and not a large-angle system like good optical microscopes. However there are some points which are worth mentioning at this stage. Firstly, the object is three-dimensional, and so the transform involved in the first stage is only a section of the full three-dimensional Fourier transform (Section III.A). Electron wavelengths of the order of 0.01 Å or less, are sufficiently smaller than crystal lattice parameters for the sphere of reflexion to become approximately planar in the region of the origin of k-space, and because we can use only small angles, the diffraction pattern observed is a plane section of the Fourier transform through the origin and normal to the axis of the instrument. As a result the image at any point is an unweighted average of the unit cell contents in a direction parallel to the axis. This planar section makes multiple scattering of the electron-beam within the sample of considerable importance, since multiply-scattered waves still satisfy conservation of momentum and energy. This point has never been fully dealt with, and we shall return to it in Section VIII.B. It has the result that the diffraction pattern is not an exact Fourier transform of the object. However, in what follows we shall assume it to be so, as is conventional.

In the diffraction-pattern plane we have a possibility of various stops or filters, the simplest being the aperture of the electron lens following it. Because of aberrations, this diameter is necessarily limited, at present, to a numerical aperture of about 0.005. The resulting degree of resolution follows from Abbe's theory. It is immediately obvious that detail of the image can be seen only if in each direction at least zero, first and second order of diffraction are transmitted by the stop. For if we only transmitted the zero and first order, we should see the periodicity; the atomic structure would always appear to be sinusoidal, whatever the reality. If we have zero and both first orders, we know from Section VI.B, that there is great danger of false detail. Only when

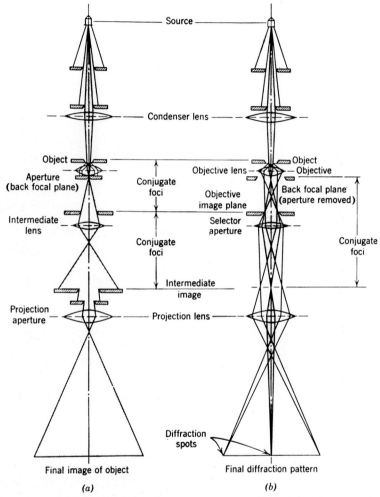

Fig. 24. Ray diagrams for double condenser, three-stage electron microscope: (a) normal image formation; (b) selected area diffraction operation. Mode (b) is obtained from (a) by decreasing the focal length of the intermediate lens so as to image the pattern on back focal plane of objective lens. (From R. D. Heidenreich, 1964.)

we use zero, first and second orders, with correct phases, can we expect to see realistic detail. This criterion has not yet been reached.

On a coarser scale one can do much more (Heidenreich, 1964). If one wants to look at deviation from the perfect crystalline structure, it is possible to concentrate attention on the ghosts around a particular order. This need not be the zero order; different information is obtained depending on the order used, the filter used being an aperture extracting it from all the others. The result of separating a single, non-zero order by means of an off-axis aperture

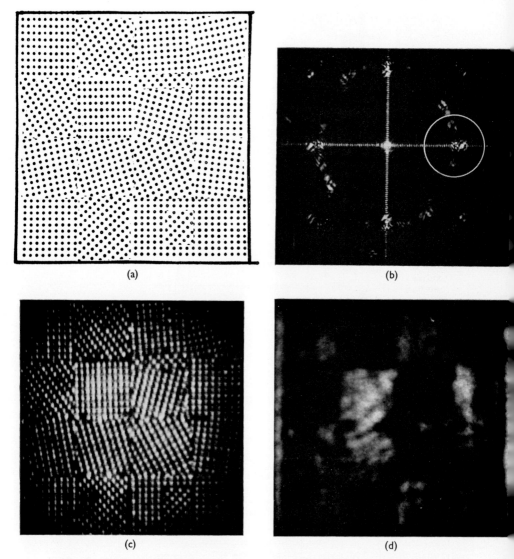

Fig. 25. Dark ground microscopy simulated in the optical experiment: (a) the object, a mosaic "crystal"; (b) its diffraction pattern; (c) image formed using the whole diffraction pattern; (d) image formed using the part of the diffraction pattern within the circle in (b).

is called dark-field microscopy. Using the apparatus described in Section VI, and an object consisting of a drawing of a mosaic crystal, the principle can be very effectively demonstrated (Fig. 25). This type of filtering is also the principle behind X-ray topography (Lang, 1959).

B. Electronic Filtering

As we saw in Section IV.A, the output from a filter is related to the input signal $u(t)$

$$v(t) = \int b(\omega) f(\omega) \exp(i\omega t) \, d\omega$$

where $b(\omega)$ is the transform of $u(t)$ and $f(\omega)$ the frequency response of the circuit. This description is analogous to that of the image-formation apparatus; $u(t)$ is the object, $v(t)$ the image and $f(\omega)$ the spatial filter. As a consequence of this analogy, one can visualize some of the properties of filters or linear amplifiers, and the way in which they modify an input signal.

Any filter must satisfy the conditions set out in Section IV.B; that is, it must obey causality and it must be symmetrical. Because of the second condition, we automatically get a real image $v(t)$ from a real input $u(t)$.

Consider, for example, the response of the L–C–R circuit to a square wave of varying frequency. The input is a row of slits with width equal to about half their spacing. The filter is described in Section IV.B. The output will be seen to consist of an oscillation of the frequency $(LC)^{1/2}$. As the object frequency is changed (by rotating the row of slits about a vertical axis) we go through resonance every time a pair of positive and negative orders is spaced by the filter spacing, and the sinusoidal nature of the output signal can also be seen. Alternatively we can look at the impulse response, or the response to other simple signals of the "1-or-0" type. Again, we can look at the effects of a low-pass filter (a single slit with continuous phase gradient from $-\pi$ to π) on signals of various sorts. We are not aware of any serious use of optical transforms for solving circuit-response problems in research, but many of the ideas can be visualized in this way.

VIII. LIMITATIONS IN THE RELATIONSHIP BETWEEN DIFFRACTION AND FOURIER TRANSFORMS

Throughout this article we have tacitly assumed that the diffraction pattern of an object is its optical transform; that is, it is the Fourier transform of a function directly representing the object, or at least a section of the three-dimensional transform of a three-dimensional object. While not wanting to dispel faith in this relationship, which is true under many conditions, it is only fair to point out that it does have its limitations. As will be seen from the short discussion below, the limitations are rarely relevant to everyday uses; but it is still wise to acknowledge their existence.

A. Polarization and the Scalar Wave Theory

One limitation arises from the scalar theory of diffraction. A basic assumption in the theory of Fraunhofer diffraction, on which most optical transform work is based, is that the light disturbance can be adequately represented by a scalar quantity, although it is well known to be a vector electromagnetic field. If one has reservations about this approximation and looks to see what can be done in the way of an exact solution of the electromagnetic theory of diffraction (Born and Wolf, 1966), one is very likely to be frightened back to unquestioning belief in the scalar theory.

There are, however, a few experiments which succeed in showing that diffraction effects are in fact polarization-dependent, and which can be understood without a very deep incursion into the electromagnetic theory. One of these is the edge-wave experiment (Lipson and Lipson, 1969) (albeit Fresnel diffraction, and properly out of the bounds of this book) in which one looks particularly at that part of the Fresnel diffraction pattern of a straight conducting edge which directly results from the discontinuity of the amplitude at the edge. If sought in polarized light, the edge wave will not be found if the light is polarized with its electric vector parallel to the edge, indicating that there is no discontinuity of amplitude for this polarization. This is a consequence of the different boundary conditions for electric field parallel and perpendicular to a conducting edge. (As a practical point, it is worth remarking that the razor blade used for this experiment should not have been shaved with!) The experiment shows that there do exist conditions under which the vector nature of light disturbances have to be taken into account.

B. Multiple Diffraction

Another limitation to the transform relationship occurs in three-dimensional diffraction, where multiple diffraction may be present. This arises when an already-diffracted beam is diffracted a second time. In X-ray diffraction, where the Ewald-sphere construction (Section I.E, Chapter 1) severely limits the observable diffracted beams, this is rarely a problem, since the probability is very remote that a second diffraction process will also satisfy the stringent conditions. However in electron diffraction the Ewald sphere becomes a plane to a good degree of approximation (Section VII.A), and then the chance of multiple diffraction is quite reasonable; it may even be that single diffraction becomes the rarity. In such a case, the diffraction pattern bears only a superficial resemblance to the Fourier transform; it has spots in the right places, but their amplitudes and phases are quite wrong. As a result, image reconstruction, which faithfully retransforms a misleading diffraction pattern, gives a false image; although it may be quite convincing in that it is built on the right lattice, the details are wrong. We should be chary of accepting electron micrographs showing anything approaching atomic detail. Fortunately the error decreases rapidly as the diffraction angle gets smaller, and electron micrographs of macroscopic detail will be quite reliable. The criterion for reliability is that $2\pi t\theta^2/\lambda$ must be much less than unity, where t is the specimen thickness (obviously in a very thin specimen multiple diffraction would not occur), θ the angle of diffraction and λ the wavelength. When $\lambda = 0\cdot01$ Å and $t \sim 100$ Å, $\theta < 10^{-3}$, implying a limit of reliable resolution of about 10 Å. This is an optimistically low value, and the much finer resolutions claimed commercially must be treated with great caution.

IX. SUMMARY

This chapter is different from others in the book in that its purpose is pedagogical, and does not aim to represent the results or direction of research work. We have tried to illustrate the way in which optical diffraction, being a visual way of producing Fourier transforms, can be used to illustrate some of the applications of these transforms in physics in general.

As regards the experimental information, it has resulted from an illustrated lecture course now being given for the third time. Some illustrations are very easy; others succeed only with perseverance; we have tried to avoid describing experiments which have only succeeded in the imagination. But it must be emphasized that any successful demonstration takes many hours in preparation and must be aimed at illustrating the point in question as simply as possible. Subjects such as pattern recognition and the differences between coherent and incoherent resolution are too difficult to demonstrate convincingly with easily-available equipment.

When talking to an experienced electronics engineer one can often get the feeling that he sees any signal intuitively as its frequency spectrum. The idea of this chapter has been to describe how optical transforms can be used to instil something of this intuitiveness into physicists.

REFERENCES

Bacon, G. E. (1962). "Neutron Diffraction," Oxford University Press, Oxford.
Bacon, G. E. (1966). "X-ray and Neutron Diffraction", Pergamon Press, Oxford.
Barrett, C. S. and Massalski, T. B. (1966). "Structure of Metals," McGraw-Hill, New York.
Born, M. and Wolf, E. (1966). "Principles of Optics," Pergamon Press, Oxford.
Furukawa, K. (1962). *Rep. Prog. Phys.*, **25**, 395.

Heidenreich, R. D. (1964). "Fundamentals of Transmission Electron Microscopy," Wiley-Interscience, New York.

Henry, N. F. M., Lipson, H. and Wooster, W. A. (1960). "The Interpretation of X-ray Diffraction Photographs," Macmillan, London.

Koppelman, G. (1969). "Progress in Optics," Vol. VII, p. 3, North-Holland Publishing Co., Amsterdam.

Lang, A. R. (1957). *Acta Met.*, **5**, 358.

Lang, A. R. (1959). *Acta Cryst.*, **12**, 249.

Lipson, S. G. and Lipson, H. (1969). "Optical Physics," Cambridge University Press, Cambridge.

Lonsdale, K. (1949). Crystals and X-rays, van Nostrand, New York.

Mills, B. Y., Little, A. G., Sheridan, K. V. and Slee, O. B. (1958). *Proc. I.R.E.*, **46**, 67.

Schiff, L. I. (1968). "Quantum Mechanics," McGraw-Hill, New York.

Smith, F. G. (1962). "Radio Astronomy," Penguin, Harmondsworth, England.

Taylor, C. A. and Lipson, H. (1964). "Optical Transforms," Bell, London.

Thompson, B. J. (1969). "Progress in Optics," Vol. VII, p. 171, North-Holland Publishing Co., Amsterdam.

CHAPTER 11

Miscellaneous Applications

J. E. Berger,[1] C. A. Taylor,[2] D. Shechtman,[3] and H. Lipson[4]

[1] *Centre for Crystallographic Research, Roswell Park Division of Health Research Inc., Buffalo, New York:* [2] *Department of Physics, University College, Cathays Park, Cardiff, Wales:* [3] *Department of Materials Engineering, Technion. Haifa, Israel:* [4] *The University of Manchester, Institute of Science and Technology, Manchester, England*

I. Analysis of Electron Micrographs (by J. E. Berger)

A. Introduction

With increasing frequency the electron microscope is revealing the presence of regularly repeating structure in biologically important systems. The list of such structures already reported includes subcellular organelles (lysosomes, mitochondria, melanosomes), fibrous constituents (collagen, actin, myosin), and viruses as well as many intracellular crystalline objects (compare Chapter 5).

Precise determination of the repeat spacings in apparently crystalline or paracrystalline arrays is the cornerstone of identification. Yet, the electron micrographic process is such that the structure as shown in the magnified image may be distorted with respect to the structure as it exists in nature. Random perturbations brought about by fixation, staining, drying and sectioning of a given specimen may, either wholly or in part, convert a regularly repeating structure into one quite lacking in apparent periodicity.

In Fig. 1(a) a number of objects have been arrayed on a hexagonal net. Figure 2(a) depicts the same net confounded by randomly placed scrawls. Even in this rather simple idealized instance, it would be difficult if not impossible to determine the periodicity of the original undistorted specimen or, in fact, even to state with certainty that any underlying periodic structure existed.

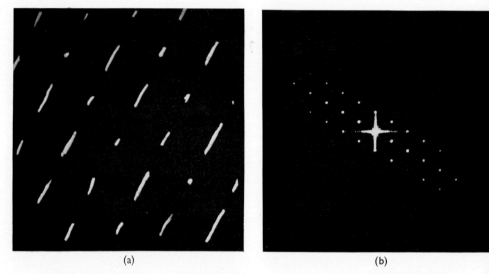

(a)   (b)

Fig. 1. (a) An array of streaks located on a centred square net; (b) optical diffraction pattern from (a).

The requirement that specimens be stained (Section III, Chapter 5) with an electron-opaque material prior to electron microscopy gives rise to additional sources of trouble. Stain may, because of microscopic local heterogeneity, fail to outline identical elements of a structure in an identical fashion. Further, the extent of penetration of stain may vary from one area to another for a variety of other reasons, including contamination of the specimen by cellular debris and local variation in the extent of pooling of the stain (Klug and Berger, 1964).

Contrast between the structure of interest and the background "noise" may, on occasion, be too low to permit clear visualization of detail, even when distortion from other sources is not present. With the problem of poor contrast added to those already outlined, it becomes even more likely that structural information will be obscured.

When the eye is called on to analyse micrographic images of periodic structures it does so by sequential scanning of the elements comprising the whole. In the process, each element contributes individually rather than co-operatively to the overall picture.

A method which permitted objective integration of the detailed data present in an electron micrograph would, of course, vastly diminish the relative importance of individual departures from a "true" position; further, the presence of a periodic array of elements not individually discernible above

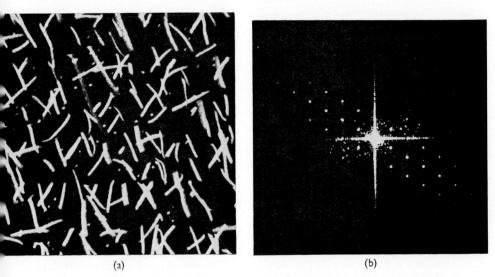

(a)   (b)

Fig. 2. (a) Net of Fig. 1(a) superimposed on a background of randomly drawn lines; (b) optical diffraction pattern from (a) showing diffraction maxima identical to those in Fig. 1(b). The presence of underlying periodicity is thus clearly revealed.

the background might be detectable if a technique were available to sum the contributions of the single units.

The required integrative evaluation of electron microscopic images can be accomplished by taking the Fourier transform of the photographic image. Application of this technique has allowed detection of periodicities quite invisible on direct inspection of the original micrograph. It has also permitted, in many instances, a detailed evaluation of structure by manoeuvres analogous to those established for X-ray diffraction analysis (Finch *et al.*, 1964; Moody, 1967; Kiselev *et al.*, 1968). Figure 2(b) shows clearly that the optical transform of the apparently quite random Fig. 2(a) does in fact derive from a periodic structure. The layer line spacing is identical with that of Fig. 1(b).

B. *Instrumentation*

Apparatus for production of Fourier transforms of pictorial data (Section IV, Chapter 1) was originally developed for X-ray crystal structure determination in the era before that of the digital computer (Taylor *et al.*, 1951). These "optical diffractometers" are commercially available and may be used with electron micrographs. However, the relatively low level of illumination produced by this instrument prolongs photographic exposures unnecessarily. A thorough discussion of the principles involved has been set forth by Taylor and Lipson (1964).

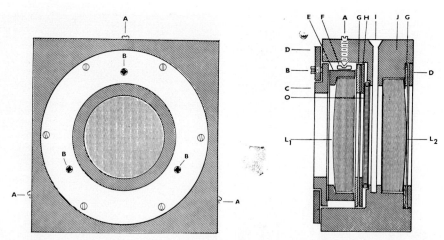

Fig. 3. The fixed lens L_1 is a snug fit in cell housing (J) and is held in place by Vellumoid gasket (G) and retaining ring (D). Because it is desirable to avoid direct application of localized forces to the lenses, tilt and translation adjustments of the movable lens are effected through screws (B) and (A) bearing on lens housing (E) through (C) and (F) respectively. An effective oil seal, which is also a yielding base for the adjustment, is obtained by the combination of O-ring (O) and gasket (G) separated by a metal washer (H). The electron micrograph is inserted at (I). (Components which would be in contact during normal operation are sometimes shown apart for clarity.) Reproduced courtesy Review of Scientific Instruments.

The availability of inexpensive, reliable helium-neon lasers operating in a single mode has permitted their substitution for the usual high-pressure mercury arc lamp. With the resulting increase in illumination, exposures can be reduced to a matter of seconds where previously hours had been required (Berger *et al.*, 1966).

Perhaps the most important single consideration in the use of optical diffractometers with electron micrographs is the elimination of optical path differences among various points on the photographic plate. These path differences arise because of undulations or asperities on the surfaces of photographic film. This factor has been emphasized previously and a means for achieving equalization of path differences has been described (Bragg and Stokes, 1945). The film under study, immersed in an oil medium of appropriate refractive index, is placed between two optical flats and the resulting sandwich

Fig. 4. Combination condensing lens and pinhole. Vertical and horizontal adjustment of the removable pinhole assembly is effected through the micrometer spindles to which the pinhole is magnetically secured. Courtesy of Jodon Engineering Associates Inc.

used as the diffracting object. Whilst this technique is effective, it is cumbersome.

The high levels of illumination achieved with laser light sources make it imperative that the total number of air-glass interfaces be kept small. Failure to do this leads frequently to the appearance of multiply-reflected images in the diffraction plane. Although in theory this problem exists irrespective of the nature of the light source, in practice it is only with the advent of laser illumination that the reflected rays have been intense enough to be troublesome.

To permit rapid examination of a number of electron micrographs with the
14

necessary oil-immersion and, simultaneously, to reduce the number of optical components in the overall diffractometer a specialized "immersion diffracto-meter" has been designed (Berger and Harker, 1967). Two plano-convex lenses are mounted in an oil-tight housing such that one lens may be tilted and translated with respect to its mate for purposes of alignment, Fig. 3. The space intervening between the lenses is filled with immersion oil of refractive index equal, as nearly as possible, to that of the photographic emulsion. The diffractometer in effect thus consists of a single very thick lens into the centre of which photographic specimens may be placed for examination. The entire

Fig. 5. Complete optical diffractometer. Components are arranged in correct sequence but without regard to proper relative spacing. Laser (S); pinhole assembly (P) see Fig. 4; lenses of diffractometer (L_1) and (L_2) see Fig. 3; recording plane for optical transform (T_1); reconstruction lens (L_3); recording plane for reconstructed image (T_2).

unit has but two air-glass interfaces instead of the usual six required by a standard diffractometer with an optical-flat immersion-oil sandwich.

The output from a helium-neon laser is condensed by a converging lens of very small focal length. The Airy disk from a pinhole placed at the focal point of this lens and at the back focal plane of the first lens of the diffracto-meter provides locally homogeneous illumination. A combination condensing lens and pinhole assembly of compact design is available commercially (Fig. 4).

An instrument capable of very high resolution may thus be assembled by mounting the three major components—laser, pinhole assembly, and im-mersion diffractometer—on a standard optical bench. Proper alignment is much facilitated by the high light intensity available from the laser (Berger and Harker, 1967). If laboratory space is limited, it is quite practicable to disassemble the entire unit for storage.

The diffractometer lenses should be of very high quality (Taylor and Thompson, 1957). Effects of spherical abberation are minimized by use only of the central portion of the lenses. A focal length of about five feet for each lens has proved generally suitable. For most precise work, the diffractometer should permit illumination of the electron micrograph by parallel light, i.e. the point source should be at the principal focus of L_1 Fig. 5 and the recording film in the back focal plane of L_2 at T_1. With slight distortion in the final optical transform, this requirement may be relaxed to require only that the pinhole and the transform plane be at conjugate foci of the diffractometer (taken as a single thick lens). This arrangement permits one to alter the resolving power of the diffractometer to meet the demands of a variety of structures and electron micrographic magnification factors. Diffraction theory dictates that optical transforms of black-and-white transparencies possess a centro-symmetric intensity distribution. Departure from this indication of good technique is most often due to use of inferior optical components, poor alignment of the diffractometer, or disregard of proper immersion methods.

C. Image Reconstruction

By adding one lens to the basic diffractometer, it becomes possible to carry out Fourier synthesis of selected portions of a Fourier transform (Buerger, 1950). One may thus often produce a clarified reproduction of an originally confused electron micrographic image.

Fourier reconstruction is accomplished by replacing the photographic recording of the transform by an opaque template in which holes have been punched to coincide with those diffraction maxima selected for synthesis. A lens, L_3, is so positioned that the Fourier transform of the diffraction pattern may be recorded photographically at T_2. This is equivalent to placing T_2 and the original electron micrograph at conjugate foci of the L_2–L_3 lens system.

Preparation of the template is facilitated by use of a special film casette (Fig. 6). In use, the casette holds film for recording the original diffraction pattern. After this is developed, holes may be punched at the sites of the selected diffraction maxima and the film rendered otherwise opaque with photographic paint. The resulting template is replaced in the casette with both caps removed. Any minor departures from perfect alignment with the diffrac-tion pattern may be corrected by use of the X–Y micrometer adjustments. The selected diffraction maxima with phase information intact now pass unhindered through the template, are resynthesized by lens L_3 and recorded at T_2. If the diffracted rays excluded by the template carry information referable *only* to artifactual distortion and background "noise," the resynthesized image will yield a clearer picture of the true structure.

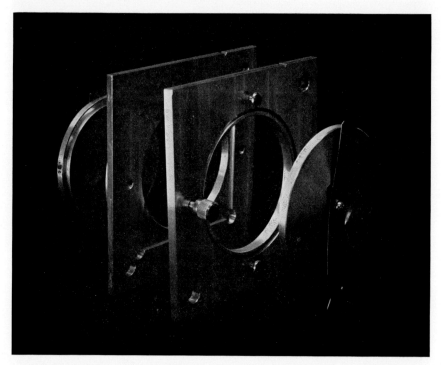

Fig. 6. Film casette, "exploded view". Note light trap at periphery of caps.

D. Photographic Considerations

Before carrying out optical diffraction on an electron micrograph it is advantageous to copy the original onto film. This avoids having to remove adherent immersion oil from the plate after finishing the study. The replicas are expendable and the original is thus maintained undamaged by cleansers.

Whilst almost any film may be used for the copy, the relatively high contrast Kodalith Ortho Type 3 has proved quite suitable in all respects. Extensive study in our laboratory has shown that the contrast and average optical density of the replica can differ substantially from that of the original plate without significant effect upon the resulting diffraction pattern. Failure to use an oil-immersion diffractometer or its equivalent can however lead to substantial loss of definition in the diffraction patterns and this effect may be aggravated by certain combinations of contrast and overall optical density in the replica.

The specific area of interest on the micrograph (or replica) should be closely framed to minimize contributions to the diffraction pattern which do not arise from the structure under study. In particular, failure to mask the border of the specimen closely will increase the prominence of those diffraction maxima related to the overall shape of the specimen but quite unrelated to the detailed structure.

It is often helpful to black out any localized regions within the specimen which appear to have poorer definition than the average for the specimen. Generally, diffraction from these areas will raise the overall "noise" level of the diffraction pattern without significant contribution to the useful signal.

Masking is readily done with photographer's opaque applied with a fine camel-hair brush. The diffraction pattern may be recorded on any red sensitive, for instance panchromatic, film. Orthochromatic emulsions are generally insensitive to the 6328 Å line of a helium-neon laser.

E. Special Techniques (Crystal Lattice Determination)

Microcrystalline inclusions are frequently found in biological materials. The appearance of these three-dimensional structures as projected onto an electron micrographic plate depends upon the direction of sectioning of the gross specimen with respect to the crystal lattice. The underlying identity of several apparently dissimilar periodic structures may thus be concealed. Conversely, closely similar one- or two-dimensional periodicities may arise from two or more quite unrelated crystals. Optical diffraction may aid in establishing the underlying three-dimensional lattice (Berger, 1969). Unit-cell dimensions can serve to characterize these structures and in addition may provide a crude estimate of the molecular weight of the crystalline substance.

The optical diffraction analysis is based on the fact (Section I.D, Chapter 1) that the Fourier transform of any projection of a crystal lattice is simply a plane section passing through the origin of the reciprocal lattice and lying parallel to the plane of projection (Fig. 7). Since electron micrographs are two-dimensional projections of structure, the optical transform of a micrograph of a crystalline specimen is a plane central section through the reciprocal lattice of that specimen. The plane of sectioning of the tissue determines the plane of the electron micrographic projection and thus the orientation of the section through the corresponding reciprocal lattice.

In general, routine electron micrographs of a given object will show a variety of random projections corresponding to the various obliquities of sectioning. Since all the resulting optical transforms are planes through the origin of reciprocal space (Section I.D, Chapter 1), any two such transforms will intersect along a line passing through that origin. Thus, when two transforms are found with a common row periodicity, ambiguity in relative orientation of these two patterns (with respect to their parent reciprocal lattice) ordinarily will be limited to one degree of rotational freedom about the common line as axis. This single degree of freedom can be fixed by analysis of additional pairs of transforms in a similar manner. Clearly, similarity of periodicity is a necessary but not sufficient condition for the identity of two

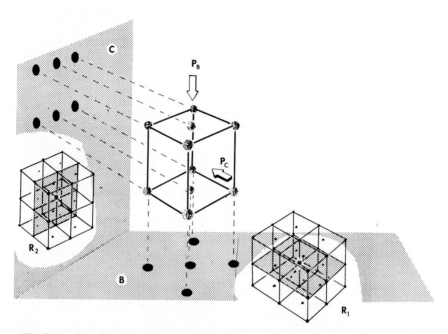

Fig. 7. Portion of a three-dimensional crystal lattice and the appearance of its projection along the directions P_B and P_C onto planes B and C, respectively. The projection pattern is indicated by solid black circles. The reciprocal lattice corresponding to the Fourier transform of the crystal lattice is shown at R_1 and again at R_2. Sections through the reciprocal lattice parallel to the planes of projection B and C are indicated by the shaded extension of those planes. These central sections through the reciprocal lattice are the Fourier transforms of the respective projections of the crystal structure. The origin of the reciprocal lattice is indicated by an asterisk. Courtesy *J. Cell Biol.*

lattice rows. However, supposition of identity for pairs of lattice rows which are in fact different will generally lead to obvious internal inconsistencies.

By making certain assumptions about the density of the crystalline substances under investigation, limits may be set on their molecular weights (Berger, 1969). Because the crystalline inclusions seen in electron micrographs are too small to yield to other analytic techniques, optical diffraction often affords the sole means of obtaining definite clues to their nature (Sternlieb and Berger, 1969).

F. Composite Micrographs

The strength of optical diffraction as described to this point rests in its ability to give a value for periodicity averaged over *one* extensive repeating structure no *part* of which is sufficiently crisp to give useful information on visual inspection. This principle may be extended to include examination of a *number* of similar periodic structures no *one* of which is well enough defined even for optical diffraction examination (Offord, 1966). Individual specimen images, e.g. tobacco mosaic virus rods, are isolated from a number of different electron micrographs. These are placed, without overlap, mutually parallel but with various translational displacements to form a composite two-dimensional array with one-dimensional order. The composite may be described as the convolution of a single "mean" virus particle with a net of more-or-less randomly disposed points. The clarity of the resulting optical transform is enhanced over that of a single particle because of the greater number of units contributing to the average. The relatively diffuse "fringe function" representing the transform of the net of points will not generally obscure important features of the transform of the particle.

In preparing composites, the magnification of all sub-units must, of course, be identical. Further, the optical density averaged over each component must not differ significantly among the several components lest components of low average density make a disproportionately large contribution to the average pattern.

G. Image Reconstruction (Benefits and Pitfalls)

Detection of periodic structure in electron micrographs by optical diffraction techniques is a quite objective operation. The method may be extended to give results of even greater utility by recombination of selected Fourier terms in the transform (compare Section III, Chapter 5 and Section II.C, Chapter 8). In carrying out this procedure, phase information in the original transform is retained. However, a certain subjectivity enters when one picks only portions of the complete transform for recombination. Ideally, those portions of the transform excluded from recombination will have derived from "noise" in the original image. To the extent that this ideal is achieved, the resynthesized image will give a clearer picture of the true structure. The accuracy of any reconstructed image must, therefore, be evaluated in relation to the probable bias introduced in selecting only some of the Fourier components for reconstruction. Figure 8 shows the result of Fourier synthesis using those diffraction maxima of a transform (Fig. 2b) which have been allowed to pass through the template shown in the insert. This result may be compared with the ideal "noise-free" image of Fig. 1(a) and the original "noisy" specimen of Fig. 2(a).

Here, the cost of eliminating most of the background confusion is mainly distortion of the shape of the individual subunits. The noteworthy feature is that a periodic array completely obscured by artifactual detail in the original image is rendered quite clearly in the reconstructed picture.

A scheme has been outlined for assigning values for both phase and amplitude to Fourier components of electron microscopic images (DeRosier and Klug, 1968). *If the phase data are correct,* the image resulting from resynthesis of the components will be an accurate representation of the true structure. This technique, while related to optical diffraction in a historical

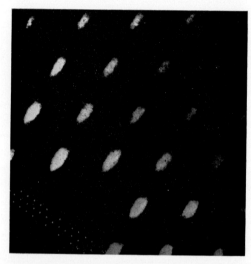

Fig. 8. Image reconstituted from those diffraction maxima of Fig. 2(b) selected by template shown (inset). Compare with Fig. 1(a).

and theoretical sense, is quite different in practice. It is noted here only for sake of completeness, and for further details the original papers should be consulted.

In summary, optical-transform techniques afford the electron microscopist a powerful tool for detecting and measuring, quite objectively, periodic structure which might otherwise remain inaccessible. By use of selective Fourier recombination, "improved" images may be derived in which background "noise" has been reduced or eliminated. There are modifications of optical diffraction methodology *sensu strictu* which attempt, by manipulation of phases or intensities of diffraction maxima, to give reconstituted images clearer than the original. The result must always be critically evaluated in light of the soundness of the premises justifying these manipulations; they may, and sometimes do, give a clear image which does not correspond to the true structure.

II. Droplet Size Determination (by C. A. Taylor)

The high-quality, high-intensity, optical diffractometer at Cardiff (Section IV.B, Chapter 4) has made possible a number of investigations outside the more obvious fields of diffraction and image formation. One such example is the work now in progress on droplet-size determination in collaboration with the National Institute for Agricultural Engineering. For various technical reasons it is necessary to know with as high precision as possible the distribution of sizes of droplets in the sprays produced by various kinds of nozzles in crop-spraying equipment.

Ideally it is desirable to measure the distribution in the "live" spray—that is, as the droplets are actually being produced and not from a photograph, or from a plate on which the drops have fallen. The object under study is thus three-dimensional and moving. Fortunately, in Fraunhofer diffraction, neither of these effects presents a problem. The three-dimensional character corresponds to the three-dimensional character of, for example, a polymer suspension being studied by low-angle X-ray scattering. The droplet size might be in the range of 50 μm–500 μm and the wavelength used is about 0·6 μm; the X-ray parallel would be particles in the range 150 Å–1500 Å being studied with an X-ray wavelength of 1·5 Å.

The movement of the particles is of no consequence because they remain in the parallel beam and translation of the object (in real space) corresponds only to a phase change, with no amplitude change, in the diffraction pattern (in reciprocal space). The objects are already randomly distributed in space and one expects only statistical information to result. The problems of interpretation are considerable unless one knows a certain amount about the scattering objects, but fortunately we can assume that all the droplets are spherical and this leads, at least theoretically, to the possibility of extracting information.

Experiments to demonstrate the feasibility of the technique have been made using commercially available glass spheres in the right size range and of known diameter. A "shower" of beads is allowed to fall through the parallel beam of the diffractometer and, fortunately, the laser provides sufficient intensity for the diffraction pattern to be recorded in a fraction of a second so that the quantity of beads involved is not great. Figure 9(a) shows a typical pattern for a shower of beads of fairly uniform diameter centred on about 600 μm. Figure 9(b) is the pattern for a similar shower of beads with their diameter centred on 800 μm. Figure 9(c) shows the diffraction pattern for a shower consisting of a mixture of equal parts of the two types of beads. Figure 9(d) shows the pattern produced by an actual spray of water droplets in which there is clearly a continuous distribution of sizes probably including both smaller and larger diameters than those of the beads. The work is still in a very early stage of development and much remains to be done before it can

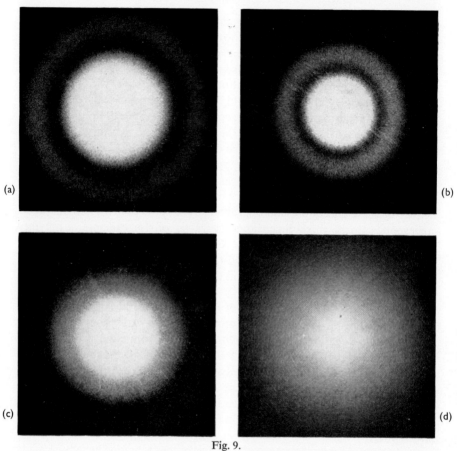

Fig. 9.

be regarded as a precise method, but initial results are promising. I am grateful to Mr. J. L. James for the preparation of Fig. 9.

III. STUDY OF WOVEN TEXTILES (BY C. A. TAYLOR)

Another application is the study of the structure of textiles. Again the technique is very much in its infancy and may prove to be of no more than academic interest; but the results are fascinating and provide excellent illustrations of some of the basic principles of diffraction and seem worth including

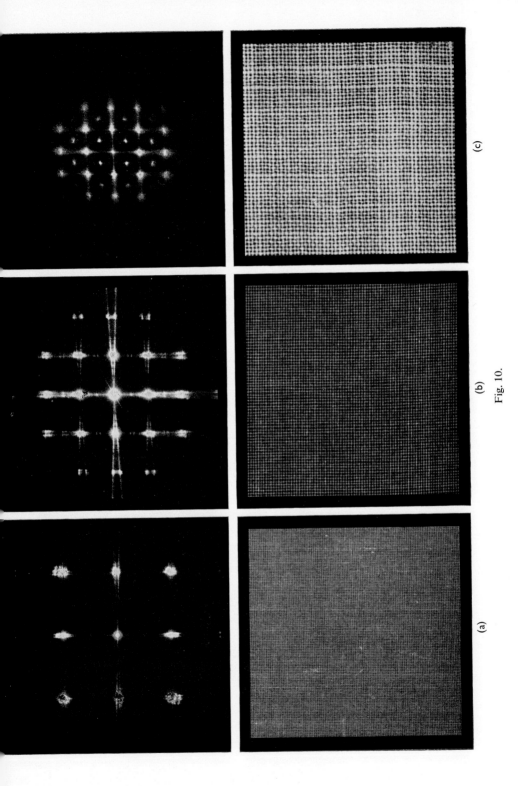

Fig. 10.

(a) (b) (c)

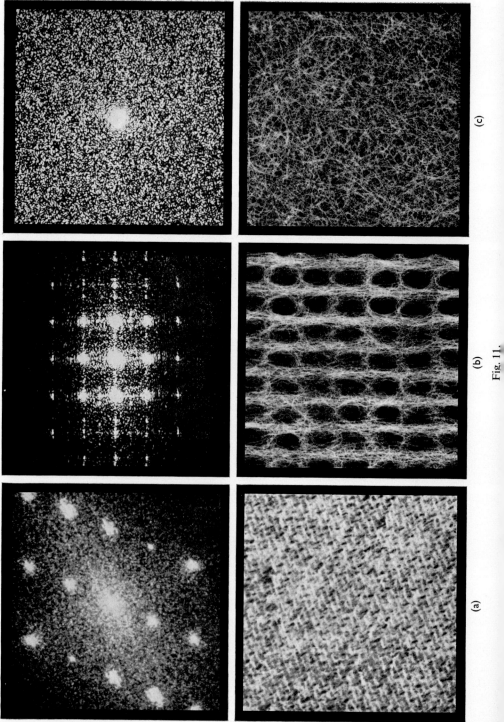

Fig. 11.

on this account alone. Again, movement is no problem and continuous monitoring of textiles moving at quite high speeds would be possible. If the periodicity of some other dimensional constant of the material changed, corresponding changes in the diffraction pattern would occur, but the diffraction pattern would, of course, remain stationary.

Figures 10 and 11 give examples of application in both woven and non-woven materials. In each example the upper photograph is the diffraction pattern and the lower photograph is a low-power micrograph of the object. Figures 10(a) and (b) are of silk gauze of two different gauges (approximately 9 threads per mm and 5 threads per mm respectively) of the type used for the manufacture of sieves. Figure 10(c) is of a handkerchief with about $3\frac{1}{2}$ threads per mm.

Figure 11(a) is for a well-known brand of paper tissue; Fig. 11(b) is for a household cleaning cloth of non-woven form and Fig. 11(c) is for a stronger non-woven material used for stiffening garments. In both Figs. 10 and 11 the objects are on the same scale and the relative enlargements of the diffraction patterns are the same. Points worthy of special interest are the "superlattice" in Fig. 10(c) which arises from the fact that the handkerchief has a conventional "over-and-under" weave which results in a "lattice" which is virtually "centred" in crystallographic terms.

The radial fanning-out of the streaks in all three patterns in Fig. 10 arises from departures from straightness and strict parallelisms of the threads.

In Figs. 11(a) and (b) the combination of regularity and randomness is interesting, but the material of Fig. 11(c) shows remarkable randomness, and the pattern is devoid of all signs of periodicities or even of preferred orientations.

A rather similar project has been carried out by Dye (1970) who has used diffraction methods to examine the perfection of field meshes used in the production of television cameras. A field mesh is a two-dimensional metal grid having a square repeat of about 20 lines per mm; it is necessary that such meshes should be of high accuracy and ordinary visual examination is not always adequate to see any blemishes that may be present. Diffraction patterns however show defects quite clearly.

Not all defects are equally important. Low-frequency errors—which give orders of diffraction with close satellites—are more important than high-frequency errors. This effect thus lends itself to completely objective testing.

By using spatial filtering (Section I, Chapter 8), it is possible to form images of the meshes showing only the defects and not the regularity; dust particles down to sizes of about 10 μm can be seen. This method also shows promise of industrial importance. I am grateful to Mr. R. Watkins for the preparation of Figs. 10 and 11.

IV. ANALYSIS OF METALLURGICAL MICROSTRUCTURES
(BY D. SHECHTMAN AND H. LIPSON)

A. Introduction

Metallography is the most important tool of the metallurgist; by polishing the surface of an alloy, etching it, and examining it under a microscope, he can usually recognize what phases are present and can see how they are distributed. This latter point is particularly important; other methods, such as X-ray diffraction, may identify the various phases more definitely, but the distribution can be found only by metallography.

Metallography is important, even for single-phase alloys, since it can tell us the sizes of the crystals—the grain size—of which the specimen is made. To put this property in quantitative form is, however, not easy. One can take a microphotograph with a given magnification, count the number of grains in a measured area, and thus work out the average size of the grains. But this process is rather lengthy. If immediate results are wanted about the grain size of a metal as soon as possible after it has solidified from the melt, the time taken to carry out these procedures would be unacceptably long.

An instrument known as the Quantimet, made by Metals Research Ltd., Cambridge, is now available for carrying out the operation automatically. The microstructure of a polished and etched specimen is exhibited on a television screen, and the intersections of the lines of the image with the grain boundaries are caused to give electrical impulses; the number of impulses is clearly a function of grain size and a small computer is programmed to give the grain size directly. This instrument clearly is as fast as one could demand, but it is expensive and is not likely to be generally available except in laboratories where grain-size determination is a general and continuing operation.

It occurred to us that optical-transform methods might provide a much simpler form of help. If a metal has a perfect uniform grain size, the transform should consist of sharp rings, the innermost being directly related to the grain size. If the grain size is not uniform the rings will not be sharp, and thus the measurement of the breadths of the rings should give an indication of the distribution of the grain sizes; this extra information could also be useful. If the grains are elongated in some particular direction—preferred orientation —this should also show in the transform of the stationary specimen.

The present paper is a preliminary account of some experiments to see if these ideas have any practical value.

B. Experimental Methods

Our first idea was to use the optical diffractometer (Section IV, Chapter 1) to produce transforms of micrographs, using negatives in which the grain boundaries show up as bright lines on a dark background. To obtain the best results, however, the plate had to be immersed in cedar-wood oil, which has the same refractive index as the emulsion of the photographic plate, in order to eliminate any variations of thickness of the emulsion (Section I.B). This procedure however turned out to be slow and not very satisfactory in its performance; waiting for the rather viscous oil to settle was rather frustrating. Moreover, the refractive index of the oil matches that of the emulsion only at a specific temperature, which can be found and kept constant only by rather laborious techniques.

We therefore tried a completely different approach—the production of the transform from the specimen directly, using a laser beam to produce a reasonable area of coherent light. The idea was to polish a specimen to optical standards, so that the laser beam would be specularly reflected; the specimen

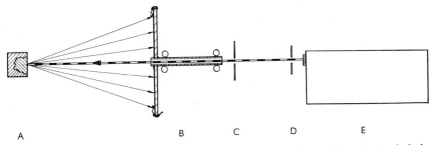

Fig. 12. Diffraction apparatus. *A*, specimen; *B*, rotating film holder; *C*, *D*, pin-hole apertures; *E*, laser.

would then be etched, and the extra scattering should then be a measure of the changes of the surface produced by the etchings. Since there may also be some changes in the crystal surfaces produced by the etching, we thought that a final polishing, which would remove any internal structure without eliminating the grain boundaries, might be necessary.

To try out these ideas, we were kindly supplied with a set of specimens from Mr. J. Hill, of the Lancashire Steel Corporation, Irlam, Lancs. The laser beam was produced from a helium-neon laser of 1 mW power. It was passed through a photographic shutter and fell on the specimen about 1 m distant (Fig. 12); here it was reflected normally and the diffraction pattern was either observed on a white card or allowed to fall on a photographic plate. To give a smoother pattern the photographic plate was rotated in its own plane; if information about preferred orientation had been sought it would have to have been kept stationary. Exposures of the order of 0·1 sec were sufficient to give satisfactory patterns.

To make the results quantitative, the photographs were measured by means of a Joyce–Loebl microdensitometer; a specimen trace is shown in Fig. 13.

C. Quantitative Analysis

For the derivation of the information about grain sizes and their distribution we require to know the variation of scattering with angle, and since the scattering at the smaller angles is spread over a smaller area, a correction has

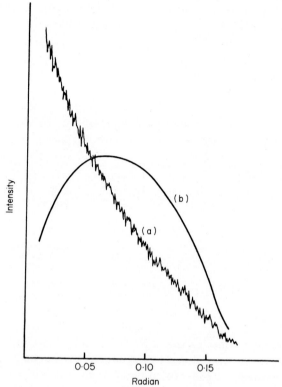

Fig. 13. Scattering results for one specimen: (a) record of scattering as function of angle of scattering; (b) smoothed curve from (a) multiplied by angle.

to be made. Clearly, the total intensity in an annulus, of radius r and width δr, is

$$I . 2\pi r \, \delta r$$

where I is the intensity at distance r from the centre of the record. In order therefore to find the total intensity we need to multiply the observed intensity by the factor r (Fig. 13).

D. Results

We have produced results from four low-carbon steel alloys, whose grain sizes were measured by means of the Quantimet (Section IV.A). The result shown in Fig. 13 is typical. The peak of the wave corresponds to an angle of diffraction of 0·066 radians, and since the wavelength of the light is 0·63 μm, the grain size is 0·63/0·066 = 9·6 μm.

The results for the four alloys are shown in the following table:

Quantimet Measurement	Diffraction Measurement
μm	μm
11·4	7·6
13·9	8·2
17·0	9·6
29·7	10·5

These results are clearly not quantitatively satisfactory and much more investigation will have to be made to find the causes of discrepancy. They are, however, of the right order of magnitude and show the correct sequence; these facts give some encouragement for continuing the work.

E. Future Possibilities

The method shows some promise of being a useful contribution to solving an important practical problem. It is, however, undesirable that photography and microdensitometry should be involved, but there are possibilities of making the analysis more direct. First, it should be easy to record the intensity directly by means of a photo-cell. Secondly, it should be possible to allow for the correction factor by means of a cam so that a recorder working directly from the photo-cell will give the required information. Scales could be made that would give the mean grain size and the variance—or whatever other measure was desired—from the recording.

Other information may also be obtained from the method. Transforms are very sensitive for showing unexpected periodicities (Section I), and it is possible that information about slight amounts of preferred orientation could be detected even if they are at the limit of the capability of a microscope; diffraction experiments can be carried out with scattered waves at angles that would hardly be acceptable to even the best designed high-powered lens systems.

It is hoped that this preliminary report will provide new approaches to these and similar problems in the field of metallurgy.

REFERENCES

Berger, J. E. (1969). *J. Cell Biol.*, **43**, 442.
Berger, J. E. and Harker, D. (1967). *Rev. Sci. Instrum.*, **38**, 292.
Berger, J. E., Zobel, C. R., and Engler, P. E. (1966). *Science*, **153**, 168.

Bragg, W. L. and Stokes, A. R. (1945). *Nature*, **156**, 332.
Buerger, M. J. (1950). *J. appl. Phys.*, **21**, 909.
DeRosier, D. J. and Klug, A. (1968). *Nature*, **217**, 130.
Dye, M. S. (1970). *J. Sci. Tech.*, **37**, 186.
Finch, J. T., Klug, A., and Stretton, A. O. W. (1964). *J. molec. Biol.*, **10**, 570.
Kiselev, N. A., DeRosier, D. J. and Klug, A. (1968). *J. molec. Biol.*, **35**, 561.
Klug, A. and Berger, J. E. (1964). *J. molec. Biol.*, **10**, 565.
Moody, M. F. (1967). *J. molec. Biol.*, **25**, 167.
Offord, R. E. (1966). *J. molec. Biol.*, **17**, 370.
Sternlieb, I. and Berger, J. E. (1969). *J. Cell Biol.*, **43**, 448.
Taylor, C. A., Hinde, R. M., and Lipson, H. (1951). *Acta Cryst.*, **4**, 261.
Taylor, C. A. and Lipson, H. (1964). "Optical Transforms, Their Preparation and Application to X-ray Diffraction Problems," G. Bell and Sons Ltd., London.
Taylor, C. A. and Thompson, B. J. (1957). *J. Sci. Instrum.*, **34**, 439.

Author Index

Subject Index

429

SUBJECT INDEX

Work function, 230
Woven textiles, 414

Y

Young's eriometer, 64
Young's fringes, 11, 14, 19, 20, 46, 93, 357

X

X-ray reflexions, 14
X-ray microscope, 12
   topography, 397

Z

Zernike image-forming process, 9
Zernike theorem, 6, 60
Zone plate, 305, 306, 337